Lecture Notes in Electrical Engineering

Volume 468

"Lecture Notes in Electrical Engineering (LNEE)" is a book series which reports the latest research and developments in Electrical Engineering, namely:

- Communication, Networks, and Information Theory
- Computer Engineering
- Signal, Image, Speech and Information Processing
- Circuits and Systems
- Bioengineering

LNEE publishes authored monographs and contributed volumes which present cutting edge research information as well as new perspectives on classical fields, while maintaining Springer's high standards of academic excellence. Also considered for publication are lecture materials, proceedings, and other related materials of exceptionally high quality and interest. The subject matter should be original and timely, reporting the latest research and developments in all areas of electrical engineering.

The audience for the books in LNEE consists of advanced level students, researchers, and industry professionals working at the forefront of their fields. Much like Springer's other Lecture Notes series, LNEE will be distributed through Springer's print and electronic publishing channels.

More information about this series at http://www.springer.com/series/7818

Gnanam Gnanagurunathan
R. G. Sangeetha · K. Usha Kiran
Editors

Optical and Microwave Technologies

Select Proceedings of ICNETS2, Volume IV

Springer

Editors
Gnanam Gnanagurunathan
Department of Electrical
and Electronic Engineering
The University of Nottingham
Malaysia Campus
Semenyih, Selangor
Malaysia

K. Usha Kiran
School of Electronics Engineering
VIT University
Chennai, Tamil Nadu
India

R. G. Sangeetha
School of Electronics Engineering
VIT University
Chennai, Tamil Nadu
India

ISSN 1876-1100 ISSN 1876-1119 (electronic)
Lecture Notes in Electrical Engineering
ISBN 978-981-10-7292-5 ISBN 978-981-10-7293-2 (eBook)
https://doi.org/10.1007/978-981-10-7293-2

Library of Congress Control Number: 2017958622

Printed on acid-free paper

This Springer imprint is published by Springer Nature
The registered company is Springer Nature Singapore Pte Ltd.
The registered company address is: 152 Beach Road, #21-01/04 Gateway East, Singapore 189721, Singapore

Contents

Enhanced Hierarchical Cluster Based Routing Protocol with Optical Sphere in FSO MANET 1
Kavitha Balamurugan, K. Chitra and A. Jawahar

Performance Improvement of Fractal Antenna with Electromagnetic Band Gap (EBG) and Defected Ground Structure for Wireless Communication 9
Shailendra Kumar Dhakad, Umesh Dwivedi, Sudeep Baudha and Tapesh Bhandari

Conventional DMTL Phase Shifter is Designed Without Meta-material and with Meta-material 21
V. Singh, G. Anitha and K. Usha Kiran

A Modified L-Slot Microstrip Antenna with Chamfered Patch Edges for UWB Applications 33
G. Jemima Nissiyah and M. Ganesh Madhan

Performance Analysis of CSRZ-DQPSK Modulator for RoF-PON-Based Wireless Access 43
C. Rimmya, M. Ganesh Madhan and S. Arshujabin

Effects of Cross-phase Modulation and Four-Wave Mixing in DWDM Optical Systems Using RZ and NRZ Signal 53
V. Sasikala and K. Chitra

Implantable Antenna for Blood Glucose Monitoring 65
Shalu Pandey and Vibha Rani Gupta

Reliability Analysis of Data Center Network 71
Abhilasha Sharma and R. G. Sangeetha

Study of Microstrip Antenna Array with EBG Structure 81
R. M. Vani, K. Prahlada Rao and P. V. Hunagund

Gain Enhancement of Compact Multiband Antenna with Metamaterial Superstrate 91
Rengasamy Rajkumar and Kommuri Usha Kiran

Cooperative Communication for Resource Sharing in Cognitive Radio Networks 99
Lekha Abraham and R. G. Sangeetha

Interdomain Traffic Engineering with BGP and MPLS VPN 105
M. Athira and R. G. Sangeetha

Analysis and Critical Parameter Extraction of an LED for Brain Implants 113
Rabinder Henry and Velmathi Guruviah

Review of Thermal Management of an LED for Brain Implants 125
Rabinder Henry and Velmathi Guruviah

Band Gap Analysis in Defectless Photonic Crystals 139
T. Sridarshini and S. Indira Gandhi

Multiband High-Gain Antenna with CPW Feed for Wi-Fi, WI-MAX and X Band Application 145
Anish Mukherjee, Abhishek Kanaujia and Ravi Prakash Dwivedi

Design and Parameter Extraction of Split Ring Resonator for Surface Crack Detection in Different Materials 153
Varun Seshadri and Ravi Prakash Dwivedi

Optical Channel Analysis of Turbo Coded MIMO-OFDM System for Visible Light Communication 161
Sabitha Gauni, C. T. Manimegalai, K. Kalimuthu, V. C. S. Kaushik and T. Rama Rao

Frequency Tuning Method for Small Profile Metamaterial Based on Tri-Ring Resonator 175
Maruti Tamrakar and K. Usha Kiran

Redesigning Mach-Zehnder Modulator with Ring Resonators 185
R. G. Jesuwanth Sugesh and A. Sivasubramanian

Study on Gain Enhancement of the Antenna Using Planar Small Metasurface Lens 193
Prakhar D. Vyas and K. Usha Kiran

Design of an Internal Multi-resonant PIFA Antenna for Mobile Telecommunication Networks 203
Subathra Thavakumar and M. Susila

Design of Substrate Integrated Waveguide Back to Back π-Shaped Slot Antenna for 60 GHz Applications 211
M. Nanda Kumar and T. Shanmuganantham

Performance Improvement of Gain in Distributed Raman Amplifier Using Forward and Backward Pumps 221
S. Shameem and A. Sivasubramanian

A Comparative Study on Asymmetric, Triangular, and Rectangular Core Large-Mode-Area PCF Designs 229
Netra Dalvi, Radhika Ramesh and Pravin Joshi

Miniaturized High-Gain UWB Monopole Antenna with Dual Band Rejection Using CSRR 237
Ravi Prakash Dwivedi and K. Usha Kiran

SOA Parameters Optimization for High Data Rate Operation 247
K. Swetha, R. Manohari and Shanthi Prince

All-Optical 3R Regenerator of Design and Simulation 255
Althi Bhavya Bindu and Shanthi Prince

Analysis of Dispersion Compensation Methods in WDM Systems 269
S. Sesha Sai Srikar and N. Subhashini

Wideband Antenna for Medical Application 279
Drishti Gandhi and Niraj Kumar

Mitigation of Cross-Phase Modulation in Multiband Radio Over Fiber Systems 289
Ch. Venkata Dharani and A. Brintha Terese

The Design of High Gain Substrate Integrated Waveguide Antennas with FR4 and RT Duroid 303
Caroline Sebastian, V. J. Amirtha Vijina and R. Ramesh

About the Editors

Gnanam Gnanagurunathan was in the cellular industry as an RF Network Engineer for almost 5+ years prior to becoming an academician in University of Nottingham Malaysia Campus. She completed her PhD in 2012 in the field of metamaterial-based planar antennas while serving the university. Currently, her research is evolving from bettering the performance of planar antennas with metamaterial to using them to harvest RF energy for IoT applications. She has a number of publications in both fields of antennas and optics. Currently, she is a Member of IET, Senior Member of IEEE /APS, and a Fellow of HEA.

R. G. Sangeetha completed her B.E. in Electronics and Communication Engineering and M.E. in Computer and Communication Engineering. She completed her Ph.D. (2012) in Optical Networks at the Indian Institute of Technology Delhi. She has a national patent and a US patent to her credit. Her research has been published in reputed journals and conferences. She has 11 years of teaching experience. Currently, she is working as an Associate Professor in VIT University, Chennai campus, and guiding M.Tech and Ph.D. scholars. She is a member of IEEE and OSA. Her main research interests are in the areas of Fiber Optic Communications, Optical Networks, and Free Space Communication. Currently, she is pursuing a Department of Science and Technology (DST)-sponsored project, titled "Test Bed for Hardware Implementation of All Optical Bi-Directional Switching Node" under the Fast-Track Young Scientist Scheme.

K. Usha Kiran completed her Ph.D. in Microwave Antennas at Gulbarga University, Karnataka, in 2007. She then joined the Microwave Lab, ECE, Indian Institute of Science (IISc), Bangalore, as Project Associate and developed several RF MEMS SPDT and SPST switches from 2007 to 2009. She served as a Project Scientist at the Indian Institute of Technology (IIT) Delhi, where she worked on an RF MEMS phase shifter from 2010 to 2012. Since 2012, she has been working at

Vellore Institute of Technology (VIT), Chennai. She has published more than 80 papers in the reputed journals, and conferences in the area of Microwave Antennas and RF MEMS. Presently, she is working on a DST-funded project on "MEM phase shifter based steerable antenna array."

Enhanced Hierarchical Cluster Based Routing Protocol with Optical Sphere in FSO MANET

Kavitha Balamurugan, K. Chitra and A. Jawahar

Abstract The existing protocols which are used for routing in MANET fail to consider FSO characteristics like the frequent physical link disruption (intermittent connectivity) and the requirement of line of sight (LOS). Hence, a hierarchical routing protocol along with spherical data propagation model is proposed. In this method, first clustering is done based on neighbor discovery algorithm and then cluster head is selected with the help of network source connector. This process provides network with reliable cluster head. After that source transmits packet based on the proposed data propagation model which helps to increase the transmission range and provides clear line of sight for smooth routing in FSO. Varying the number of transceivers as 2, 4, 6, and 8 it is seen that the performance characteristics such as throughput, delivery ratio, drop, and delay is enhanced. Proposed protocol has the advantage of providing high throughput and clear line of sight.

Keywords Free space optics · Mobile ad hoc networks · Routing protocol Transceivers

---

K. Balamurugan (✉)
Department of Electronics and Communication Engineering, KCG College of Technology, Karapakkam, Chennai 600097, India
e-mail: kavitha@kcgcollege.com

K. Chitra
School of Electronics Engineering, VIT University, Vandalur Kelambakkam Road, Chennai 600127, India
e-mail: chitra.krishnan@vit.ac.in

A. Jawahar
Faculty of Electronics and Communication Engineering, SSN College of Engineering, Kalavakkam, Chennai 603110, India
e-mail: jawahara@ssn.edu.in

G. Gnanagurunathan et al. (eds.), *Optical and Microwave Technologies*,
Lecture Notes in Electrical Engineering 468,
https://doi.org/10.1007/978-981-10-7293-2_1

1 Introduction (Free Space Optical Wireless Networks)

FSO provides relatively good bandwidth for point-to-point communication involving last mile applications and also for indoor wireless communication. FSO has many important features like [1]

- Spatial reuse,
- Usage of low-power per bit transmitted,
- Licensed-free band, and
- Higher bandwidth.

Though the above characteristics are advantageous, it is not used practically for metropolitan area networks (MANs) or multi-hop ad hoc networks, which are presently based on radio frequency (RF) communication technologies. The reasons are [1]:

- LOS is needed for alignment between communicating nodes.
- LOS has to be continuously adjusted.
- Transmission quality reduction in adverse weather conditions.

FSO systems are slowly gaining acceptance in the marketplace as a solution to replace expensive optical fibers [2].

Though FSO is useful in solving the wireless capacity problem, it faces new challenges, like physical link disruption frequently and the LOS requirements [3].

For the FSO network performance to get improved through design of networks, the important issues are design of topology and routing. The few routing algorithm [4–7] proposed for a FSO network are delay-constrained minimum hop [4] (DCMH), an optimal diverse routing algorithm, all hops shortest paths (AHSP) algorithms [5], a load balanced scheme for routing, and a minimum hop count with load-balancing (MHLB) algorithm for routing [6].

The problems in routing protocols like DSDV and AODV are; it uses the reverse path technique, but links which are unidirectional are not considered. Therefore in FSO MANET, these protocols cannot be directly applied [8–15].

Local directionality utilization as a property to route packets itself is an important issue [16].

Sensor readings of routing to the base station and with nodes-to-base station, base station-to nodes, and neighborhood communication are also some of the problems [5].

The paper is scheduled as follows: Literature review is given in Sect. 2. Problem identification and solution in Sect. 3, which explains the data propagation model including clustering process, network source connector discovery process, and multielement routing protocol. Simulation results are presented in Sect. 4. Finally, conclusion with a brief summary is given in Sect. 5.

2 Literature Review

In [3], the author has proposed a spherical model of FSO as a basic building block and verified the influence of such FSO models to upper layers. The main problems shown here are the line of sight (LOS) and the suffering of FSO from beam spread with distance and unreliability during bad weather (especially fog) conditions.

In this work [17], a strategy is given for providing backup instantaneously in a RF/FSO hybrid mesh network to traffic. In this paper [18], author has defined neighborhood and base station discovery algorithms for routing in FSO sensor networks. The paper addresses the problem of finding the neighbor node and a base station taking into account the problem of line of sight. In this paper [8], the author has shown that the least cost path (LCP) routing algorithm minimizes the end-to-end delay and minimum hop count with load-balancing (MHLB) routing algorithm based on the number of hops is used between the source and the destination to route the traffic. The metrics taken are traffic matrix, link utilization matrix, departure rate, link traffic load, total traffic load present internally, demand for total external traffic, average delay for a packet moving through the network. In this paper [19], a hierarchical protocol for routing with multiple transceivers for FSO MANET is proposed. The paper also introduces how formation of cluster happens through neighborhood discovery algorithm.

The performance metrics taken into account are average end-to-end delay, average packet delivery ratio, drop, and throughput.

3 Problem Identification and Solution

In previous papers, we found that there are many problems related to routing in FSO networks. So if routing has to be smooth between source and destination, a hierarchical routing protocol [19] with multielements comprising of 2, 4, 6, and 8 transceivers of FSO multielement structures to solve maximum routing problems [3] is proposed.

In hierarchical routing protocol, first clusters are formed based on neighborhood discovery algorithm and then with the help of network source connector a routing table is built by gathering information from cluster head. Now when a node, i.e., source wants to send information to the other node, i.e., destination node, it has to use FSO MANET routing protocols. The direct connection of network source connectors and routing algorithms provides end-to-end throughput and low delay. But for reliable communication, clear line of sight is needed. Hence, hierarchical routing protocol is enhanced with data propagation model which provides a spherical structure for propagation of data with 2, 4, 6, or 8 transceivers so that it can cover large transmission area. Also, it considers several factors such as beam intensity, angle of divergence, geometric attenuation factor, and atmospheric attenuation factor to provide a clear LOS for communication.

If we could use hierarchical routing protocol for the connectivity with multiple transceivers we can reduce the delay and drop while improving the delivery ratio.

3.1 Data Propagation Model

This section describes different metrics used in the proposed protocol in order to provide better routing path between source and destination. Figure 1 shows the data propagation model [20] with conditions for coverage area and uncovered area. The various parameters of interest are the intensity of the beam, geometric attenuation factor (A_F), and atmospheric attenuation factor (A_P). The received power is denoted as B, and the source power is denoted as S_p in Fig. 1.

3.2 Phases of proposed Hierarchical Routing protocol with Multielements

The different phases of proposed multielement hierarchical routing protocol are described in the following section:

- Clustering process,
- Network source connector discovery process,
- Multielement routing protocol with 2, 4, 6, and 8 number of transceivers.

(1) *Clustering Process*

The clustering process depends on the neighbor discovery algorithm (NDA) [18, 19] which defines that every node consists of a table that has information about the identity of the neighboring nodes and position of the link.

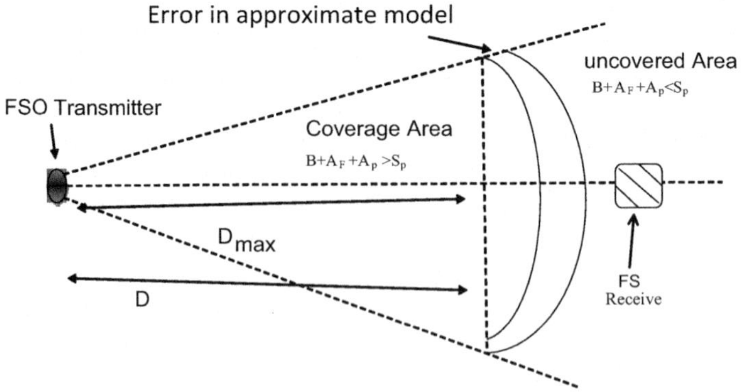

Fig. 1 Data propagation model

(2) *Network Source Connector Discovery Process*

Once clustering is done, all the information about the cluster member is maintained by the network source connector [18, 19] in the routing table.

(3) *Multielement Routing Protocol*

After the clustering process, election of CH and creation of routing table by each network source connector is done. The next step includes transmission of this information from source node to destination node.

The proposed protocol considers various factors that provide better communication between source and destination node. The FSO spherical structure [3, 9, 10, 21] is used for the transmission of data from source to destination so that we can achieve angular diversity using multiple, i.e., 2, 4, 6, or 8 transceivers.

4 Simulation Results

4.1 Simulation Results

The hierarchical routing protocol with optical sphere for smooth routing (HROS) performance is verified through NS2 [22] simulation. A network is deployed randomly in an area of 2200 × 2200 m. The nodes can be varied as 25, 50, 75, and 100. In this work, keeping number of nodes as 100, number of transceivers were varied as 2, 4, 6, and 8 and various parameters like throughput, delivery ratio, delay, and drop were verified. In simulation, the nodes were placed randomly in the given area. Same value is set for the channel capacity of mobile hosts.

Table 1 lists the simulation parameters used.

Table 1 Simulation parameters

No. of nodes	100
No. of transceivers	2, 4, 6, and 8
Area size	2200 × 2200
Mac	802.11
Time of simulation	50 s
Traffic source	CBR
Size of packet	512
Coverage	300 m
Rate	250 Kb
Propagation model	FSO sphere
Type of antenna	FSO antenna
Type of modulation	BPSK/FSO

4.2 Performance Metrics

The proposed HROS performance is done with 2, 4, 6, and 8 transceivers. The performance is verified according to the following parameters: average end-to-end delay, average packet delivery ratio, drop, throughput.

5 Results

The no. of transceivers is varied as 2, 4, 6, and 8 in the simulation environment.

Figures 2, 3, 4, and 5 present the graphical representations of the results of delay, delivery ratio, packet drop, and throughput, respectively.

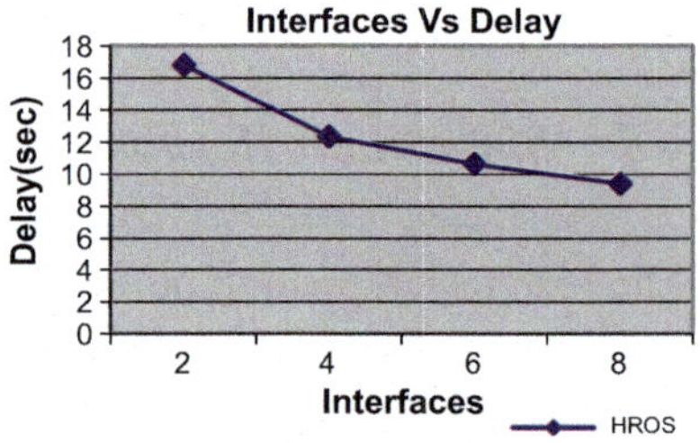

Fig. 2 Interfaces versus delay

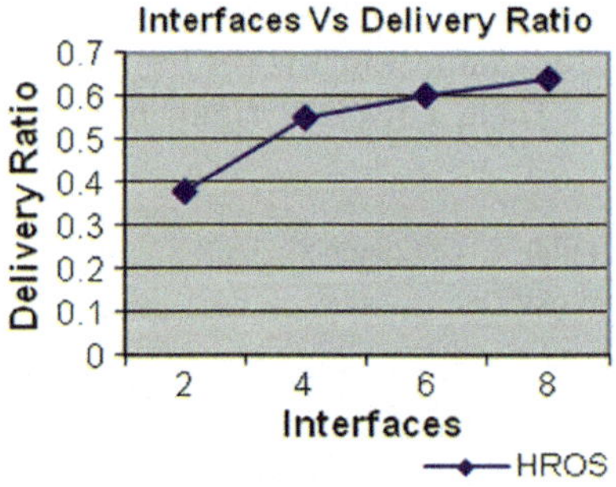

Fig. 3 Interfaces versus delivery ratio

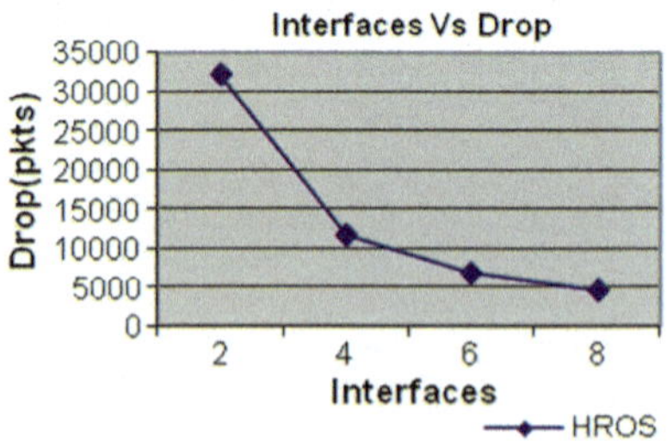

Fig. 4 Interfaces versus drop

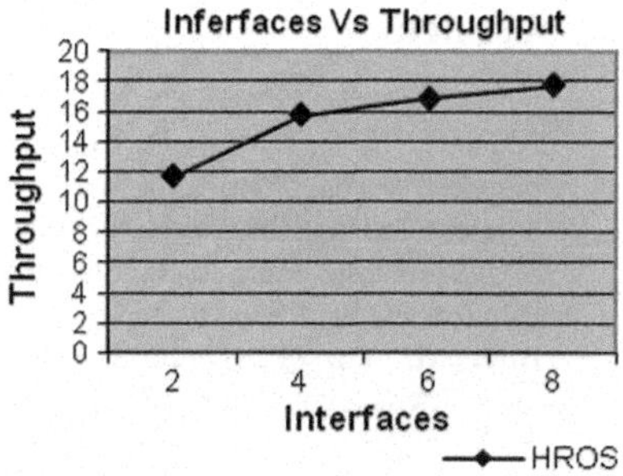

Fig. 5 Interfaces versus throughput

Figure 2 shows the end-to-end delay in the network by varying the number of transceivers. From the figure, we can see that the delay decreases as the number of transceivers increase. This is due to the fact that with more number of transceivers, establishing the alignment is quicker, the delay involved in the route discovery process will be lesser.

Figure 3 shows the packet delivery ratio of HROS by varying the number of transceivers. From the figure, we can see that the delivery ratio begins to increase as number of transceivers increase as 2, 4, 6, and 8.

From Fig. 4, it can be seen that the packet drop of the HROS decreases as number of transceivers increases.

From Fig. 5, it can be seen that the throughput of HROS increases with the number of transceivers.

6 Conclusion

In this paper, hierarchical routing protocol along with data propagation model varying the number of transceivers is proposed to provide smooth routing protocol in FSO MANET. Here, first clustering process is done according to neighbor discovery algorithm and then CH is selected by network source connector. For a clear LOS, proposed model considers parameters such as intensity of beam, divergence angle, geometric attenuation factor, and atmospheric attenuation factor that helps to provide consistent and reliable communication without any interruption in the network. The advantage of proposed protocol by varying the number of transceivers is that it provides high throughput, better delivery ratio, drop, delay, and clear line of sight. The comparison with the existing hierarchical cluster-based routing protocol by varying the number of nodes is our future study.

References

1. Akella J, Liu C, Partyka D, Yuksel M, Kalyanaraman S, Dutta P (2005) Building blocks for mobile free-space-optical networks. In: Proceedings of IFIP/IEEE international conference on wireless and optical communications networks (WOCN), IEEE, pp 164–168
2. SANS Institute InfoSec Reading Room (2002) Free-space optics: a viable, secure last-mile solution? © SANS Institute
3. Bilgi M, Yuksel M (2008) Multi-element free-space-optical spherical structures with intermittent connectivity patterns. In: Proceedings of the IEEE Infocom Student Workshop, Phoenix, AZ
4. Mohd Shariff AR, Woodward ME (2007) A delay constrained minimum hop distributed routing algorithm using adaptive path prediction. J Netw 2(3):46–57
5. Cheng G, Ansari N (2004) Finding all hops shortest paths. IEEE Commun Lett 8(2):122–124
6. Hu Z, Verma P, Sluss Jr J (2009) Routing in degree-constrained FSO mesh networks. Int J Hybrid Inf Technol 2(2)
7. Mohd Shariff AR, Woodward ME (2007) A delay constrained minimum hop distributed routing algorithm using adaptive path prediction. J Netw 2(3):46–57
8. Kishore Kumar D, Murthy YSSR (2013) Hybrid cluster based routing protocol for free-space optical mobile ad hoc networks. Springer, New York
9. Bilgi M, Yuksel M (2010) Multi-transceiver simulation modules for free-space optical mobile ad hoc networks. In: Proceedings of SPIE 7705. [doi:https://doi.org/10.1117/12.851756]
10. Nakhkoob B, Bilgi M, Yuksel M, Hella M (2009) Multi-transceiver optical wireless spherical structures for MANETs. IEEE J Sel Areas Commun 27(9)
11. Yuksel M, Akella J, Kalyanaraman S, Dutta P (2009) Free-space-optical mobile ad hoc networks: auto-configurable building blocks. ACM Springer Wireless Netw 15:295–312
12. Bilgi M, Yuksel M (2014) Capacity scaling in free-space-optical mobile ad hoc networks. Ad Hoc Netw (Elsevier) 12:150–164
13. Di W, Alhussein A (2011) Throughput and delay analysis for hybrid radiofrequency and free-space-optical (RF/FSO) networks. Springer Wireless Netw 17:877–892
14. Derenick J, Thorne C, Spletzer J (2005) On the deployment of a hybrid free-space optic/radio frequency (FSO/RF) mobile ad-hoc network. In: IEEE/RSJ international conference on intelligent robots and systems
15. George D, Sankar SP (2016) Stability routing in FSO-MANET. Int J Res Eng Technol 05
16. Cheng BN, Yuksel M, Kalyanaraman S (2008) Using directionality in mobile routing. In: IEEE international conference on mobile ad-hoc and sensor systems (MASS)
17. Kashyap A, Rawat A, Shayman M (2006) Integrated backup topology control and routing of obscured traffic in hybrid RF/FSO networks. IEEE Globecom
18. Okorafor UN, Kundur D (2005) Efficient routing protocols for a free space optical sensor network. In: IEEE international conference on mobile adhoc and sensor systems (MASS), pp 251–258, IEEE, Washington DC (2005). [https://doi.org/10.1109/MAHSS.2005.1542807]
19. Kumar DK, Murthy YSSR (2012) Hierarchical routing protocol for free-space optical mobile ad hoc networks (FSO/RF MANET). Int J Comput Appl 60(10):0975–8887
20. Bilgi M, Yuksel M (2010) Packet-based simulation for optical wireless communication. In: Proceedings of IEEE workshop on local and metropolitan area networks, IEEE
21. Asaad Kaadan, Hazem H. Refai, Peter G. LoPresti (2014) Multielement FSO transceivers alignment for inter-UAV communications. J Lightwave Technol 32(24)
22. Network simulator. http://www.isi.edu/nsnam/ns

Performance Improvement of Fractal Antenna with Electromagnetic Band Gap (EBG) and Defected Ground Structure for Wireless Communication

Shailendra Kumar Dhakad, Umesh Dwivedi, Sudeep Baudha and Tapesh Bhandari

Abstract This paper is aimed at developing a Sierpinski carpet fractal antenna with electromagnetic band gap (EBG) in the ground plane. This is designed using Rogers Duroid 5880 substrate, and validation of model is done using extensive simulations in CST MWS 2011. Several iterations are done and optimized to give the best result. The simulation results showed return loss of −22.25 dB at 8.592 GHz for third iteration. There is an enhancement of bandwidth by 209% from 75 MHz (conventional patch antenna) to 234.4 MHz using SCFA with EBG and directivity of 4.4 dB at resonant frequency. It shows reduction in size of antenna, and the proposed antenna size is 17.2×20.5 mm^2. This antenna works well in X-band communication (8–12 GHz) for ultra-wideband imaging for medical application and weather monitoring radars in satellite communication.

Keywords Sierpinski carpet fractal antenna · Return loss · Voltage standing wave ratio · EBG structures · Microstrip patch antenna

1 Introduction

Communication system is a rapidly growing field in the world right now, and Antennas are a very important part of them. These days, designing of the microstrip antennas has become a great field of research as they are light in weight, easy to manufacture, and conformable to both planar as well as non-planar surfaces.

S. K. Dhakad · U. Dwivedi · S. Baudha (✉) · T. Bhandari
Department of EEE&I, BITS Pilani, KK Birla Goa Campus, Goa, India
e-mail: sudeepb@goa.bits-pilani.ac.in

S. K. Dhakad
e-mail: skdhakad@goa.bits-pilani.ac.in

U. Dwivedi
e-mail: ukdwivedi@jpr.amity.edu

T. Bhandari
e-mail: tapesh.bhandari@gmail.com

G. Gnanagurunathan et al. (eds.), *Optical and Microwave Technologies*,
Lecture Notes in Electrical Engineering 468,
https://doi.org/10.1007/978-981-10-7293-2_2

One popular kind of antenna is a fractal antenna. Fractal shapes generally are the geometrical shapes which repeat themselves with respect to their previous iteration design by modifying the dimension of each new iteration. They are chosen over conventional shapes for antenna design because they increase the electrical length while keeping the surface area and volume same.

Electromagnetic band gap structures help in creating a gap in the band around the operating frequency of the antenna. Popular EBG structures are mushroom-like EBG, polygonal, circular, and spiral. However, it is found that circular-shaped EBG exhibits higher directivity and gain as compared to other types of EBG structures.

In this paper, a third-iteration Sierpinski fractal antenna is designed with circular EBG. Analysis of the antenna is done using various modifications. Analyzing the results of this antenna with the number of iterations and studying the effect of EBG and non-EBG structures in the ground plane are done in the following paper. The final antenna is found to be working in the X-band radar application including single polarization, dual polarization, Synthetic Aperture Radar and also for weather monitoring by meteorological department.

2 Methodology

In this work, the third iteration of a Sierpinski fractal antenna with a circular-shaped electromagnetic band gap structure on its ground plane was analyzed. This antenna was designed using Rogers RT Duroid 5880 substrate. Table 1 shows the substrate properties.

2.1 Antenna Design

Firstly, in the zeroth iteration, a conventional rectangular patch antenna is designed with the dimension shown in Fig. 1. Table 2 shows the values of the variables used in Fig. 1. In the first iteration, a rectangle is cut from the middle of the zeroth iteration, the dimension of which is 1/3rd the dimension of the rectangle in the zeroth iteration. In the second iteration, seven more rectangles are cut from the first iteration, the dimension of each, again being 1/9th of the zeroth iteration. In the third iteration, 50 more rectangles are cut symmetrically from the second iteration, the dimensions of which are 1/27th as compared to the zeroth iteration (Fig. 2).

Table 1 Properties of the substrate

Properties	Value
Relative permittivity, ε_r	2.2
Thickness	0.38 mm

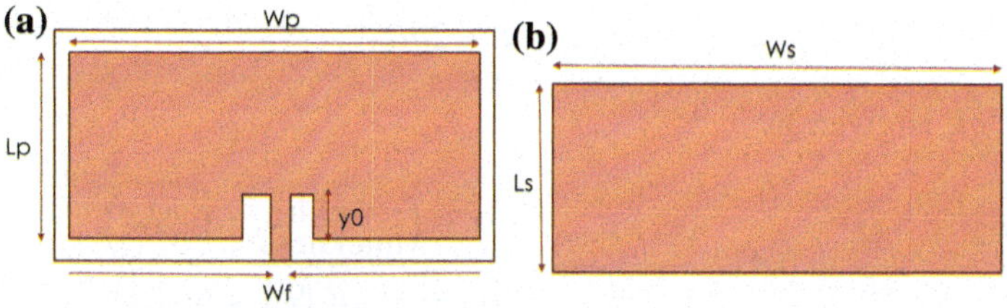

Fig. 1 Patch antenna **a** Front view. **b** Back view

Table 2 Proposed antenna design parameters

L_p	L_s	W_p	W_s	W_f	Y_o	r/a
17.27	19.6	20.5	22.5	1.18	5	0.5

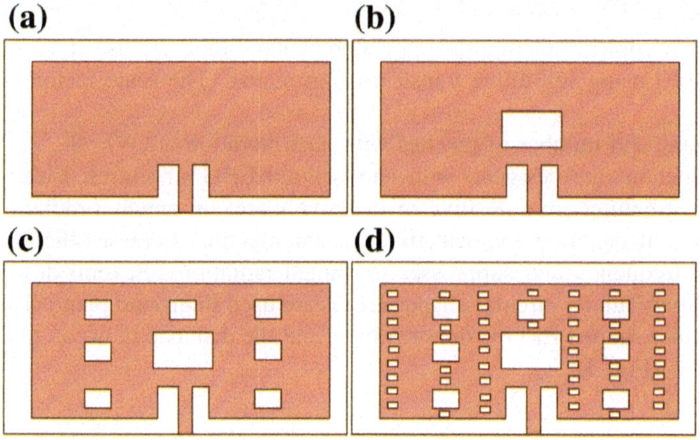

Fig. 2 **a** Zeroth iteration. **b** First iteration. **c** Second iteration. **d** Third iteration

The antenna was designed using equations [1] to calculate dimensions and resonant frequency.

(1) Effective dielectric constant for $E_r = 2.2$

$$\varepsilon_{r,\text{effective}} = \frac{\varepsilon r + 1}{2} + \frac{\varepsilon r - 1}{2}\left[1 + 12\frac{\text{thickness}}{\text{Width}}\right]^{\frac{-1}{2}}$$

where ‘thickness’ is the thickness of substrate and ‘width’ is the width of the patch antenna. Due to fringing, effective length is more than physical length. The effective length is L_n

$$L_n = L + 2\delta L$$

While δL is given by

$$\frac{\delta L}{h} = 0.412 \frac{\left(\in_{r,\text{effective}} + 0.3\right)\left(\frac{\text{width}}{\text{thickness}} + 0.264\right)}{\left(\in_{r,\text{effective}} - 0.258\right)\left(\frac{\text{width}}{\text{thickness}} + 0.8\right)}$$

The actual length L can be calculated using following formula

$$L = \frac{1}{2f_r\sqrt{\in_{r,\text{effective}}\, \mu_o \in_o}} - 2\,\Delta L$$

Width W is given by $W = \frac{c}{2f}\sqrt{\frac{2}{\in_r + 1}}$, where c is the speed of light and f is the frequency.

In the Sierpinski antenna design different iterations, we use iteration function system (IFS) using self-affine transformation matrix. The scale factor of length of each side.

$L_n = L/3^n$ and number of patches with each iteration are $N = 8^n$.

The antennas are designed with and without EBG structures. Electromagnetic band gap structures give positive or negative effect on antenna radiation at some frequencies. It increases the selectivity of antenna and increases the directivity at particular frequency and suppresses unwanted radiations. Several designs can be used, and particularly circular-shaped EBGs are used in ground plane. The radius of these circular shaped EBG is denoted by r, and the distance between them is given by 0.45 * r (Fig. 3).

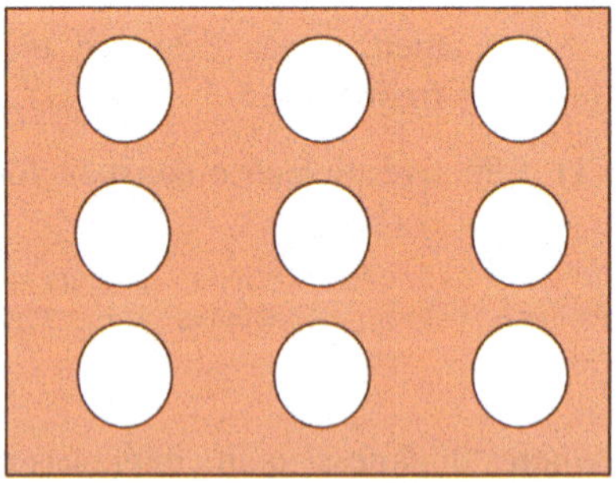

Fig. 3 EBG structure in ground plane

3 Results and Discussions

3.1 Analysis with Respect to Number of Iterations and EBG Structure in Ground Plane

The following Fig. 4 shows the return loss plot of conventional patch antenna. From the graph, we infer that the conventional patch antenna resonates at 9.204 GHz with a bandwidth of 75.7 MHz.

After applying the EBG structure in the ground plane, the resonant frequency is now found to be 9.708 GHz and the bandwidth reduced to 40.6 MHz. Figure 5 shows the return loss plot for the conventional patch antenna with and without EBG structure in the ground plane.

After the first iteration, the resonant frequency without EBG is 8.988 GHz and the bandwidth for the same increased to 169.9 MHz. After applying the EBG structure to this antenna, the resonant frequency was found to be 8.604 GHz and the bandwidth further increased to 206.6 MHz. Figure 6 shows the return loss plot for conventional patch antenna, first iteration and first iteration with EBG with the same frequency axis.

Figure 7 shows the conventional patch antenna, second iteration and second iteration with EBG with the same frequency axis. We infer from the figure that second iteration without EBG has a resonant frequency of 9.036 GHz and a bandwidth of 212.3 MHz. After applying the EBG, the resonant frequency is decreased to 8.604 GHz and bandwidth reduced to 165.6 MHz.

Finally, Fig. 8 shows the conventional patch antenna along with third iteration without EBG and third iteration with EBG. We see that the resonant frequency for third iteration is 9.036 GHz and the bandwidth increased to 220.7 MHz. After

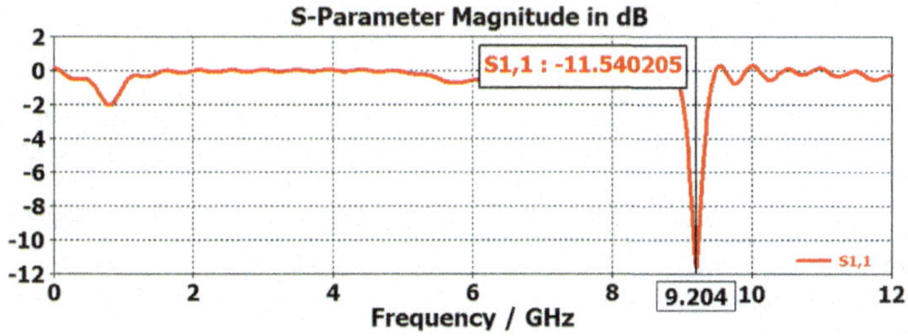

Fig. 4 Return loss plot for conventional patch antenna

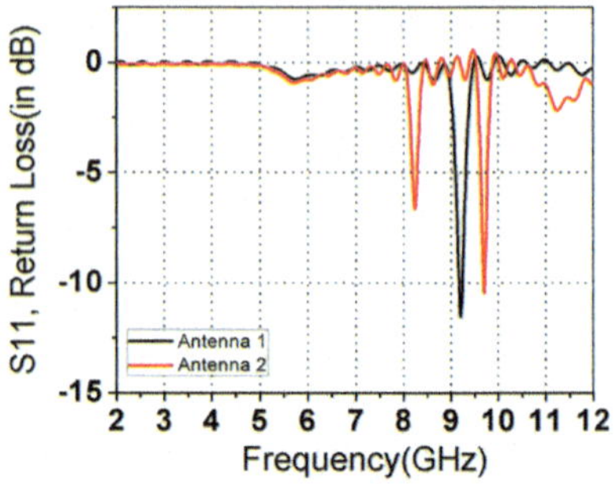

Fig. 5 Antenna 1 represents conventional patch antenna and antenna 2 represents conventional patch antenna with EBG

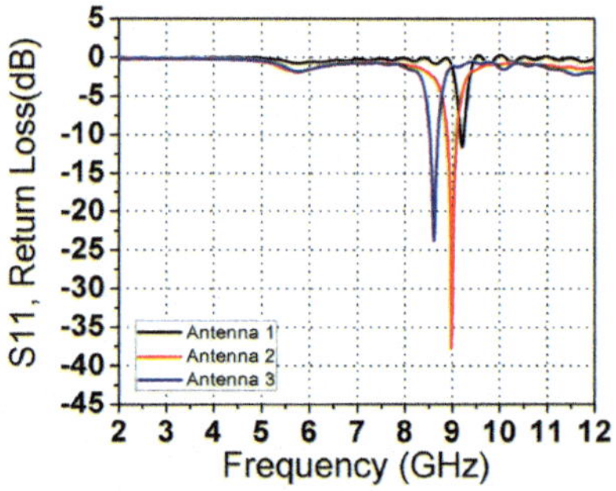

Fig. 6 Antenna 1 represents conventional patch antenna, antenna 2 represents first iteration without EBG and antenna 3 represents first iteration with EBG

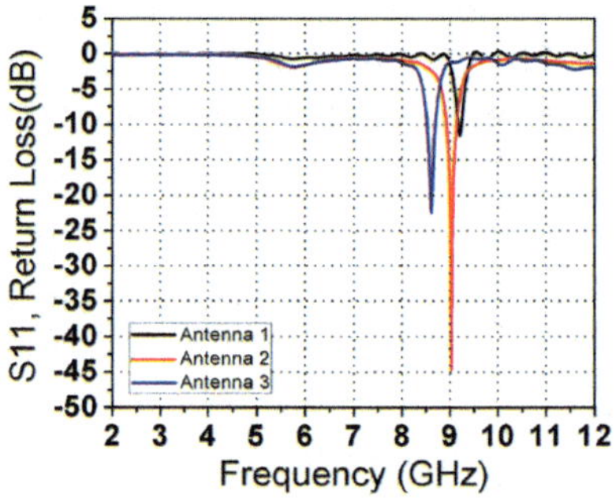

Fig. 7 Antenna 1 represents conventional patch antenna, antenna 2 represents second iteration without EBG and antenna 3 represents second iteration with EBG

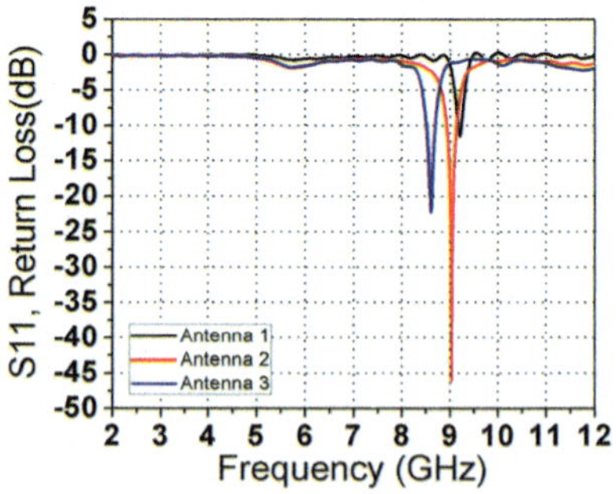

Fig. 8 Antenna 1 represents conventional patch antenna, antenna 2 represents third iteration without EBG and antenna 3 represents third iteration with EBG

applying the EBG in third iteration, we find that the resonant frequency decreased to 8.592 GHz and bandwidth increased drastically to 234.4 MHz. We see that there is an increment of 209.6433% in the bandwidth as compared to conventional patch antenna.

To summarize and tabulate all the results as the number of iterations increase, we compare the parameters such as resonant frequency, radiation efficiency at resonant frequency, total efficiency at resonant frequency, directivity at resonant frequency, and bandwidth. This work is done in Table 3.

3.2 Proposed Antenna

The return loss plot for the final antenna after third iteration and EBG is shown in Fig. 9. The resonant frequency for the same is 8.592 GHz.

The voltage standing wave ratio (VSWR) plot shows good matching at frequency 8.592 GHz and is shown in Fig. 10.

The plot for directivity (3D, radiation efficiency) vs frequency is shown in Fig. 11.

The plot for gain (in dB) with respect to frequency is given in Fig. 12.

The plot for polar directivity (2D) at resonant frequency (8.592 GHz) with respect to phi is shown in Fig. 13.

Figure 14 represents the directivity of the proposed antenna. We see that the directivity is 4.595 dB at the resonant frequency which is much higher as compared to that of conventional patch antenna, i.e., 4.034 dB.

Figure 15 shows the gain (IEEE) pattern of the proposed antenna. We see that the gain at the resonant frequency is 4.064 dB which is very large as compared to that of conventional patch antenna, i.e., 1.443 dB.

After performing the analysis on the antenna by changing the number of iterations, it is clear that antenna improves its performance as we increase the number of iterations. The operational frequency of the antenna is found to be lowered when the number of iterations is increased. Also, it is found that EBG is responsible for the increment in bandwidth from 75 MHz to 234 MHz. The same trend is followed in radiation efficiency, directivity and gain. Hence, the proposed antenna is the most optimal antenna among all of them.

Table 3 Performance analysis

	Conventional Patch	Conventional patch with EBG	First iteration	First iteration with EBG	Second iteration	Second iteration with EBG	Third iteration	Third iteration with EBG
Return loss resonant frequency (GHz)	9.204	9.708	8.988	8.604	9.036	8.604	9.036	8.592
Radiation efficiency at resonant frequency (%)	55.07	31.16	93.6	89.38	93.39	89.53	93.08	88.49
Total efficiency at resonant frequency (%)	51.21	28.31	93.58	89	93.39	89.01	93.08	87.96
Directivity at resonant frequency (%)	4.034	4.712	4.025	4.584	4.009	4.588	3.994	4.595
Bandwidth (MHz)	75.7	40.6	169.9	206.8	212.3	165.6	220.7	234.4

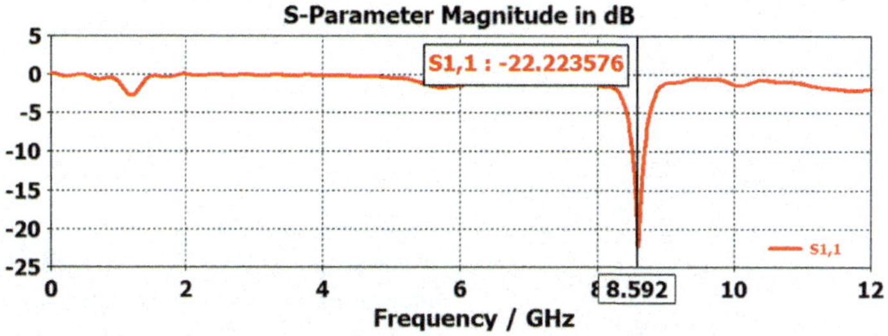

Fig. 9 Return loss plot of the proposed antenna

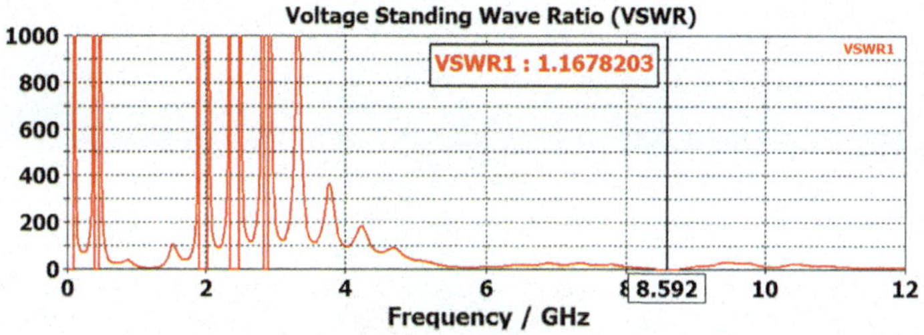

Fig. 10 VSWR plot of the proposed antenna

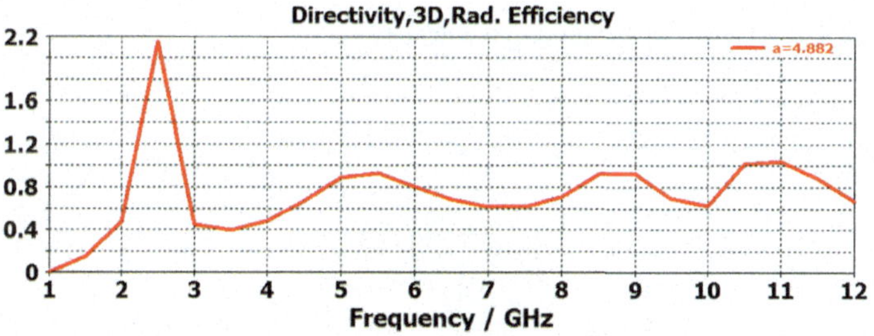

Fig. 11 Directivity versus frequency

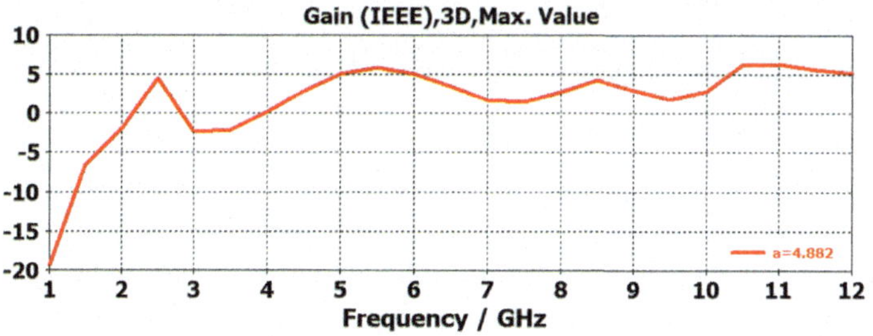

Fig. 12 Gain versus frequency curve

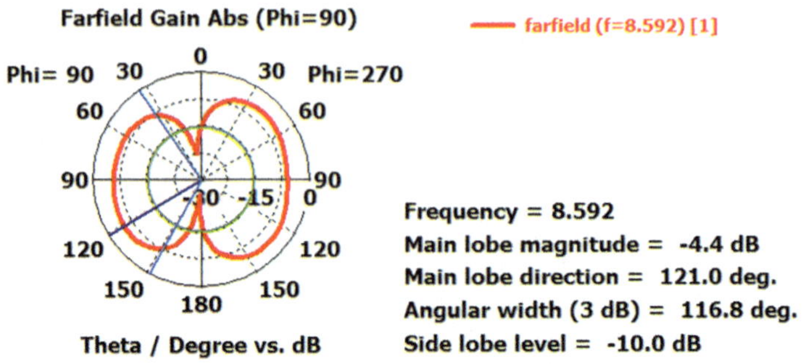

Fig. 13 Polar directivity (2D) at resonant frequency 8.59 GHz

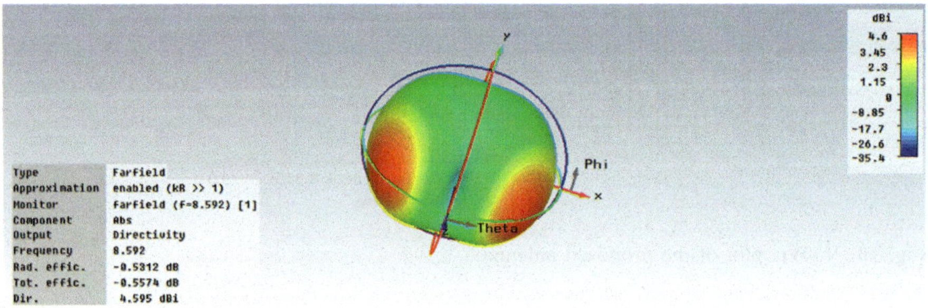

Fig. 14 Directivity pattern of the proposed antenna

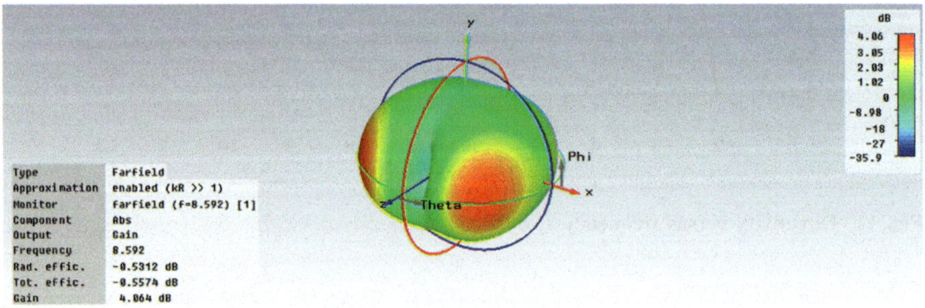

Fig. 15 Gain (IEEE) radiation pattern of the proposed antenna

4 Conclusion

The proposed antenna shows better results in terms of return loss, bandwidth, radiation efficiency, and directivity. The analysis of the antenna with respect to number of iterations and changing ground plane is done extensively, and the results are verified by CST Microwave Studio 2011. The geometry lowers the frequency of operation and increases the bandwidth with good axial ratio which is well suited for X-band applications including high-imaging radars and ultra-wideband (UWB) biomedical applications.

Reference

1. Balanis CA. Antenna theory—analysis and design, 2nd edn. Wiley, New Yoek

Conventional DMTL Phase Shifter is Designed Without Meta-material and with Meta-material

V. Singh, G. Anitha and K. Usha Kiran

Abstract Miniaturized antenna and phase shifter are designed for 10 GHz. The antenna is composed of a silicon substrate having high dielectric constant which reduces the size of the antenna. In this design, micro-machining concept is used to reduce the losses by removing some part of the silicon underneath the rectangular patch. DMTL phase shifter is designed and later stacked with meta-material complementary split ring resonator (CSRR) in the transmission line and ground plane of the coplanar waveguide to diminish the effective length of the CPW and increases phase shift, and both the results are discussed in this paper. Single-bit phase shifter for 90° is demonstrated, and the effect of loading is studied.

Keywords Meta-material · Miniaturized low-profile antenna · Capacitive coupled feed · Miniaturized phase shifter · Radar antenna · Phased array

1 Introduction

Microwave phase shifters are square measure basic components in phased-array antennas for telecommunication, radio detection, ranging applications. Nowadays, most of the micro-electro-mechanical systems phase shifters are created based on setup styles except that the solid-state switch is substituted by a MEMS switch. RF MEMS switches end in low loss in the frequency range of 8–120 GHz. The insertion loss of progressive 3-b MEMS phase shifters in the frequency range 10–14 GHz is 0.9 dB, which might be a 3–4 dB change compared to on-wafer styles exploitation transistor switches. This interprets to the improvement of 6–8 dB

V. Singh · G. Anitha (✉) · K. Usha Kiran
School of Electronics Engineering, VIT University, Chennai 600127, Tamil Nadu, India
e-mail: anitha.g2014phd1176@vit.ac.in

V. Singh
e-mail: vineeta.singh2015@vit.ac.in

K. Usha Kiran
e-mail: usha.kiran@vit.ac.in

G. Gnanagurunathan et al. (eds.), *Optical and Microwave Technologies*,
Lecture Notes in Electrical Engineering 468,
https://doi.org/10.1007/978-981-10-7293-2_3

during radio detection and ranging or two-way communication equipment. MEMS section shifters can even be needed to diminish 50–75% of the T/R modules in monster phased arrays. The precise variety relies on the transmit control and fluctuate required from the microwave radar system and on the gain-over-noise (G/T) request in communication systems. For low-loss segment shifters, it is going to be attainable to nourish two to four components from one T/R module, accordingly decreasing system complexness [1].

Distributed techniques are utilized for exceptionally wideband devices. The main objective is based on periodical stacking of a transmission line with passive components such as capacitors or stubs, or Schottky diodes, to obtain wide-band amplifiers, mixer, oscillators, pulse-shaping circuits, and multipliers. The DMTL comprises of a >50 Ω transmission line which is capacitive stacked by placing the MEMS bridges in periodical manner. This could be finished by utilizing a wide range of sorts of t-lines; be that as it may, it is anything but difficult to actualize utilizing CPW. The MEMS bridge has a thickness t, a width w, and a length l [2].

The following sections are discussed in this paper: In Sect. 2, literature and proposed work are presented; in Sect. 3, the design of 10 GHz antenna concept and results are presented, followed by optimization of effect of coupling gap; in Sect. 4, conventional phase shifter is designed without meta-material and with meta-material; in Sect. 5, results are presented, and finally, in Sect. 6, conclusion is presented.

2 Literature and Proposed Work

"MEMS" is innovation used to make small coordinated gadgets or frameworks to join mechanical and electrical parts. The thought is to expel every existent capacitance of an inductor, as it is in the planar conveyor, and fuse controllable capacitor in shunt with the ideal inductor to create good for nothing disseminated CPW phase shifter utilizing holding wire as an inductor. The resonant state of the holding wire inductor and included controllable capacitors makes extensive variety of working zone to accomplish huge stage moving per unit cell. A spacious range of MEMS devices have substantial size diverged from solid-state devices. The benefits of diminished size are less demanding in fixed bundling, ease and low misfortune. Utilizing this strategy, the quantity of cells is diminished and subsequently the size and misfortunes are diminished.

The simple DMTL phase shifter was produced to limit addition hardship and size, utilizing high tuning RF capacitors. Spring constant because of uniform power connected over the whole bar is given by:

$$k_a = 32Ew\left(\frac{t}{l}\right)^3 \tag{1}$$

where Young's modulus is represented by E; width of the beam is represented by w; thickness of the beam is represented by t, and length of the beam is represented by l. At the point when a voltage is connected between altered settled bar and the draw down terminal, an electrostatic power is affected on bar. Pull down voltage is calculated by:

$$V_p = \sqrt{\frac{8kg_o^3}{27\varepsilon_o Ww}} \quad (2)$$

where k is spring constant, is permittivity; W is center electrode width, and w is beam width. Theoretical calculation of electrical and mechanical parameters to design the MEMS switch has been taken into consideration [3, 4].

3 Design of Antenna

Miniaturized antenna for phase shifter application is designed at 10 GHz. The antenna is made up of a silicon material having high dielectric constant ($\mathcal{E}_r$) which reduces the size of the antenna. The bandwidth performance of the fix reception apparatus can be essentially enhanced by choosing a coupled nourish arrangement and optimizing the measurements to get the maximal change in transfer speed. The capacitive nourishing system comprising of a tiny probe-fed for rectangular patch, which is capacitive coupled with the transmitting element, can be utilized in wide-band application. In order to obtain maximum benefits, the antenna design parameters are given Table 5, for improving gain and decreasing the substrate losses. In this design, micro-machining concept is used to reduce the losses by removing some part of the Si underneath the rectangular patch. This method can viably lessen the viable dielectric constant of the substrate by specifically expelling material and along these lines enhance its bandwidth and radiation effectiveness. The effective dielectric constant, propagation constant, and the characteristic impedance depends upon the patch width and furthermore on the augmentation of the cavity past the microstrip patch [4, 5] (Fig. 1).

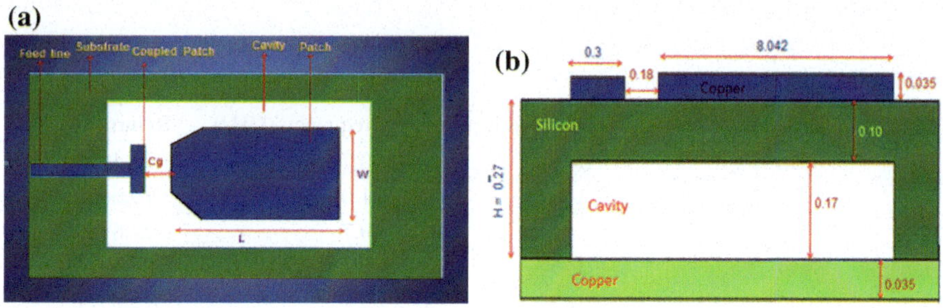

Fig. 1 **a** Top-view and **b** Lateral of the antenna

The straight most strategy for enhancing the transmission capacity of the microstrip is to use a low and thick $\mathcal{E}_r$ (dielectric constant) substrate. Micro-machining can results in notable improvement in bandwidth. The main feature of capacitive bolstering mechanism is a tiny probe-fed for rectangular patch which is capacitive coupled with the emanating component could be utilized in wideband application [6].

(a) **Design Optimization**

The design optimization results were observed that the rectangular patch of the length L and the coupling gap Cg between the rectangular strip and the patch. The input impedance of the antenna as appeared in Tables 1, 2, 3, and 4 is controlled by the key parameters, such as rectangular strip and the patch. Each of these parameters for the most part influences the resistive part of the input impedance, which diminishes with expanding Cg. It can be proficient by decreasing of the sides of the patch. The design is initialized with feed method known as inset, which is matched with 50 Ω characteristic input impedance [5] (Table 5).

Table 1 Coupling gap = 0.1 mm

S. No.	Patch length (mm)	Patch width (mm)	Frequency (GHz)	Return loss (dB)
1	7.146	7.146	9.40	−3.8518
2	7.546	7.146	12.7	−16.9636
3	8.549	7.146	12.50	−16.99
4	8.646	7.146	9.60	−3.8654
5	15	7.146	6.200	−20.4766

Table 2 Coupling gap = 0.2 mm

S. No.	Patch length (mm)	Patch width (mm)	Frequency (GHz)	Return loss (dB)
1	7.146	7.146	12.40	−10.6287
2	7.546	7.146	12.5	−21.6711
3	8.646	7.146	10.80	−13.9865
4	13.646	10	6.600	−6.3824

Table 3 Coupling gap = 0.18 mm

S. No.	Patch length (mm)	Patch width (mm)	Frequency (GHz)	Return loss (dB)
1	8.646	7.146	10.70	−15.4854
2	8.950	7.146	11.00	−19.6463
3	9.464	7.146	9.7	−27.8788
4	9.646	7.146	9.10	−26.9172
5	9.646	7.1855	9.2	18.0571
6	10	7.146	9.00	−15.31

Table 4 Comparison of different coupling gap

S. No.	C_g (mm)	Frequency (GHz)	Return loss (dB)
1	0.1	10.70	−18.588
2	0.18	10.70	−15.8454
3	0.2	10.80	−13.9865

Table 5 Antenna parameters

Geometrical parameters	Dimensions (mm)
Patch	7.2645 × 9.646 × 0.035
Coupled patch length	4
Coupled patch width	0.3
Feed line length	3
Feed line width	0.26
Coupling gap	0.18
Cavity dimensions	8.876 × 10.558 × 0.17
Substrate	15.8055 × 16.126 × 0.27

4 Conventional DMTL Phase Shifter Is Designed Without Meta-material and with Meta-material

First started with the design of conventional phase shifter using two MEMS bridges without using meta-material, substrate material is GaAs of height 275 μm as shown in Fig. 2, when switch is ON $g_0 = 5$ μm and in OFF condition $g_0 = 0.1$ μm given in Table 6.

Simulation Result For Simple Phase Shifter:

Phase angle (ON) = −18.2581
Phase angle (OFF) = 70.8018
Total phase angle = 89.0599

Two MEMS bridges are used in phase shifter design, and these switches are loaded with meta-material. The height of the substrate (GaAs) is 275 μm, when the

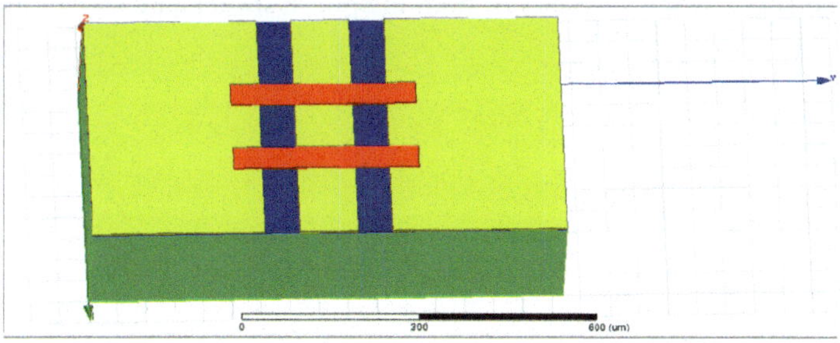

Fig. 2 Simulation design of conventional phase shifter

Table 6 Switch parameters

Geometrical parameters	Dimensions (μm)
Substrate length	500
Substrate width	820
Substrate height	275
CPW (G/S/G)	300/100/300
Bridge length	320
Bridge width	50
Bridge height	0.8
$g_{\circ}$	5
SiN_3	0.1
SiO_2	0.1

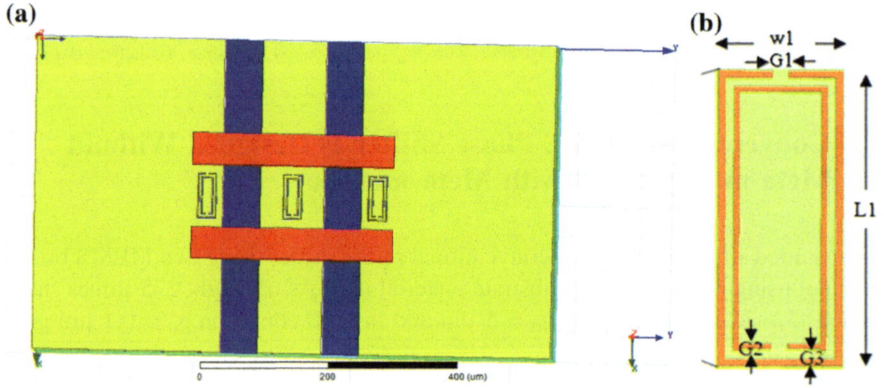

Fig. 3 **a** Simulation using meta-material and **b** CSRR structure

switch is ON $g_0 = 5$ μm and in OFF condition $g_0 = 0.1$ μm. Meta-material dimensions used in this design are $L1 = 70$ μm, $G1 = 8$ μm, $w1 = 30$ μm, $G2 = 2$ μm, and $G3 = 5$ μm shown in Fig. 3.

MEMS switch demonstrate for conservativeness and enhanced isolation have been proposed which are stacked with CSRR for upgrade in isolation qualities of CPW based shunt switch. Results of both the phase shifters are shown in Figs. 6, 7 and 8.

Simulation Result For Simple Phase Shifter:

Phase angle (ON) = −17.7167
Phase angle (OFF) = 73.56.10
Total phase angle = 91.2777

5 Result

(a) **Result of Antenna**

A micro-machined antenna along with capacitive feed operating at 10 GHz has been shown in Fig. 4. Micro-machining has been used to achieve a low $\mathcal{E}_r$ (dielectric constant) locale under the patch. The antenna parameters are obtained using Ansoft HFSS.

The return loss of the proposed antenna is -14.42 dB at the frequency range of 10GHz. The positive gain is 3.057 dB and bandwidth of the S11 is 2000MHz.

(b) **Result of Phase Shifters**
(i) Fig. 5 shows the return loss = −23.7793 dB, phase angle = −18.2581°
(ii) Fig. 6 shows the isolation = −16.7793 dB, phase angle = 70.8018°
(iii) Fig. 7 shows the return loss = −25.4545 dB, phase angle = −17.7167°
(vi) Fig. 8 shows the isolation = −43.6022 dB, phase angle = 73.5610°

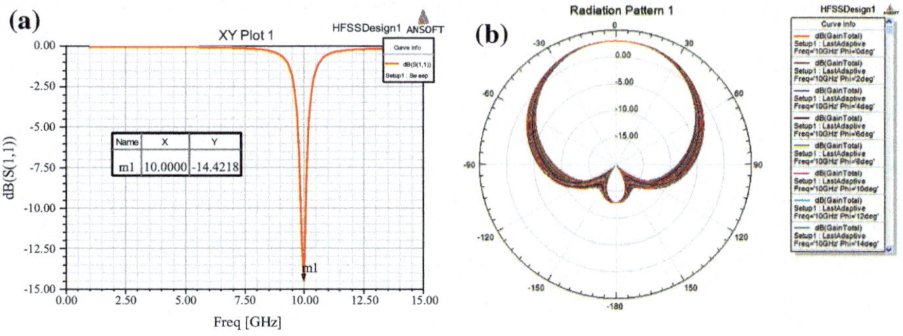

Fig. 4 **a** Return loss of antenna design resonating at 10 GHz and **b** Radiation pattern of antenna design resonating at 10 GHz

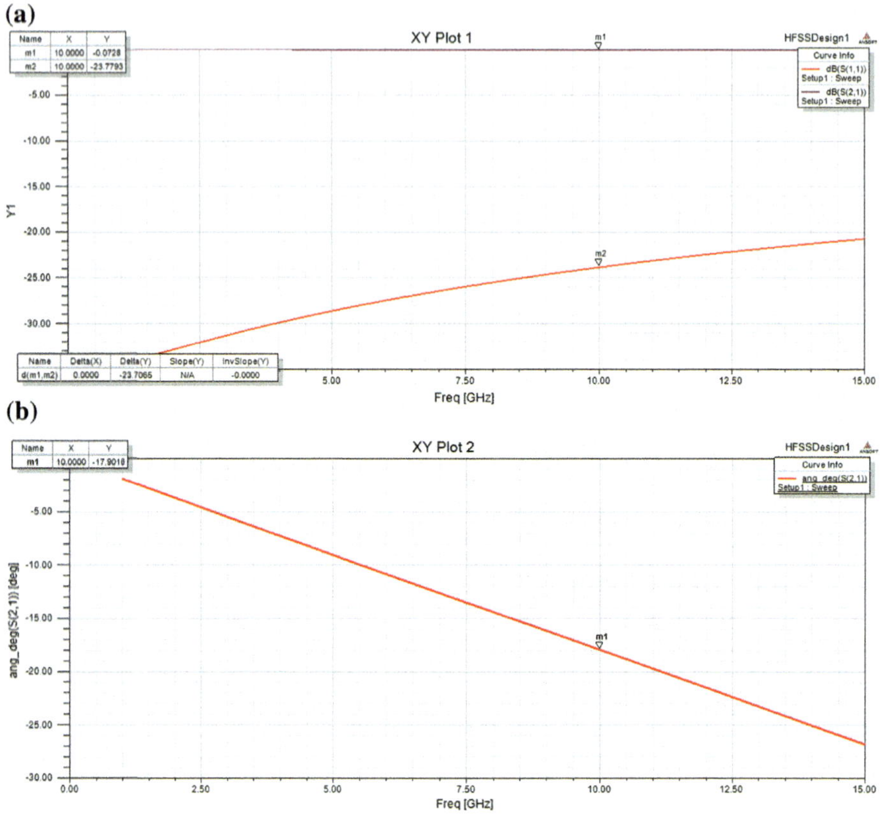

Fig. 5 **a** Return loss and **b** Phase angle of ON state without meta-material

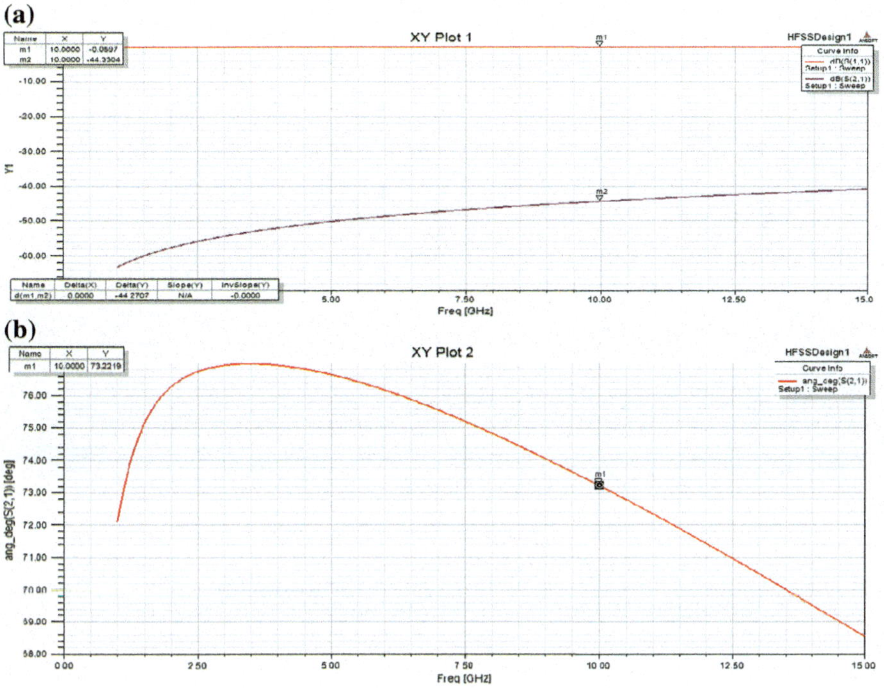

Fig. 6 **a** Isolation and **b** Phase angle of OFF state without meta-material

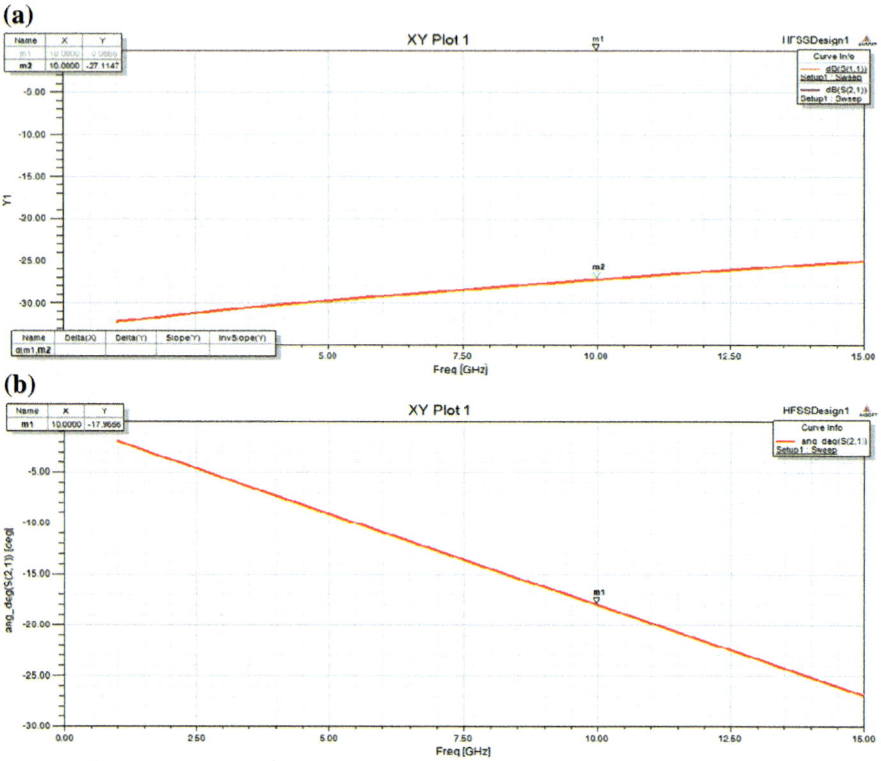

Fig. 7 **a** Return loss and phase angle of ON state with meta-material

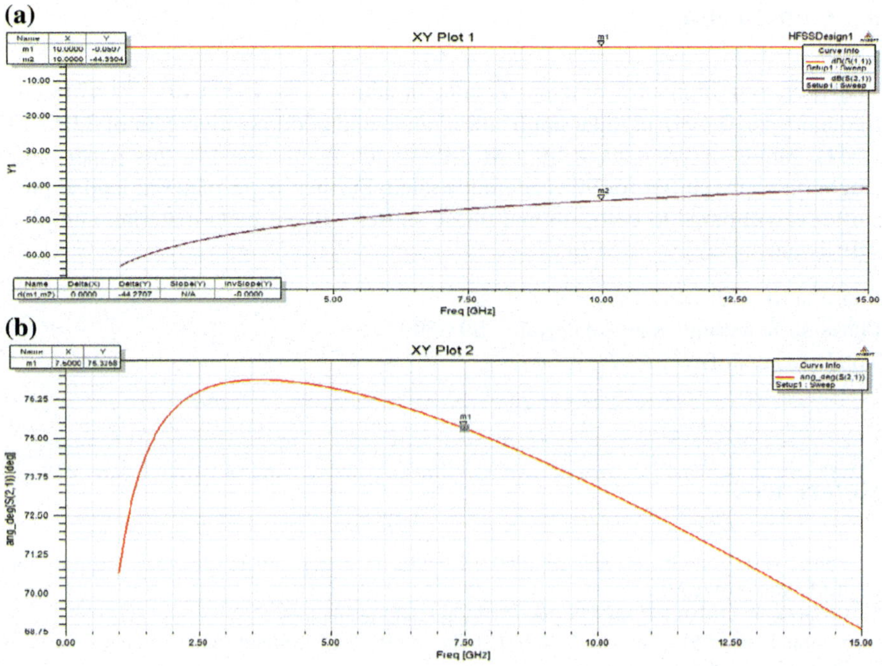

Fig. 8 **a** Isolation and **b** Phase angle of OFF state with meta-material

6 Conclusion

The proposed miniature antenna results in return loss (S11) of −14.4218, and frequency is 10 GHz in radar application range and good positive gain of 3.057 dB. DMTL phase shifter stacked by a MEMS switch is discussed in this paper, and a single-bit phase shifter for 90° is demonstrated. Further, meta-material loaded phase shifter is compared to conventional phase shifter without meta-material, and phase shift is enhanced by +2.2178°.

Phase shift with meta-material = 91.2777°
Phase shift without meta-material = 89.0599°
Total phase shift = 2.217°.

References

1. Rebeiz GM, Tan GL, Hayden JS (2002) RF MEMS phase shifters: design and applications. IEEE Microwave Mag
2. Rebeiz GM (2010) RF MEMS theory, design, and technology
3. Sharma P, Koul SK, Chandra S (2007) Studies on RF MEMS shunt switch. Indian J Pure Appl Phys 45:387–394
4. Jose S, Vinoy KJ (2006) Design and optimization of broadband micromachined antenna. In: 2006 IEEE. 1-4244-0370-7/06
5. Sathanur AV, Vinoy KJ (2005) A design procedure for micromachined antennas on semiconductor substrates. In: 2005 IEEE. 0-7803-9503-4/05
6. Ridgers GM, Odendaal JW, Joubert J (2003) Single-layer capacitive feed for wideband probe-fed microstrip antenna elements. IEEE Trans Antennas Propag 51:1405–1407

A Modified L-Slot Microstrip Antenna with Chamfered Patch Edges for UWB Applications

G. Jemima Nissiyah and M. Ganesh Madhan

Abstract In this paper, an improved rectangular patch antenna with L-shaped slot in the ground plane with rotated patch, fed by a microstrip line, is reported. The rotated patch has been chamfered at the top edges, which leads to an improvement in efficiency. This antenna design provides better return loss characteristics in 3.1–10.6 GHz bands, making it a suitable candidate for UWB systems. The antenna is implemented in FR4 substrate, and it is found that the measured results and the simulated results are in good agreement.

Keywords Microstrip · Patch · Slot antennas · Wide-band antennas
Ultra-wideband (UWB)

1 Introduction

The Ultra-wideband (UWB) antenna design has received considerable attention in recent days due to the widespread use of the UWB systems. As the lower frequency bands are already over-crowded, large amount of data transmission requires higher carrier frequencies. Hence, the ultra-wideband antenna technology, covering the frequency band of 3.1–10.6 GHz, provides an excellent solution for short distance, high data rate wireless transmissions [1]. In the past, several antenna designs have been proposed and developed for UWB and wideband applications. They include vivaldi antenna designs, log-periodic, bowtie, and spiral types. Alternate method to achieve broadband characteristics is by using slot antennas. They are advantageous due to wide bandwidth with low cost, light weight, and ease of fabrication [2]. In order for the UWB band to have a stable radiation pattern, suitable feed and slot

G. Jemima Nissiyah (✉) · M. Ganesh Madhan
Department of Electronics Engineering, MIT Campus, Anna University,
Chennai 600044, India
e-mail: jemigeorge23@gmail.com

M. Ganesh Madhan
e-mail: mganesh@annauniv.edu

G. Gnanagurunathan et al. (eds.), *Optical and Microwave Technologies*,
Lecture Notes in Electrical Engineering 468,
https://doi.org/10.1007/978-981-10-7293-2_4

shapes must be incorporated. By tuning the slot sizes, an optimum impedance bandwidth can be obtained. Impedance matching can be controlled by altering the feed shape or slot shape, thereby changing the coupling property. For an optimum impedance matching, the antenna feed and the slot must be of similar shapes. Using a large slot in the antenna results in a high-electromagnetic coupling to the feed line. To obtain a good impedance matching, coupling between the slot and feed must be increased. However, the impedance matching would deteriorate if the coupling is increased beyond an optimum value. Thus, over coupling would degrade impedance matching [3].

Printed wide-slot antenna operating at a bandwidth of over 120% for wideband applications has been designed (S11 < −10 dB). Stable radiation pattern and low cross-polarization can be obtained by choosing an appropriate combination of feed and slot shapes [4]. In [5], a wideband antenna having a rotated slot, which is fed by a microstrip line, is proposed. By proper selection of rotation angle with respect to the center of the square patch, other resonant modes are created which has similar polarization planes, thereby enhancing the bandwidth of the wide-slot antenna.

Bandwidth enhancement can also be achieved by using Z-like slot shapes. Two or three slots are introduced near the two slots of Z-like slot antenna, to generate additional resonances over the operating frequency bands. These additional resonances can be combined together to create a wideband characteristic [6]. A rhombus-shaped slot was created for the antenna [7]. Along with this, a pair of parasitic strips were created which excites additional resonant modes close to 4 GHz. Operating bandwidth of the antenna was improved by various other slot shapes like L and inverted T shapes, which provide impedance bandwidth of about 60 and 84%, respectively [8].

In this paper, an improvement in return loss is realized by chamfering the edges of the rotated square patch, along with L-shaped slot in the ground plane. The rotated square patch is asymmetrically fed by a microstrip line. The simulations were implemented by using CST Microwave Studio, which is based on finite integral techniques. The entire design is simple, cost-effective and results in a compact form. The results obtained show an improvement in performance over the similar structure reported in the literature.

2 Antenna Design and Simulation

The antenna design is similar to the approach carried out by Song et al. [2], and the same geometry is considered. The geometry of the antenna proposed in [2] is shown in Fig. 1. In this paper, further improvement is made by optimizing the dimensions of the antenna and by chamfering the top edges of the rotated patch with a chamfer width of 0.5 mm. This reduces the return loss well below −20 dB. The bandwidth of the antenna is improved and covers the complete UWB band (2.8–12 GHz). The design of the antenna proposed by Song et al. is shown in Fig. 1, and the modified

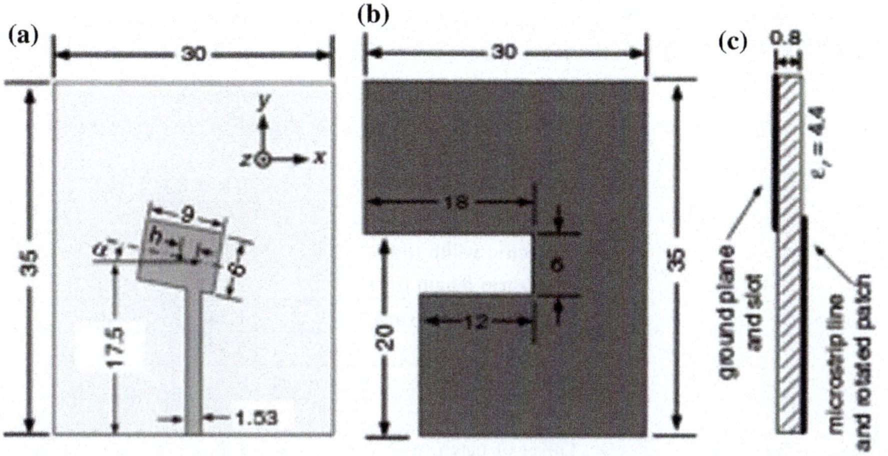

Fig. 1 Geometry and dimensions of L-slot microstrip antenna with rotated patch [2], **a** top view, **b** bottom view, **c** side

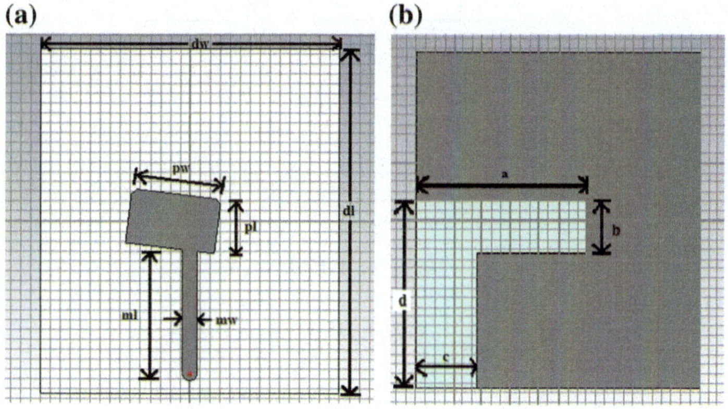

Fig. 2 Design of the proposed antenna, **a** top view, **b** bottom view

antenna is shown in Fig. 2. The optimized dimensions of the proposed antenna are provided in Table 1.

2.1 Design Equations

The design equations for calculating the parameters of the antenna are given by [1]. They are,

Table 1 Dimensions of the modified antenna design

Parameters	Dimensions (mm)
Slot 1 width (*a*)	6.55
Slot 1 length (*b*)	19.5
Slot 2 width (*c*)	18
Slot 2 length (*d*)	5.45
Dielectric height (dh)	0.8
Dielectric length (dl)	35
Dielectric width (dw)	30
Microstrip length (ml)	20.4
Microstrip width (mw)	1.542
Patch length (pl)	5.5
Patch width (pw)	9
Rotation angle (α)	353°
Offset of patch (tr)	1.2
Chamfer width (cha)	0.5

$$W = \frac{1}{2f_r\sqrt{\mu_0\varepsilon_0}}\sqrt{\frac{2}{\varepsilon_r+1}} = \frac{\upsilon_0}{2f_r}\sqrt{\frac{2}{\varepsilon_r+1}} \tag{1}$$

$$W = 30\ \text{mm}$$

$$\varepsilon_{\text{reff}} = \frac{\varepsilon_r+1}{2} + \frac{\varepsilon_r-1}{2}\left[1+12\frac{h}{w}\right]^{-1/2} \tag{2}$$

$$\varepsilon_{\text{reff}} = 4.179$$

$$\frac{\Delta L}{h} = 0.412\frac{(\varepsilon_{\text{reff}}+0.3)\left(\frac{W}{h}+0.264\right)}{(\varepsilon_{\text{reff}}-0.258)\left(\frac{W}{h}+0.8\right)} \tag{3}$$

$$L = \frac{1}{2f_r\sqrt{\varepsilon_{\text{reff}}}\sqrt{\mu_0\varepsilon_0}} = -2\Delta L \tag{4}$$

$$L = 6.56\ \text{mm}$$

The width (*W*), effective dielectric constant (ε_reff), and the length (*L*) of the patch are given by Eqs. (1), (2), and (4). ΔL is a function of the effective dielectric constant (ε_reff) and also the width to height ratio (*W*/*h*). The width, effective dielectric constant, and the length are calculated as 30 mm, 4.179, and 6.56 mm, respectively. The dielectric length and width of the antenna are 35 and 30 mm, respectively.

The antenna designed and fabricated with FR4 substrate is shown in Fig. 3. A rectangular patch (9 × 6 mm), rotated at an angle of $\alpha = 353°$, is fed

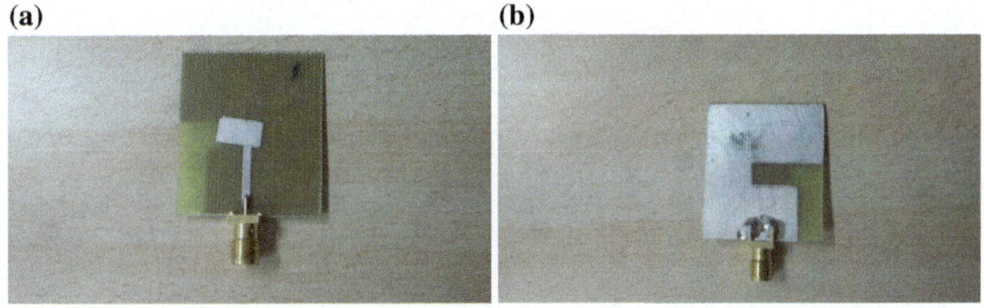

Fig. 3 Fabricated antenna, **a** top view, **b** bottom view

asymmetrically by a 50 Ω microstrip line of width 1.542 mm. The patch is shifted at a distance of 1.2 mm along the $-x$ direction.

2.2 Chamfering and Optimization of Parameters

Chamfering of patch edges improves the bandwidth of the system. Before chamfering, the return loss achieved is comparable to the result obtained in [2], covering the UWB band from 3.6 to 10.6 GHz. Chamfering the edges of the patch improves the bandwidth by covering the bands from 2.8 to 12 GHz.

The value of offset (tr) has been increased from 0.5 to 1.2 mm. This increase in offset value provides a better response at 6–7.4 GHz. The simulated return loss for different values of 'tr' is shown in Fig. 4a. The length of the slot 1 (b) is reduced from 20 to 19.5 mm yielding a better return loss in 2.3–3.2 GHz band. The return loss for different values of the slot 1 length 'b' is shown in Fig. 4b. As the slot 2 length (d) is reduced from 6 to 5.45 mm, the S11 characteristics are found to improve in the frequency range of 4.3–8 GHz. This is shown in Fig. 4c. The 50 Ω microstrip length (ml) and the microstrip width (mw) are increased and decreased, from 17.5 to 20.4 mm, and 1.56–1.542 mm, respectively. The chamfering of patch edges has a strong influence on the slot antenna performance [9]. The return loss of the antenna is compared with [2] and is depicted in Fig. 5a, b.

Comparison of the simulated result with [2] is shown in Fig. 5a, and comparison of the measured result with [2] is shown in Fig. 5b. We infer that there is an improvement in return loss of 10 dB at 4 GHz, 5 dB at 8 GHz, and 3 dB at 10 GHz, respectively. Thus, it is observed that there is a significant improvement in the return loss, over the previous report by Song et al. [2]. By optimizing the values of the patch offset parameter 'tr', the length of the L-shaped slots in the ground plane, the microstrip length and width and by chamfering the patch edges, we obtain a bandwidth of 9.2 GHz. The improvement in return loss leads to an

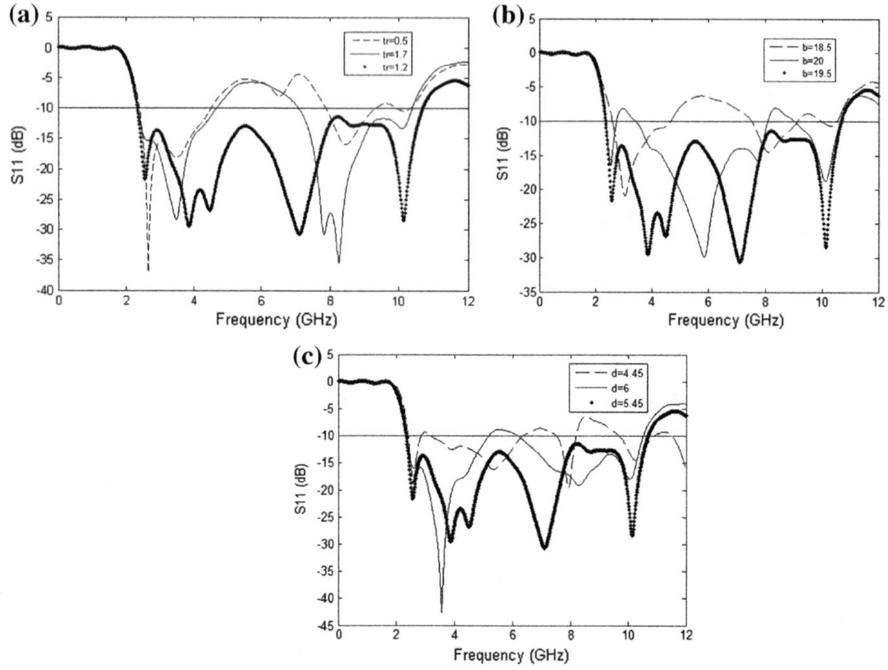

Fig. 4 Simulated return loss for different values of: **a** tr, **b** slot 1 length '*b*', **c** slot 2 length '*d*'

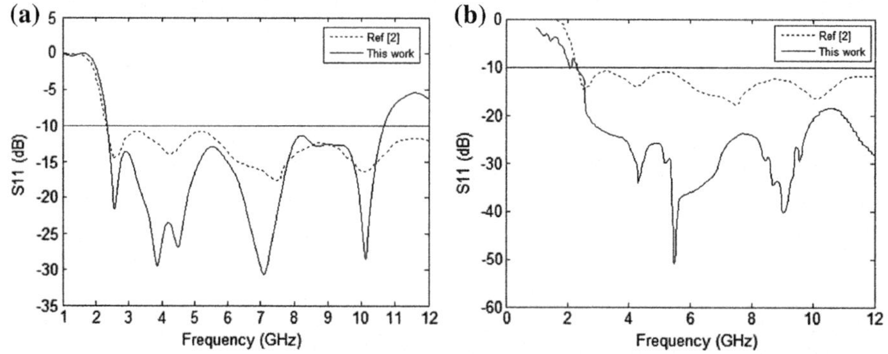

Fig. 5 Comparison of return loss, **a** simulated output with [2], **b** measured output with [2]

improved efficiency of 70–94% over the frequency band of 3.1–10.6 GHz and a gain of 2.775 dB at 8 GHz.

The current distributions for the antenna are observed at three frequencies 3.1, 8, and 10.6 GHz and are illustrated in Fig. 6. The current is distributed at the feed

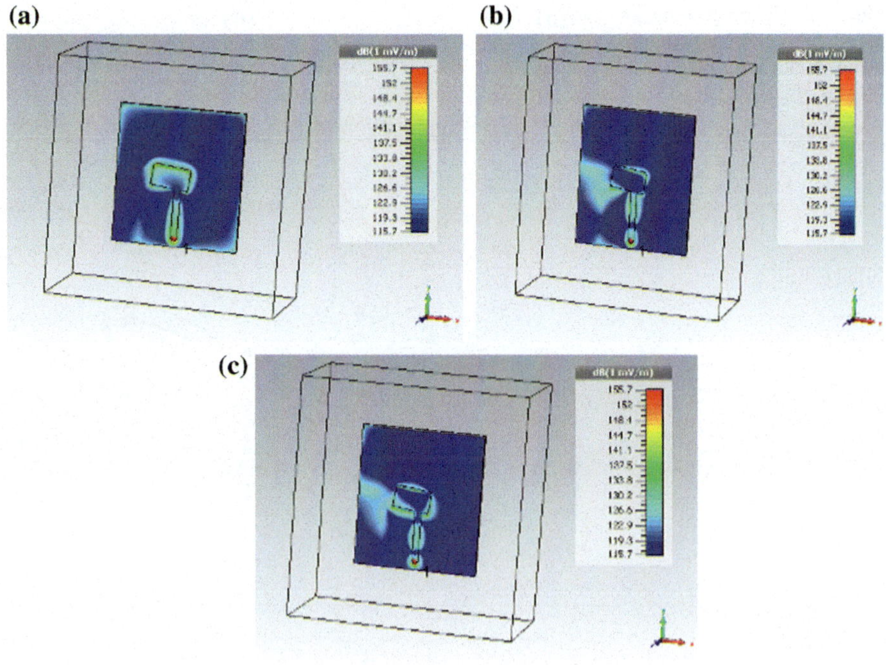

Fig. 6 Current distributions at **a** 3.1 GHz, **b** 8 GHz, **c** 10.6 GHz

location and the top region of the patch and indicates a better radiation. The 2D radiation patterns measured at 3.1, 8, and 10 GHz, in azimuth and elevation planes, are shown in Fig. 7. It is seen that the antenna shows an omnidirectional radiation pattern on both the azimuth and elevation planes. The radiation pattern in 3D view, with azimuth and elevation plane, is shown in Fig. 8. From the 3D radiation pattern, it is observed that at 8 GHz, the antenna exhibits a gain of 2.758 dB. There is an improvement in gain of the antenna in this design, when compared with literature [1], which is 1.32 dB. The main radiation pattern exists along the positive z axis.

The VSWR of the antenna is shown in Fig. 9. In this plot, it is seen that the measured and simulated outputs are close to each other. The VSWR lies between 1.2 and 1.5 in the entire frequency range of 2.8–12 GHz, which covers the complete UWB band. The overall area of the antenna is 35×30 mm^2. This antenna provides a better bandwidth of 9.2 GHz than those provided by [1, 3, 4] which are 8.28, 6.06, and 8.6 GHz, respectively. As the return loss of the antenna is improved, the antenna shows a good efficiency of 94% at 3.5 GHz and 70% at 10.6 GHz, better than [1] which is 54.7% at 3.5 GHz. The efficiency of the antenna is depicted in Fig. 10.

Fig. 7 Radiation patterns at 3.1, 8, 10.6 GHz, **a** azimuth plane, **b** elevation plane

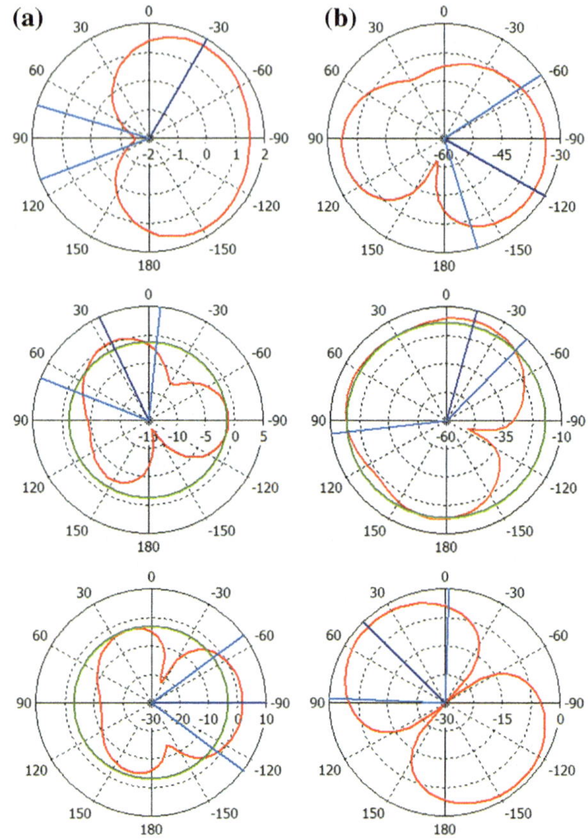

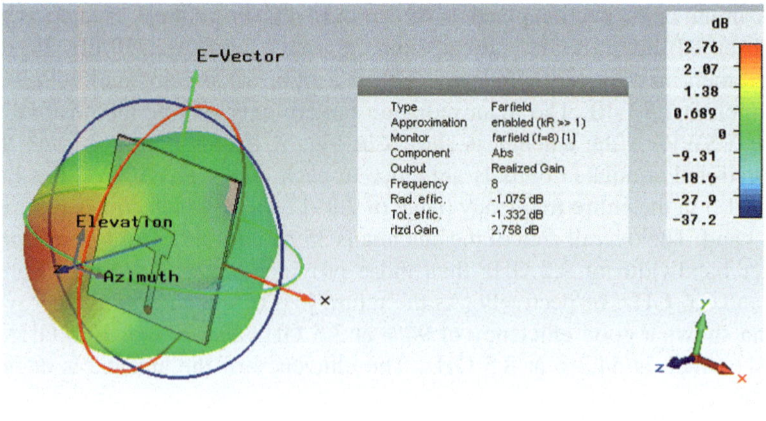

Fig. 8 Radiation pattern of the antenna in 3D

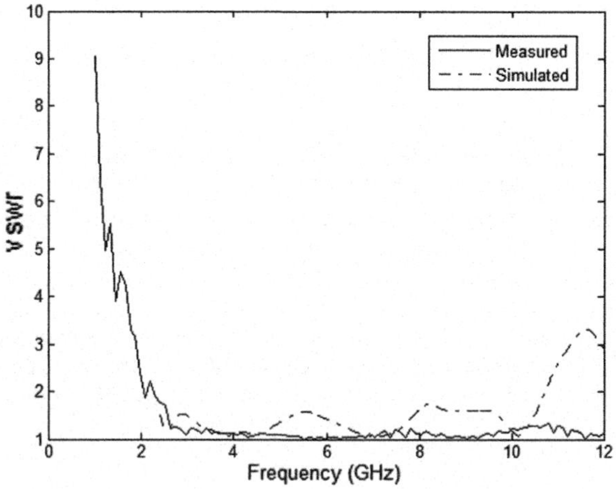

Fig. 9 VSWR of the antenna

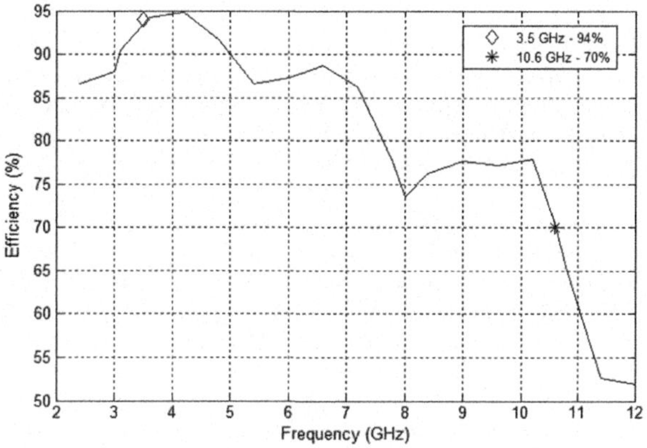

Fig. 10 Efficiency of the antenna

3 Conclusion

A simple modification in the UWB antenna design reported earlier is proposed. The basic structure was taken, and additional changes in the design dimensions and chamfering of the patch edges were done to obtain better results compared to the existing literature. The fabricated antenna was measured using network analyzer, and the variation from the simulated values might be due to connector, fabrication, or soldering effects. The antenna provides a good return loss less than −10 dB and it covers the entire UWB band (2.8–12 GHz), thus making it useful for Ultra-wideband systems. A minimum efficiency of 70% is observed at 10.6 GHz, whereas in the other frequency bands, efficiency is better than 70% and even reaches a maximum of 95% at 4 GHz. In the case of Ref. [1], the maximum efficiency is only 54.7%. Further, a bandwidth of 9.2 GHz is achieved in this design, which is better in comparison with [1] which provides a bandwidth of

8.28 GHz. The bandwidth obtained was also found to be better than [3, 4], which are 6.06 and 8.6 GHz, respectively. A gain of 2.758 dB has been achieved at 8 GHz. The measured results were found to be consistent with the simulated results. The UWB antenna finds application in short range, high-speed wireless communication systems, vehicle radar system, sensing/monitoring systems. Communication can be established in indoor mobile wireless connection, between laptops and devices such as printers and digital cameras.

Acknowledgements The authors would like to thank 'Visvesvaraya Ph.D. Scheme for Electronics and IT'—DeitY, for providing the financial support to carry out this research work. The authors would also like to thank Mr. Ganesh Balasubramanian, Amphenol Omniconnect India Pvt. Ltd. for providing measurement setup and Mrs. Kamalaveni Ganesh, Research Scholar, and Mr. J. Logeshwaran, Teaching Fellow, MIT, for support throughout this work.

References

1. Padmavathy AP, Ganesh Madhan M (2015) An improved UWB patch antenna design using multiple notches and finite ground plane. J Microw Optoelectron Electromagn Appl 14(1)
2. Song K, Yin YZ, Fan ST, Wand, Zhang L (2009) Open L slot antenna with rotated rectangular patch for bandwidth enhancement. Electron Lett 45(25)
3. Liu YF, Lau KL, Xue Q, Chan CH (2004) Experimental studies of printed wide slot antenna for wide-band applications. IEEE Antennas Wirel Propag Lett 3:273–275
4. Dastranj A, Imani A, Moghaddasi M-N (2008) Printed wideslot antenna for wideband applications. IEEE Trans Antennas Propag 56(10):3097–3102
5. Foster I, Kesselman C, Nick J, Tuecke S (2002) The physiology of the grid: an open grid services architecture for distributed systems integration. Technical report, Global Grid Forum
6. Wang CJ, Chang SW (2008) A technique of bandwidth enhancement for the slot antenna. IEEE Trans Antennas Propag 56(10):3321–3324
7. Jan JY, Wang LC (2009) Printed wideband rhombus slot antenna with a pair of parasitic strips for multiband applications. IEEE Trans Antennas Propag 57(4):1267–1270
8. Latif SI, Shafai L, Sharma SK (2005) Bandwidth enhancement and size reduction of microstrip slot antennas. IEEE Trans Antennas Propag 53(3):994–1003
9. Noghabaei SM et al (2013) A dual-band circularly-polarized patch antenna with a novel asymmetric slot for WiMax Application. Radioengineering 22(1)

Performance Analysis of CSRZ-DQPSK Modulator for RoF-PON-Based Wireless Access

C. Rimmya, M. Ganesh Madhan and S. Arshujabin

Abstract Conventional schemes use three Mach–Zehnder modulators in the transmitter for the generation of CSRZ-DQPSK signal in passive optical networks. It also simultaneously provides wireline connectivity till ONU and wireless access from ONU, without the need of a separate RF source at the ONU. In this paper, a design of CSRZ-DQPSK transmitter with reduced complexity using two Mach–Zehnder modulators (MZM) and performance analysis of the wireless access from the ONU is analyzed. For wireless transmission of high data rate, a 20 GHz RF carrier signal is generated at the ONU from the received 20 Gbps CSRZ-DQPSK signal. Further, QAM modulation technique is used for 5 Gbps wireless data transmission from the recovered 20 GHz carrier signal. System simulation of the entire fiber link is carried out in Optisystem, and the performance of 2-MZM-based transmitter is analyzed for the wireless transmission from the ONU to the user end, using the eye diagram and received power. Results indicate that 2-MZM-scheme achieves similar performance compared to conventional 3-MZM scheme.

Keywords Mach–zehnder · Modulator (MZM) · Carrier-suppressed return-to-zero (CSRZ) · Differential quadrature phase shift keying (DQPSK) · Quadrature amplitude modulation (QAM) · Radio over Fiber (RoF)

1 Introduction

Recently, broadband signals have been transported through passive optical network (PON) in a cost-effective manner. In a PON, optical line terminal (OLT) transmits the signal from the central office, which is sent through the fiber to optical network unit (ONU) after splitter. The ONU is normally kept near the end-user equipment. As a demand for wireless access and longer reach increases, Radio over Fiber

C. Rimmya (✉) · M. Ganesh Madhan · S. Arshujabin
Department of Electronics Engineering, MIT Campus, Anna University, Chennai 600044, India
e-mail: rimmyac@gmail.com; rimmya@annauniv.edu

G. Gnanagurunathan et al. (eds.), *Optical and Microwave Technologies*, Lecture Notes in Electrical Engineering 468,
https://doi.org/10.1007/978-981-10-7293-2_5

(RoF) becomes a viable solution. RoF can be realized by placing antenna ports at the ONUs [1]. RoF is advantageous for radio access networks as it requires simple antenna, low cost and easy deployment. Wireless and wireline access can be achieved simultaneously without additional radio system in the optical network unit [2].

On-Off Keying (OOK) modulation is normally preferred in optical networks as it is simple. But at high data rates, OOK performance degrades significantly. Bandwidth and spectral efficiency are also not optimum in OOK. Beyond 20 Gbps, material and polarization mode dispersion are detrimental for data transmission [3]. These factors limit the transmission distance in OOK system. Differential quadrature phase shift keying (DQPSK) provides a critical advantage of 3 dB improvement in the optical signal-to-noise ratio and spectral efficiency when compared to OOK [4].

In [5] a single MZM-based RZ/CSRZ-DPSK (return-to-zero/carrier-suppressed return-to-zero—differential phase shift keying), scheme using standard digital integrated circuits is reported. It also requires a clock pulse along with XOR gate to generate the pulse which is modulated in a single dual-drive MZM. However, this structure can be used for low data rate only.

Future networks require 60 GHz carrier for the transmission of high data rates. Though it covers a short distance, the cost of implementing the antenna and generating high carrier frequency is expected to be very high. Generating carrier frequencies from the received signal itself will be cost-effective for reaching the end-user. The method of recovering the carrier signal using precise filtering is also available in the literature [1].

By making use of quadrature amplitude modulation (QAM) modulation technique along with Radio over Fiber, a high data rate wireless access can be provided at remote user side from the ONU. This work focuses on the generation of the 20 GHz RF signal under 2- and 3-MZM-based CSRZ-DQPSK transmitters and the wireless transmission from the ONU. The performance of QAM modulation for wireless transmission along with CSRZ-DQPSK link from OLT to ONU is investigated using Optisystem tool-based simulations.

The outline of the paper is as follows. Section 2 describes the overall system simulation blocks, and in Sect. 3, the results are discussed along with a comparison of conventional and proposed schemes. Section 4 concludes the work.

2 System Model

The transmitter of optical line terminal unit is shown in Fig. 1. A 20 Gbps data is generated using PRBS and data is DQPSK precoded to provide 10 Gbps I and Q components. Optical signal at a wavelength of 193.1 THz is phase modulated with I component of precoded data using first MZM. The second MZM completes the remaining process of generating CSRZ-DQPSK signal

The mechanism of CSRZ pulse carver which modifies the shape of Q channel signal is shown in Fig. 2. An XOR logic gate performs XOR operation of Q

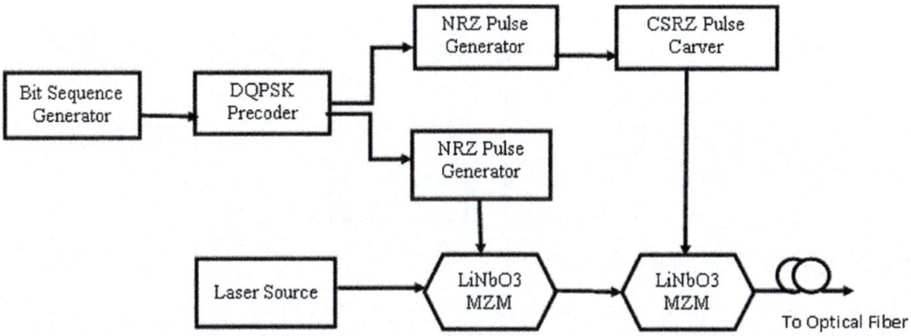

Fig. 1 Block diagram of 2-MZM-based CSRZ-DQPSK transmitter at optical line terminal unit

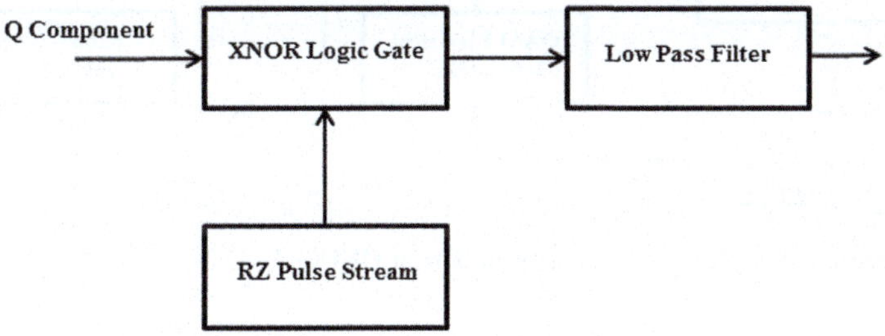

Fig. 2 CSRZ pulse-carver block

component and half of the bit rate of RZ pulse stream with duty cycle of 0.5. A sinusoidal-carved signal is generated by allowing the output of XOR gate to pass through a low-pass filter [5].

The CSRZ-DQPSK-modulated optical signal is transmitted via a single-mode fiber. At an optical network unit, the signal is split into two branches. The 20 Gbps data is decoded from the one of the branches and the other branch is used for the RF carrier extraction. The electrical signal is passed through a bandpass Bessel filter with a center frequency of 20 GHz and then amplified to produce RF carrier. Figure 3 illustrates the block diagram of the receiver at ONU.

A 5 Gbps data is QAM-modulated using the extracted 20 GHz RF carrier as shown in Fig. 4. The downstream wireless data is transmitted over the free space. The propagation loss resulting from free space is calculated as from the standard equation,

$$\mathrm{FSL} = 32.44 + 20\log f(\mathrm{MHz}) + 20\log d(\mathrm{km}) \tag{1}$$

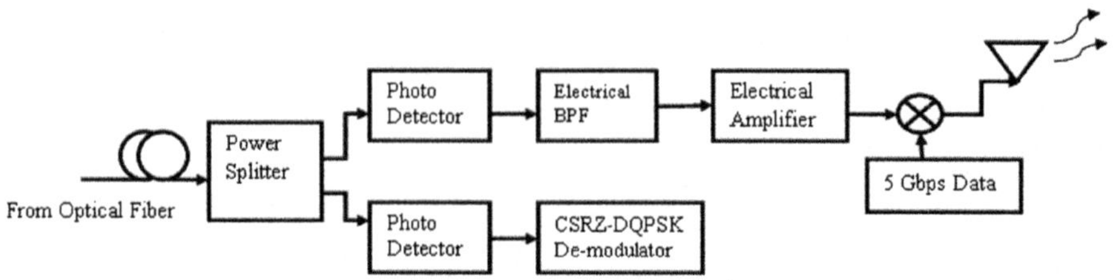

Fig. 3 Block diagram of receiver at ONU

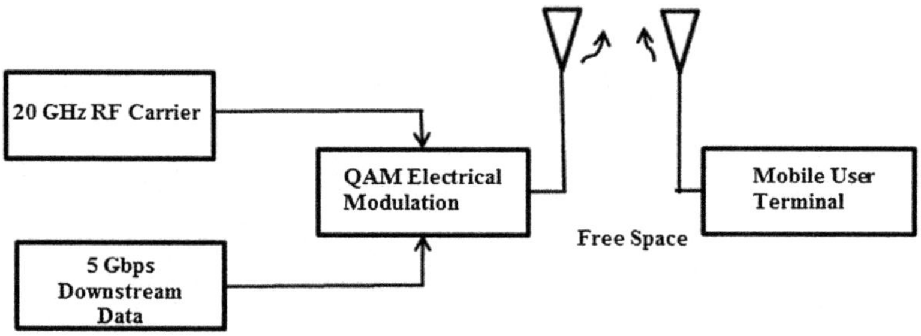

Fig. 4 Electrical modulation of downstream data at ONU

3 Results and Discussions

The design is implemented in Optisystem tool for simulation. The parameters used for the link analysis is listed in Table 1.

The CSRZ-DQPSK transmitter is simulated for 20 Gbps transmission in the OLT unit, and the ONU provides both the received data as well as 20 GHz RF carrier. For simulating the wireless transmission from the ONU to the user terminal,

Table 1 Simulation parameters

S. No.	Parameter	Value
1	Laser linewidth	1 MHz
2	Operating wavelength	1552.4 nm/193.1 THz
3	Laser source power	0 dBm
4	Attenuation in fiber	0.25 dB/km
5	Dispersion in fiber	17 ps/nm/km
6	Pulse-carving signal	Bit rate/4
7	PIN responsivity	1 A/W
8	PIN dark current	10 nW
9	Downstream data rate	5 Gbps
10	Baseband modulation	QAM
11	Bits per symbol	6

binary bits are generated using pseudo random bit stream generator. Six bits are combined to form a symbol in the QAM sequence generator. The generated sequence is pulse shaped using M-ary pulse generator. For wireless transmission, the extracted 20 GHz RF carrier is QAM-modulated.

Direct optical to electrical conversion using PIN photo diode followed by electrical filter, an amplifier and QAM modulator finishes RF signal transmission at the ONU. In Optisystem tool, RF spectrum analyzer indicates the modulated radio signal spectrum, and the constellation visualizer helps to identify the performance of 64 QAM. The wireless channel propagation loss is calculated for 100 m distance and implemented in Optisystem as an equivalent electrical attenuator and is shown in Fig. 5.

Figure 6 shows the schematic of the wireless receiver structure. The extracted 20 GHz RF carrier signal at ONU for wireless downstream data transmission is shown in Fig. 7.

A small variation in the amplitude with time is observed in both the cases. However, the 2-MZM-basedtransmitter configuration needs a more narrow filter than the 3-MZM-based transmitter configuration. The corresponding frequency spectrum is shown in Fig. 8, where the amplitude of 20 GHz signal is similar in both the cases, i.e., −50 and −52 dBm for 3- and 2-MZM-based transmitters, respectively.

The filtered 20 GHz carrier signal is used for QAM modulation. The modulation of downstream data in both amplitude and phase can be observed form Fig. 9a and the corresponding downstream frequency spectrum at ONU is shown in Fig. 9b.

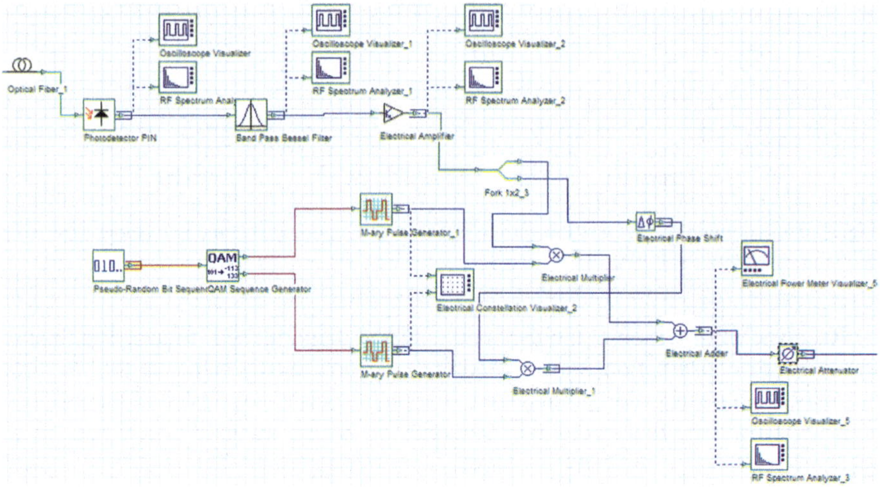

Fig. 5 Schematic of ONU—20 GHz RF carrier signal generation and QAM modulation

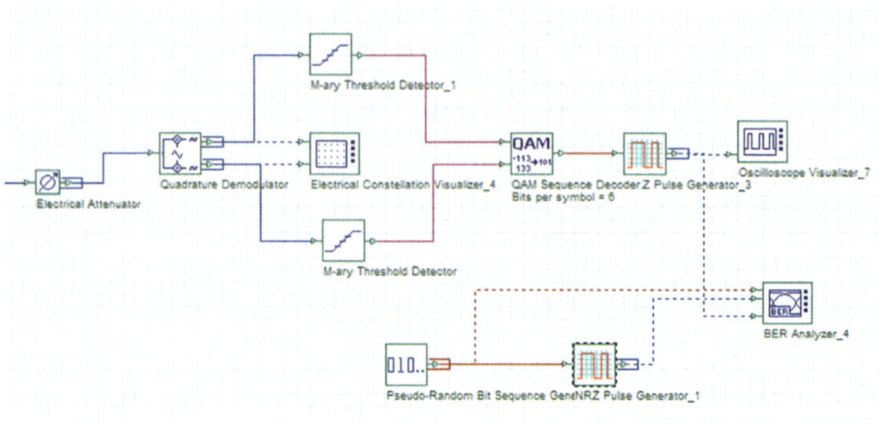

Fig. 6 Schematic of QAM demodulation at the mobile-user terminal

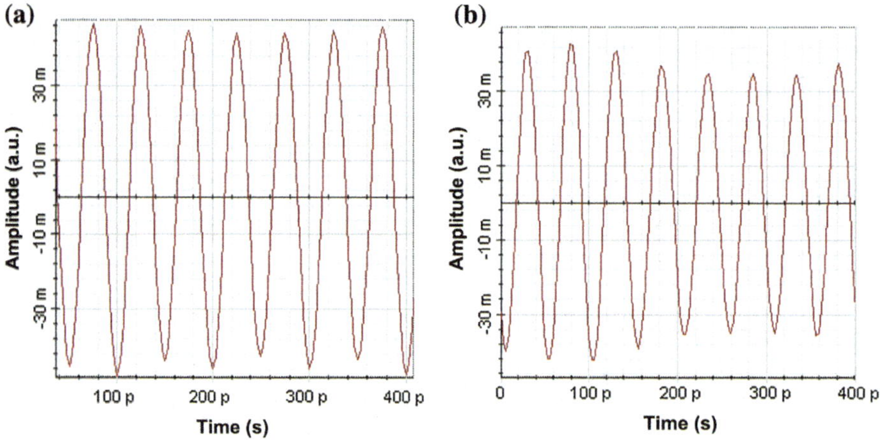

Fig. 7 Time domain of 20 GHz RF carrier signal: **a** conventional scheme with three MZMs and **b** proposed scheme with two MZMs

Received information at the user end is QAM demodulated. Figure 10a, b show the received constellation for conventional and proposed scheme at the user end. A rotation of constellation points due to the phase noise is observed in the proposed scheme. However, this could be overcome by using standard algorithms in the receiver [1].

The sensitivity of the system is determined by evaluating the eye diagram. Figure 11 indicates the eye-opening for 100 m wireless transmission using 2-MZM case. A comparison with the conventional scheme shows similar eye-opening for the same operating conditions.

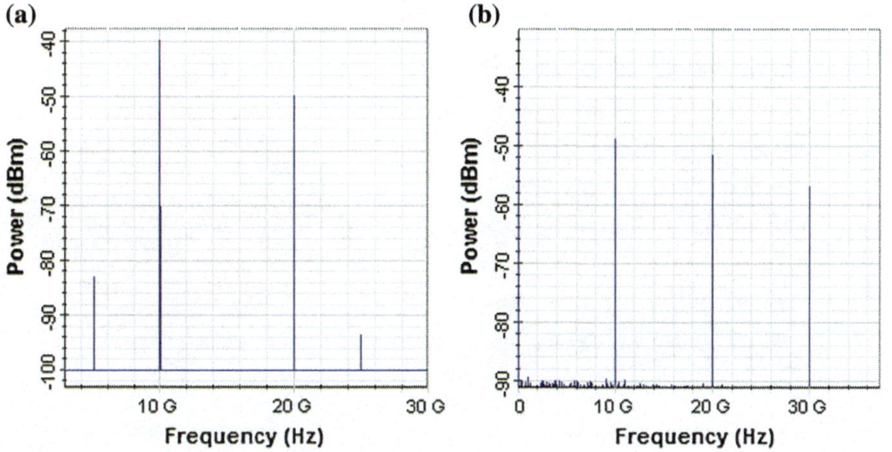

Fig. 8 Spectrum of recovered 20 GHz RF carrier signal: **a** conventional scheme with three MZMs and **b** Proposed scheme with two MZMs

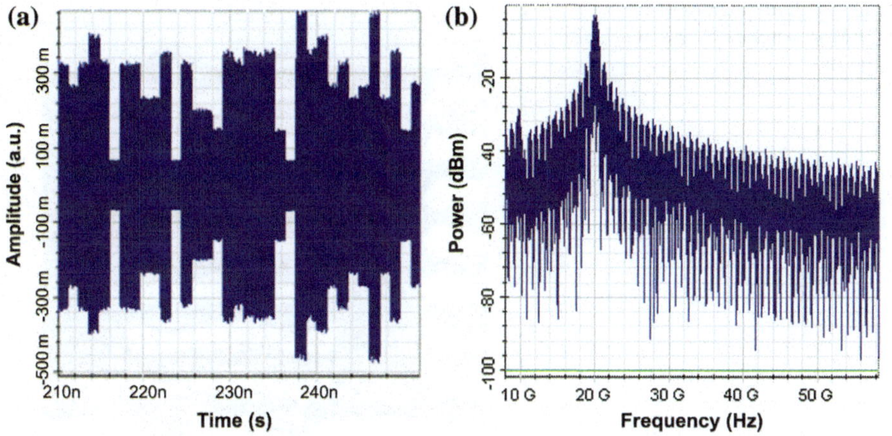

Fig. 9 QAM-modulated downstream data: **a** time-domain signal and **b** frequency spectrum

The user's wireless receiver decodes the received QAM signal provided by the ONU. In Fig. 12, the impact of free-space path loss over received power at user terminal is observed.

From the above graph, it can be observed that the wireless access in both 2-MZM-based and 3-MZM-based transmitters gives similar performance. The received power for the conventional scheme after free-space path loss is −62.5 dBm, where as for proposed scheme, the received power is observed as −63.8 dBm for a distance of 10 m.

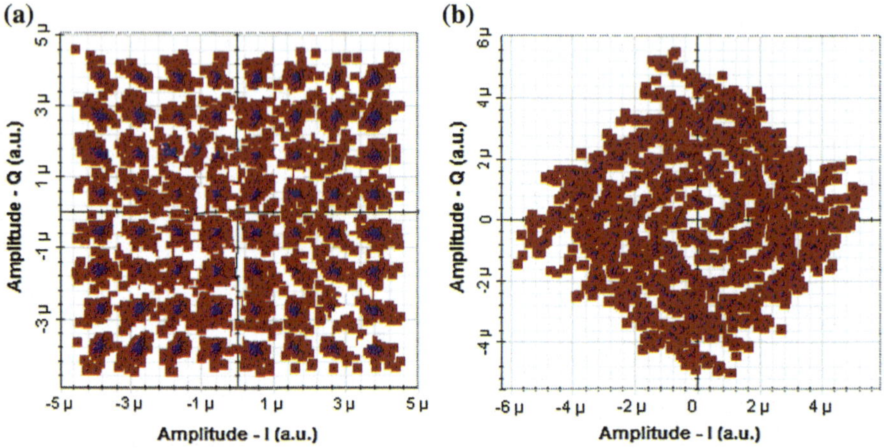

Fig. 10 Received constellation diagram of 64 QAM in wireless transmission at 100 m from ONU: **a** Conventional scheme with three MZMs and **b** proposed scheme with two MZMs

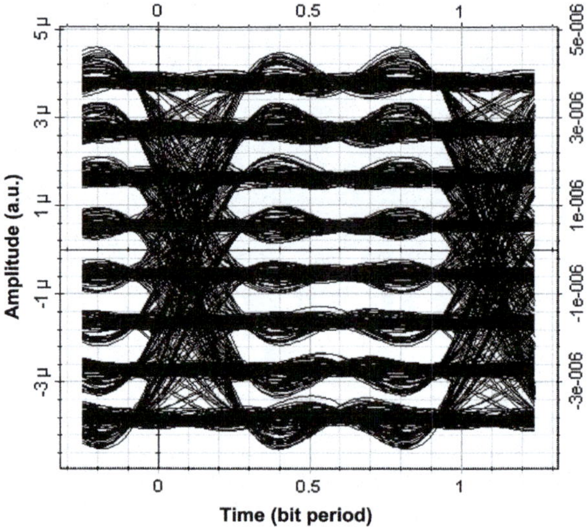

Fig. 11 Eye diagram of 64 QAM in wireless access at 100 m

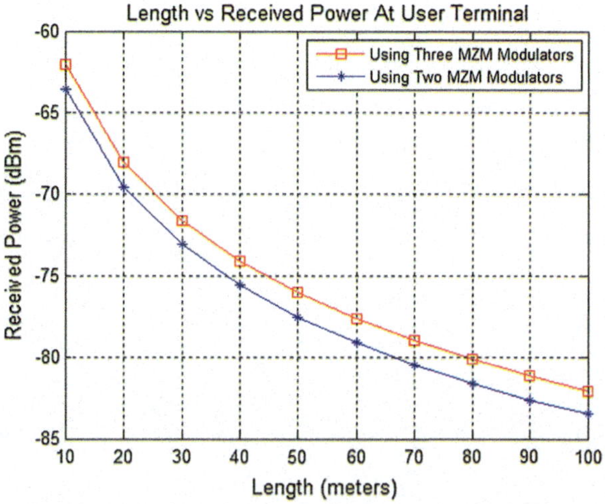

Fig. 12 Free-space path loss impact on received signal strength

4 Conclusion

We have demonstrated a 20 GHz PON system for wireless transmission based on CSRZ-QDPSK modulation at OLT. The performance of the designed system is evaluated in terms of received constellation diagram, eye diagram and received power at end-user terminal for both 3-MZM and 2-MZM cases. The 2-MZM-based CSRZ-DQPSK transmitter shows similar performance as that of the 3-MZM-based CSRZ-DQPSK transmitter.

References

1. AI-Raweshidy H, Komaki S (2002) Radio over fiber technologies for mobile communication network. Artech House Publishers, Norwood
2. Ji W, Chang J (2013) IEEE Trans Opt Commun Netw 5
3. Agarwal GP (2010) Fiber optic communication systems. Wiley, New York
4. Ho K-P (2005) Phase-modulated optical communication systems. Springer series in Optics and Laser, New York
5. Dong Y, He H, Su Y, Hu W (2006) RZ/CSRZ-DPSK signal generation using only one Mach-Zehnder modulator. In: Proceedings of the optical fiber communication conference, Mar 5, Anaheim, California United States

Effects of Cross-phase Modulation and Four-Wave Mixing in DWDM Optical Systems Using RZ and NRZ Signal

V. Sasikala and K. Chitra

Abstract Nonlinear effects play a major role in optical fibers and are occurred due to the changes in refractive index of the optical medium with respect to the intensity of light propagation. Four-wave mixing (FWM) and cross-phase modulation (XPM) are the major nonlinear effects that affect the performance of the dense wavelength division multiplexing (DWDM) system. In FWM, three signals are combined to form a fourth spurious mixing component results in waste of power from wanted signals to unwanted spurious signals which lead to cross talk and degrade the system performance. Nonlinear phase changing due to power variations in co-propagating light is called XPM which causes pulse distortion and affects the system transmission capacity. This paper shows the effect of XPM and FWM in return-to-zero (RZ) and non-return-to-zero (NRZ) signals for various ranges of dispersion. The effect has been analyzed based on eye diagrams, Q values, BER values, and FWM power for two, four, and eight channels at 10 Gbps in DWDM systems.

Keywords Four-wave mixing · Cross-phase modulation · DWDM
Dispersion · Nonlinearity

1 Introduction

Dense wavelength division multiplexing (DWDM), is a key technology, enables a very large-capacity photonic networks by multiplexing number of optical carrier signals on to a single fiber in the wavelength range of 1550 nm. It leverages the use

V. Sasikala (✉)
Department of Electronics and Communication Engineering,
Sri Sairam Engineering College, Chennai, India
e-mail: sasikalajp@gmail.com

V. Sasikala · K. Chitra
School of Electronics Engineering, VIT University, Vandalur Kelambakkam Road,
Chennai 600127, India
e-mail: chitra.krishnan@vit.ac.in

G. Gnanagurunathan et al. (eds.), *Optical and Microwave Technologies*,
Lecture Notes in Electrical Engineering 468,
https://doi.org/10.1007/978-981-10-7293-2_6

of erbium-doped fiber amplifiers (EDFA) and reduces the cost of the system [1]. The advancement in the optical communications has been progressed in part by the betterment of efficient and powerful optical amplifiers such that the need for costly conversions from optical to electrical signal and vice versa is eliminated [2]. A favorable candidate among the all optical amplifiers is the semiconductor optical amplifier (SOA) having the practical advantages of very strong nonlinearity, short latency, high stability, strong compactness, and low power consumption. In general, the nonlinear design based on SOA is classified into two categories. The nonlinear properties of SOA such as cross-phase modulation (XPM), cross-polarization modulation (CPM), cross-gain modulation (XGM), and four-wave mixing (FWM) are in the first category, whereas the designs consist of SOA-based fiber Sagnac interferometer, ultra-fast nonlinear interferometer (UNI), SOA-based Mach–Zehnder interferometers (SOA-MZIs), and an SOA-based Michelson interferometer (SOA-MI) are in the second category. The limitation of the formal bulk SOA is the operation speed due to the slow secular response of gain and phase recovery. On the other hand, the gain and phase recovery response are significantly faster in Quantum dot SOA (QD-SOA) compared to Quantum well (QW) SOA, which provide high-speed performance in logic systems based on QD-SOA [3].

Nonlinear optical effects are classified into two sets of effects, one resulting from the light propagation in a single channel and the other caused due to interactions between numbers of channels in DWDM system. Single-channel nonlinear effects are evidenced mainly through self-phase modulation (SPM), whereby each signal propagates in a channel modulates its own phase. DWDM nonlinear effects are classified into cross-phase modulation (XPM) and four-wave mixing (FWM). In XPM, the power of the co-propagating channels modifies the phase of each channel. Three channels interacted together to transfer a fraction of their energy to a fourth one is called FWM. We realize that the DWDM systems, which have already become entirely common all over the world; therefore, the reason and complete understanding of nonlinear effects is needful for DWDM systems.

In this paper, we have simulated the effect of FWM products and XPM effect in DWDM environment by switching the dispersion level from −4 to 10 ps/nm/km for two, four, and eight channels. Introduction to DWDM systems and nonlinear effects is presented in Sect. 1. Sections 2 and 3 discuss the theory of FWM and XPM. Section 4 discusses the transmitter section for return-to-zero (RZ) and non-return-to-zero (NRZ). Section 5 discusses the simulation setup for XPM and FWM. The simulation output representing the effects of FWM and XPM is presented in Sect. 6, and finally, the conclusion and future outline are made in Sect. 7.

2 Four-Wave Mixing

The origination of FWM process resides in the nonlinear response of bound electrons of a material to the given optical field.

The linearity property of the optical signal remains as long as the power of the launched optical signal is low. When a high-power optical signal is launched into an optical fiber, the refractive index of the medium depends on the intensity of light in the medium, and hence, it lost the property of linearity. The FWM process emerges from third-order nonlinear susceptibility $\chi_{(3)}$ called as optical Kerr effect [4] and occurs when more than one light signal of different wavelengths are launched into an optical fiber. If any three optical fields with carrier frequencies ω_1, ω_2, and ω_3 propagate simultaneously inside the fiber, then the third-order nonlinear susceptibility $\chi_{(3)}$ develops a fourth optical field with a new frequency ω_4 as shown in Fig. 1 and is related to other original frequencies by a relation $\omega_4 = \omega_1 \pm \omega_2 \pm \omega_3$.

The performance of the system is degraded due to these nonlinear effects by introducing interference and excess attenuation in the optical signals. The nonlinear effects tend apparently visible themselves when the input optical power is very high, and thereby, understanding its effect is gaining importance in DWDM systems.

3 Cross-phase Modulation

Fiber nonlinearities caused by optical Kerr effect are a restricting factor in fiber optic communication systems. In DWDM systems, the most influencing effect is cross-phase modulation (XPM). The nonlinearity in phase changes with respect to power variations in neighboring channels can strongly influence the performances of the system [5]. The cross-phase modulation effects are pointed below:

- The optical intensity measurement on one signal can be possible by monitoring the phase variations in the co-propagating signal.
- Synchronization of two mode-locked lasers using the medium with same gain experiences pulse overlap and cross-phase modulation.
- Cross talk due to interchannel effects degrades the system performance in optical fiber communications.

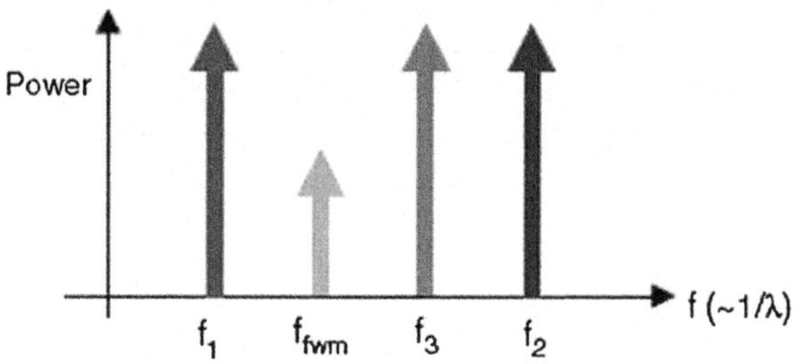

Fig. 1 Three adjacent optical frequencies ω_1, ω_2, and ω_3 interact to produce a fourth frequency ω_4 as frequency of four-wave mixing

4 Transmitter Section

Optical modulation format is the method used to modulate the optical carrier with respect to the logical information provided as input to the system. The simplest modulation technique is on-off keying (OOK). It is in two forms namely return-to-zero (RZ) and non-return-to-zero (NRZ). The NRZ signals are easy to generate, and its signal bandwidth is 50% smaller compared to RZ signals. The reasonable cost of NRZ systems makes it as most beneficial for WDM systems. The RZ signals require wider bandwidth, and it is more susceptible to intersymbol interference (ISI). The RZ signals are self-synchronized and achieve better performance compared to NRZ-based systems. RZ signals are used where the cost is not the driving factor and performance is the major goal.

The transmitter section to generate FWM and XPM effects consists of modules of optical source, logical signal generator, electrical modulator driver, and optical modulator. In this paper, we have analyzed the nonlinear effects for RZ and NRZ signals by varying the signal type in the electrical modulator driver and are discussed in the following section.

4.1 NRZ Optical Transmitter

The NRZ rectangular driver shown in simulation setup of Fig. 2 as electrical generator has two different electrical levels based on the transmitted bit. When an input “1 or 0” is fed into the electrical driver, the output signal is switched between the **Low Level and High Level** during the entire bit period. Figure 2 also shows the output signal waveform and the optical spectrum of NRZ transmitter. Switching between the two electrical levels is occurring without any delay, and the time slope may be zero or the desired level.

4.2 RZ Optical Transmitter

The RZ rectangular driver shown in Fig. 3 has an output signal that also has two different electrical levels. When an input “1” is fed into the electrical driver, the output signal is at the **High Level** for a period equal to the product of the duty cycle by the bit time. Then, it remains in **Low Level** for the rest of the time. When a “0” is transmitted, the output is continuously constant at the low level for the entire bit duration.

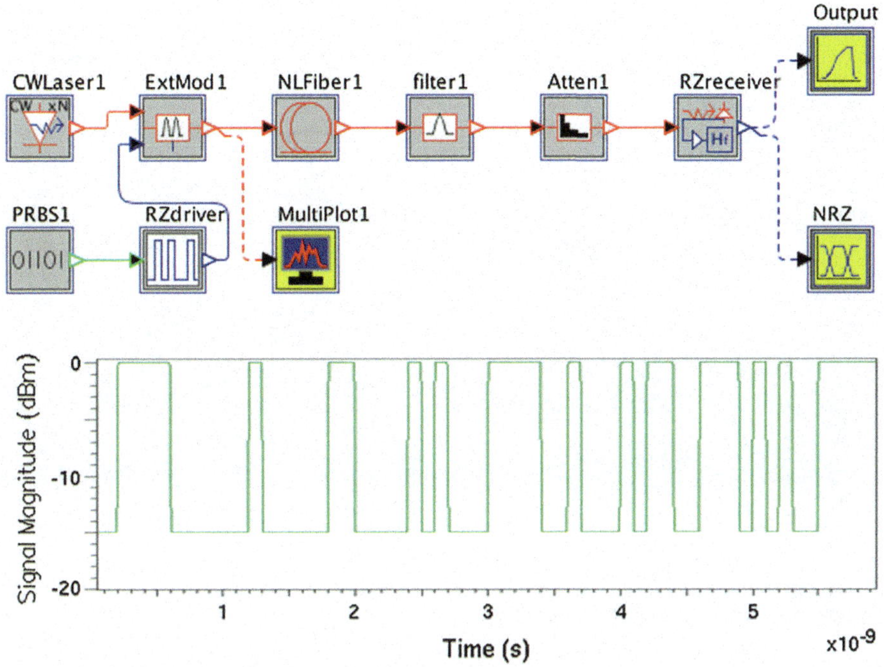

Fig. 2 NRZ transmitter-simulation setup and output waveform

5 Simulation Setup

The schematic of simulation setup shown in Fig. 4 consists of eight-channel DWDM system at 10 Gbps to analyze the effects of XPM and FWM. It consists of laser sources, electrical pulse generators of type NRZ/RZ signals, and MZM modulator driver in the transmitter section. Optical scope is also connected to observe the optical spectrum, and electroscope is connected to view the eye diagrams. The measurement components such as BER and Q estimator are also attached to analyze the respective values for NRZ and RZ signals by switching the dispersion from −4 to 10 ps/nm/km through parametric scan in OptSim 2016.06 software. The parameters assigned during the simulation setup are shown in Table 1.

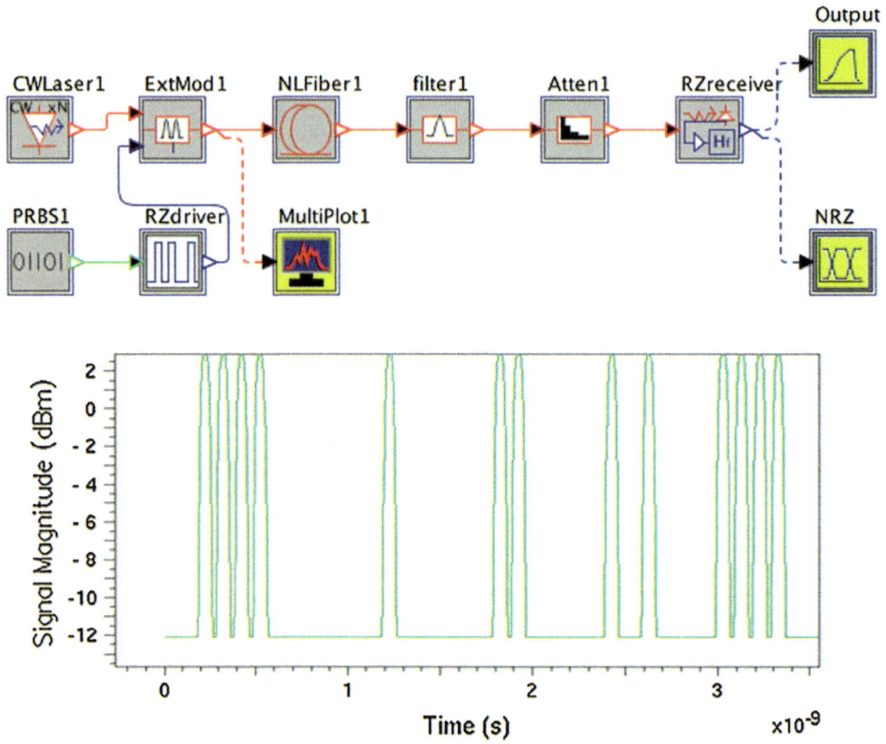

Fig. 3 RZ transmitter-simulation setup and output waveform

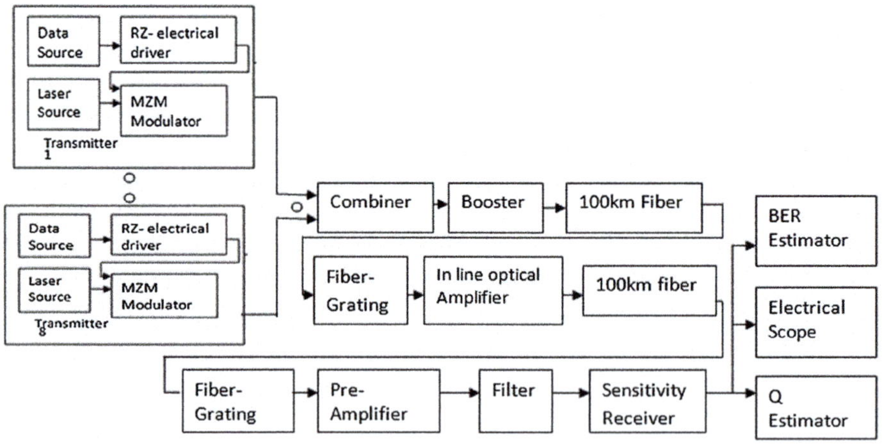

Fig. 4 Simulation setup for eight-channel RZ transmitter to produce FWM output

Table 1 Simulation parameter

S. No.	Parametric values		
	Parameter name	Value	Unit
1	Samples per bit	31	
2	Bit rate	10.0	Gb/s
3	Center frequency	193.05	THz
4	Extinction ratio	20.0	dB
5	Chirp factor	0.0	
6	V pi	5.0	V
7	Noise figure	10.0	dB
8	−3 dB bandwidth	10.0	GHz
9	First filter notch	22.57609	GHz
10	CW pump channel power	−10	dBm
11	CW probe channel power	−30	dBm

6 Results and Discussions

6.1 Effects Due to Cross-phase Modulation

Figures 5, 6, 7, and 8 show the effects of XPM for two-, four-, and eight-channel DWDM system and the corresponding eye diagrams, Q and BER values plot for the signal types of NRZ and RZ in the transmitter section.

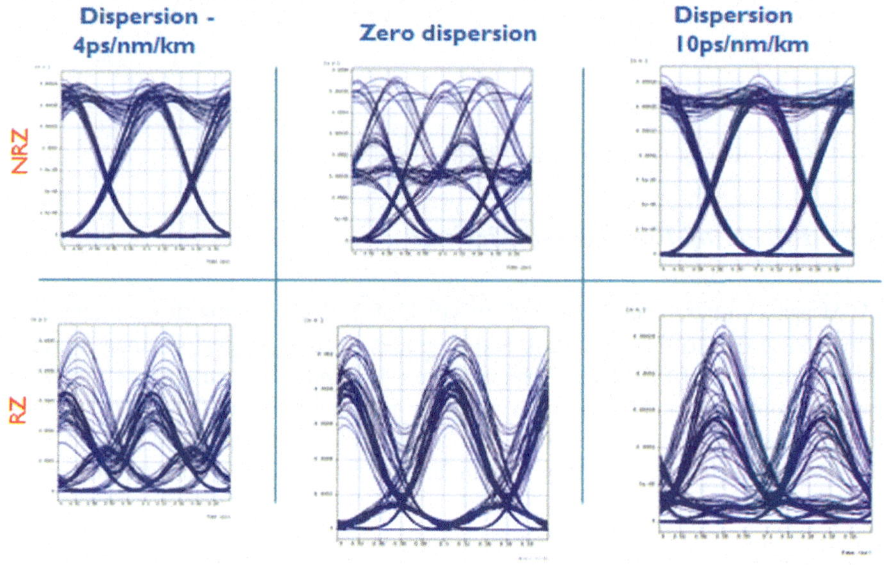

Fig. 5 Eye diagrams represent XPM effect for NRZ and RZ signal by varying dispersion in two-channel systems

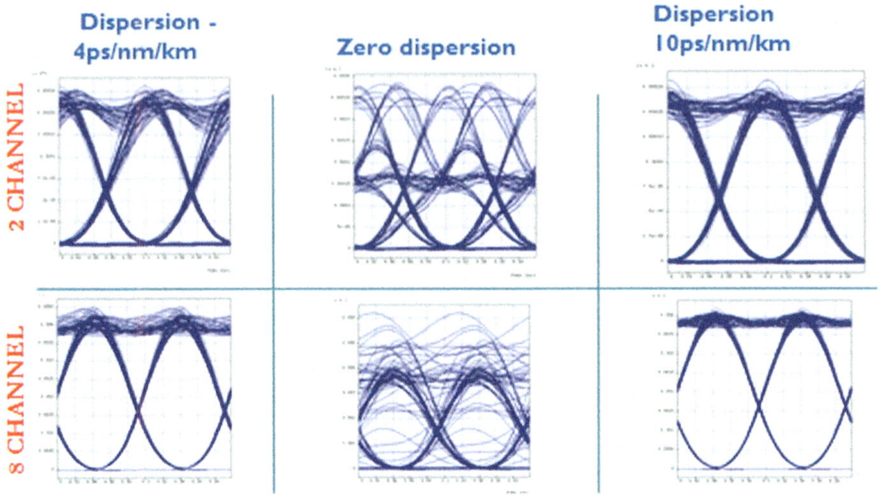

Fig. 6 Eye diagrams represent XPM effect for NRZ signal by varying dispersion in two-channel and eight-channel systems

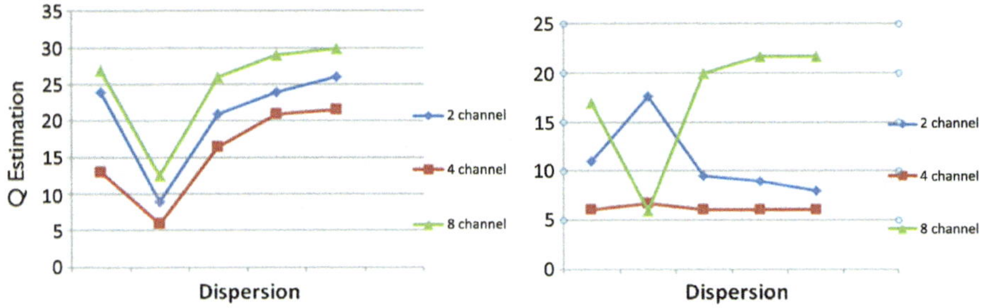

Fig. 7 Comparison of Q estimation for NRZ (left) and RZ signals

6.2 *Effects of Four-Wave Mixing*

FWM causes interchannel crosstalk in DWDM optical systems. It generates additional noise due to this cross talk and degrades the system performance. Figure 9 shows the effect of FWM for 2 channels and 4 channels by varying the dispersion ranges from −4 to 10 ps/nm/km through parametric scan feature in OptSim. Figure 10 shows the comparison of four-channel FWM effects in DWDM system at the data rate of 10 and 40 Gbps to validate the results for higher data rate.

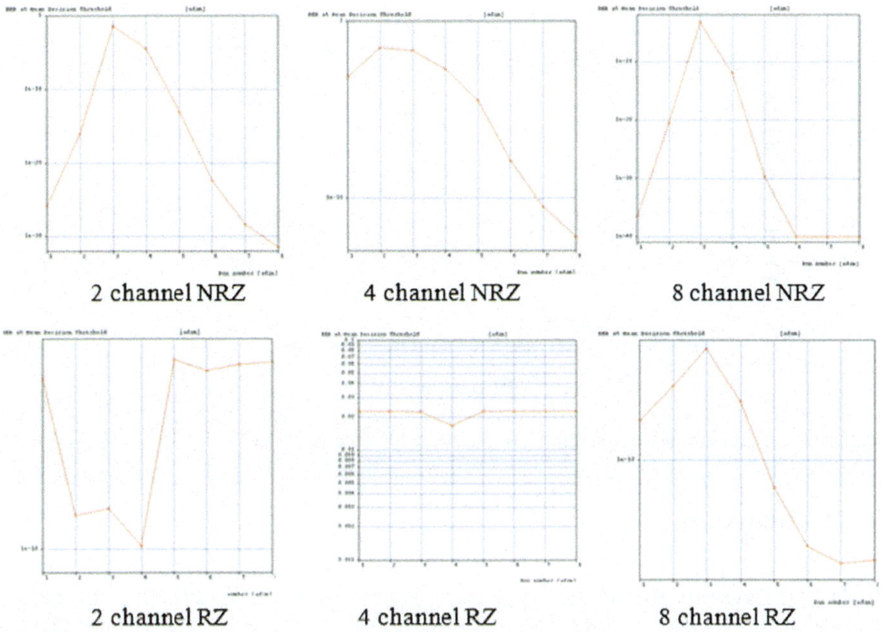

Fig. 8 Comparison of bit error rate for NRZ and RZ signals

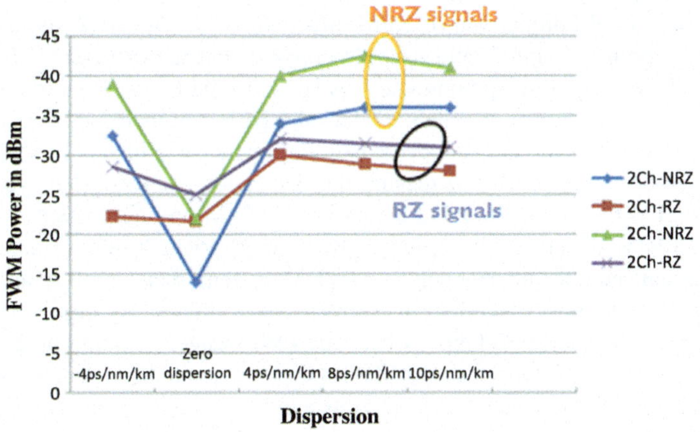

Fig. 9 Comparison of FWM power for two-channel and four-channel systems

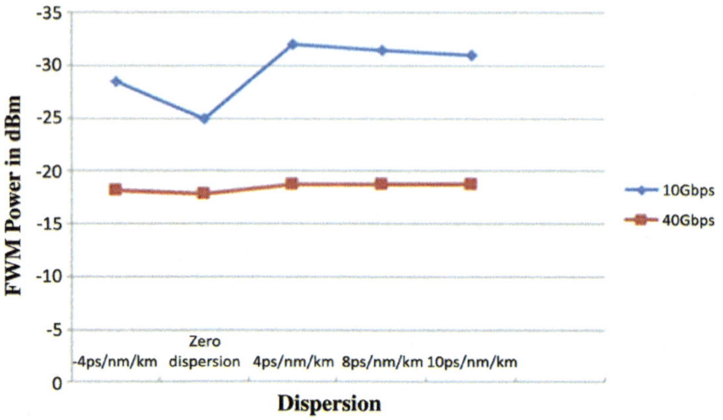

Fig. 10 Comparison of four-channel FWM power for 10 and 40 Gbps

7 Conclusion

In DWDM system, the nonlinear effects play a major role in affecting the performance of the system. In this paper, we analyzed the effects of FWM and XPM at 10 Gbps for two-, four-, and eight-channel DWDM system. This analysis is done exclusively for NRZ and RZ signals by varying the dispersion ranges from −4 to 10 ps/nm/km through parametric scan feature in OptSim 2016.06 software. The eye diagrams and BER comparison of XPM effect show that the NRZ signals perform much better than RZ signal, and it is also proved that the dispersion shifted fiber (DSF) produces large eye opening and less BER compared to zero dispersion fiber. The increase in number of channels reduces the BER and increases the Q value for NRZ signal, whereas the Q value decreases with increase in channel for RZ signals. This can be compensated by increasing the dispersion for RZ-based systems. The analysis on FWM effect shows that the power of FWM product reduces when using the DSF. FWM effect can be reduced to an appropriate level by increasing the dispersion and number of channels in DWDM system. This result is analyzed for the bit rate of 10 Gbps. The effect of FWM is also validated for higher data rate by increasing the bit rate to 40 Gbps, where the FWM effect is more compared to the same parameters at 10 Gbps. The analysis can be extended for higher data rate by varying the modulators, channel space between the channels, fiber length, and laser power, and this is the topic of research and challenge in the optical field to improve the performance of DWDM systems.

References

1. Dewra S, Kaler RS (2013) Performance evaluations of optical add drop multiplexers with Mach-Zehnder interferometer techniques for dense wavelength division multiplexed System. J Opt Technol 80(9):526–531
2. Singh S, Kaler RS (2013) Comparison of pre-, post- and symmetrical compensation for 96 channel DWDM system Using PDCF and PSMF. Optik 124:1808–1813
3. Kotb A (2016) Simulation of high quality factor all-optical logic gates based on quantum-dot semiconductor optical amplifier at 1 Tb/S. Optik 127:320–325
4. Nazemosadat E, Pourbeyram H, Mafi A (2016) Phase matching for spontaneous frequency conversion via four-wave mixing in graded-index multimode optical fibers. J Opt Soc Am B 33(2):144–150
5. Kashyap K, Singh H, Singh P, Kumar S (2013) Compensation of dispersion in optical fiber using Fiber Bragg Grating (FBG). Int J Adv Res Sci Eng 02(04):124–132. ISSN 2319-8354

Implantable Antenna for Blood Glucose Monitoring

Shalu Pandey and Vibha Rani Gupta

Abstract In this paper, a miniaturized implantable antenna for ISM (industrial, scientific and medical) band 2.4–2.4835 GHz has been proposed. The designed antenna can be used for biomedical application called blood glucose monitoring system. The proposed antenna will be placed under the skin tissue. To make the antenna biocompatible, it is coated with the PMMA (polymethylmethacrylate) biocompatible layer. The designed antenna is tested for single-layer skin tissue as well as multilayer muscle and skin tissue.

Keywords Implantable antenna · ISM (industrial, scientific and medical) Biomedical · Biocompatible

1 Introduction

Implanted devices placed inside the human body to diagnose serious health problem are the major revolution in the biomedical field. These devices can work as stimulators and sensors depending upon types of applications such as monitoring body temperature, pressure, blood glucose level, pacemakers. The data collected from the devices can be transmitted to the outside monitoring devices through wireless telemetry [1]. In such implanted devices, the transmitting and receiving role is performed by the implanted antenna. The design of such antenna is a challenging task as the antenna has to be planted inside human body so the biocompatibility and miniaturization of size is an important area to be taken into consideration. The difficulty lies in the design of antenna to work in human tissues as the medium is lossy. The consideration of the designing factors such as minia-

S. Pandey (✉) · V. R. Gupta
Electronics and Communication Engineering, Birla Institute of Technology, Mesra, Ranchi, Jharkhand, India
e-mail: shalupandey970@gmail.com

V. R. Gupta
e-mail: vrgupta@bitmesra.ac.in

G. Gnanagurunathan et al. (eds.), *Optical and Microwave Technologies*,
Lecture Notes in Electrical Engineering 468,
https://doi.org/10.1007/978-981-10-7293-2_7

turization of antenna size, biocompatibility, high conductivity of tissue, power requirement, and impedance matching is an important task.

In this work, the aim is to design a miniaturized single-band implantable antenna for 2.4–2.4835 GHz industrial, scientific and medical (ISM) band. The intended design is for one of the medical application called continuous glucose monitoring (CGM). The CGM system consists of a sensor, transmitter, and an external monitoring unit. The sensor is inserted under the skin to check glucose level in tissue fluid. The sensed data are transmitted through the lightweight and miniaturized transmitter to the outside monitoring unit. In this work, the higher frequency of operation is chosen so as to lower the size of antenna and to get higher bandwidth so that large amount of data can be transmitted.

2 Antenna Design

The antenna proposed is CPW fed slot antenna as shown in Fig. 1, which provides miniaturization of size and improvement in bandwidth. Firstly, the antenna is designed in free space, and then to make it biocompatible, the whole structure is embedded inside the biocompatible material called PMMA (polymethylmethacrylate), whose electrical properties are $\varepsilon = 2.16$ and $\tan\delta = 0.01$. The simulation tool used for antenna design is the full-wave EM solver IE3D. It is based on the method of moment in which the Maxwell's equations are solved in an integral form through the use of Green's functions. Here, the substrate used is Rogers RO3210 for antenna design having permittivity $\varepsilon = 10.2$ and loss tangent = 0.003. Rogers RO 3210 has been selected because of its similar electrical properties to biocompatible ceramic alumina ($\varepsilon = 9.6$, $\tan\delta = 3 \times 10^{-5}$) [2, 3], which are being used in medical implants since long. The side view of antenna with substrate RO3210 of thickness 1.27 mm is shown in Fig. 2.

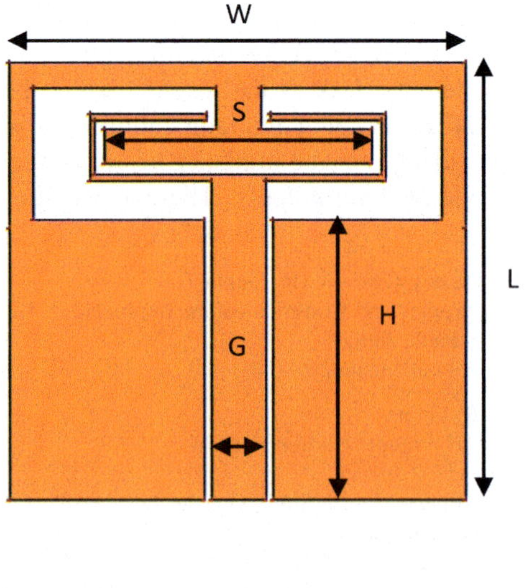

Fig. 1 Top view of antenna with CPW feed

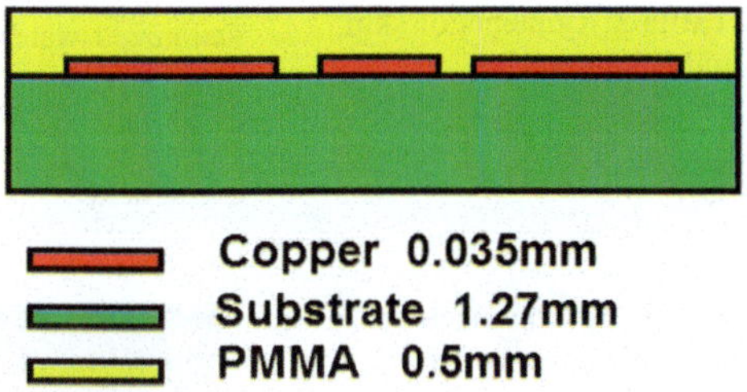

Fig. 2 Side view of antenna with biocompatible PMMA layer

Table 1 The dimensions of the proposed antenna

L	21 mm
W	23.9 mm
H	13.5 mm
S	11 mm
G	2.3 mm

A PMMA layer of thickness 0.5 mm is placed over the metallic portion of the designed antenna to make it completely biocompatible. The coplanar waveguide (CPW) feed is applied for 50 Ω impedance. Once the antenna is designed for free space, it is coated with superstrate layer of biocompatible material to make it biocompatible. In the next step, biocompatible antenna has been re-optimized with the layer of human tissues above the superstrate and below the substrate. As per requirement of the application of blood glucose monitoring system, the chosen layer of tissue is skin tissue. The skin mimicking gel has been prepared, and its electrical properties at 2.45 GHz have been observed [4]. The designed antenna is miniaturized in size as compared to the literature [4, 5].

The dimensions of the proposed antenna are listed in Table 1.

3 Results

The designed antenna is tested under the following conditions:

(i) Antenna under single layer of skin tissue and
(ii) Antenna embedded in the muscles and skin tissue

(i) The first analysis is performed with the single layer of skin tissue. The electrical properties of the skin gel at ISM band 2.4–2.4835 GHz are $\varepsilon = 38.0639$ and $\sigma = 2.39$ S/m [4]. The recipe for the preparation of skin mimicking gel consists of deionized water, sugar, agarose as shown in Table 2.

The 100 mL of mixture is heated at a temperature of 80 °C till the clear solution is formed. The prepared skin tissue is further tested at network analyzer to obtain the required electrical property as shown in Fig. 3.

Table 2 Contents of skin mimicking gel

Deionized water	43%
Sugar	57%
Agar	1.2 gm in 100 mL
Citric acid	1 gm in 100 mL

Fig. 3 Testing electrical properties of skin tissue

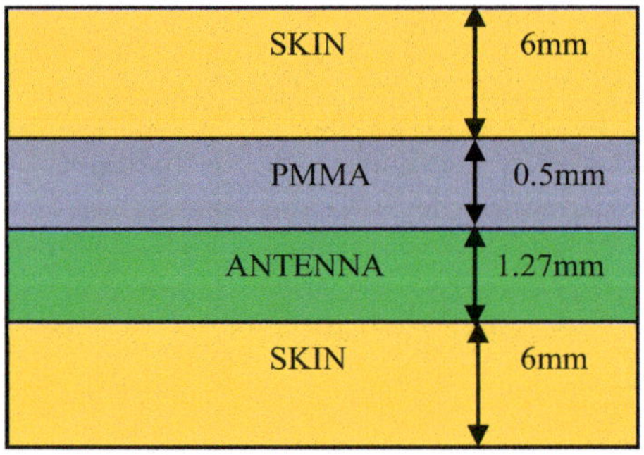

Fig. 4 Single-layer arrangement of antenna

The layout of single-layer position of antenna placed in between human skin tissues is shown in Fig. 4.

The simulation result of antenna in single-layer skin tissue is shown in Fig. 5. The value of reflection coefficient value S_{11} is −23.5 dB at 2.4 GHz, and bandwidth obtained is 440 MHz ranging from 2.3 to 2.74 GHz which covers the complete ISM band. The obtained antenna bandwidth is 40.9% higher than reported in the literature [5] for ISM band.

For performing the multilayer test, muscle tissue is prepared with content of deionized water (62%), sugar (36%), and salt (2%). The 100 mL solution is then tested at network analyzer. The multilayer test consists of muscle and skin tissue with antenna implanted inside (Fig. 6).

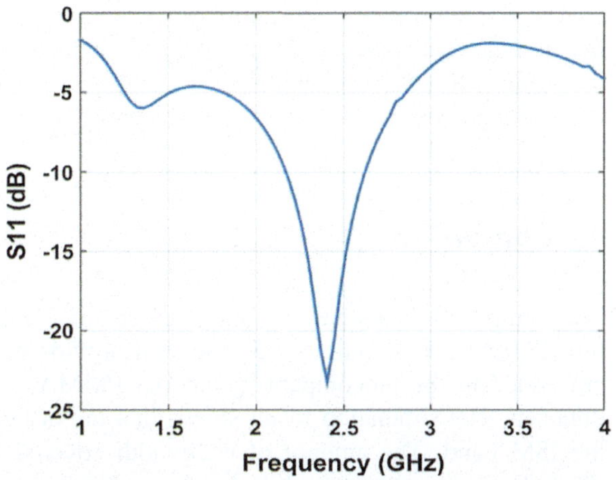

Fig. 5 Simulation result S_{11} of antenna in single layer

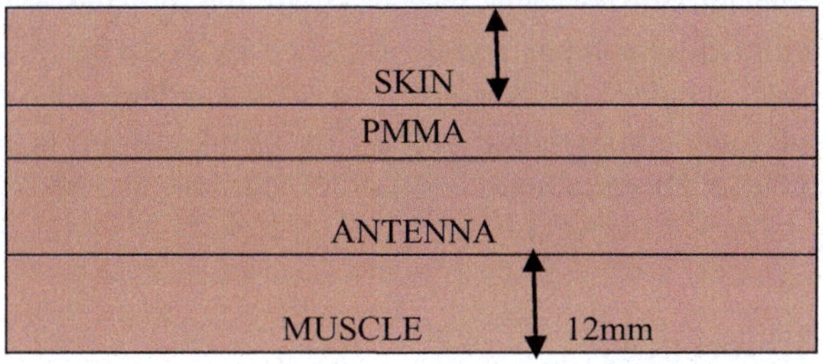

Fig. 6 Multilayer structure for antenna simulation

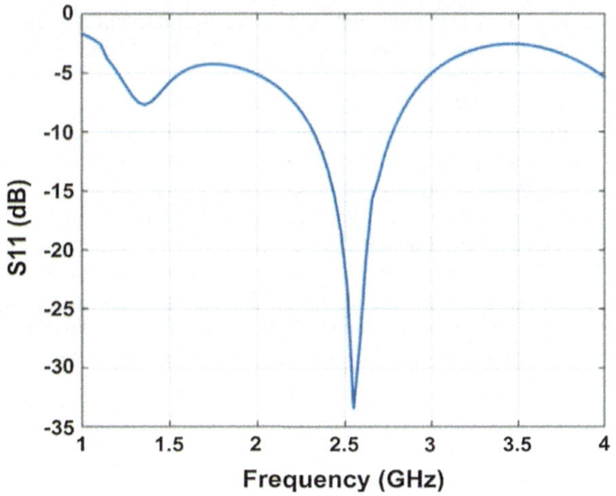

Fig. 7 Simulation result S_{11} of antenna in multilayer tissue

The muscle tissue prepared has $\varepsilon = 52.73$ and $\sigma = 2.79$ S/m. The simulations of antenna with multilayer phantom have been performed as shown in Fig. 7. The obtained result gives S_{11} of −33.42 dB at 2.51 GHz, and bandwidth is 350 MHz from 2.395 to 2.745 GHz.

The single-layer test and multilayer test have been performed which show that the designed antenna is suitable to be implanted inside human body with other interfacing circuits containing sensors, amplifiers, rectifiers to work for sensing the change in glucose level of body and sending the data to outside monitoring device.

4 Conclusion

The proposed antenna is miniaturized implanted antenna of cross section 23.9 mm × 21 mm × 1.57 mm. To provide biocompatibility, the antenna is embedded in the biocompatible material PMMA, whose electrical properties are obtained. The simulation result shows that the designed antenna performs well in the ISM band. The amount of bandwidth covered is 440 and 350 MHz for the single-layer tissue and multilayer tissue, respectively, and can be used for the transmission of data wirelessly to the outside receiving system. The radiation emitted by antenna can be absorbed by body up to certain limit. It is calculated by SAR standard [6]. According to SAR standard, 1.6 W/kg for 1 gm tissue is the limit of American Standard ANSI C95.1 and 2 W/kg in 10 gm tissue is the safety limit of ICNIRP guidelines.

References

1. Merli F (2011) Implantable antennas for biomedical applications. Ph.D. dissertation, Sept 2011
2. Kim J, Rahmat Samii Y (2004) Implanted antenna inside the human body: simulation design and characterization. IEEE Trans Microw Theory Tech 52(8):1934–1943
3. Scarpelb ML, Kurup D, Rogier H, Ginste DR, Axisa F, Vanfleteren J, Joseph W, Martens L, Venneeren G (2011) Design of an implantable slot dipole conformal flexible antenna for biomedical applications. IEEE Trans Antennas Propag 59(10):3556–3564
4. Karacolak T, Hood AZ, Topsakal E (2008) Design of a dual-band implantable antenna and development of skin mimicking gels for continuous glucose monitoring. IEEE Trans Antennas Propag 60(8):3568–3575
5. Xia W, Saito K, Takahashi M, Ito K (2009) Performances of an implanted cavity slot antenna embedded in the human arm. IEEE Trans Antennas Propag 57(4):894–899
6. IEEE Standard for Safety Levels With Respect to Human Exposure to Radio frequency Electromagnetic Fields, 3 kHz to 300 GHz, IEEE Standard C95.1 (2005)

Reliability Analysis of Data Center Network

Abhilasha Sharma and R. G. Sangeetha

Abstract High computational systems consisting of multiple processors need interconnection networks to provide communication between them. Torus and Benes networks are highly scalable data center network and found to be the suitable candidates to work in this environment. Fault tolerance and reliability are the crucial issues for any scalable network. This paper analyzes the network reliability of torus and Benes networks of size 8×8 and 16×16. In this study, we have evaluated the network reliability for both the networks using an accurate analytical method. This analysis is performed for the network of size 8×8 and 16×16 two-dimensional torus and Benes networks. For this analysis, the confidence levels from 0.99 (high switch reliability) to 0.90 (low switch reliability) have been considered. The overall system reliability has been evaluated considering both node and link failures. The results are validated and compared with existing Benes network using MATLAB. The analysis shows that the reliability of torus network is better than Benes network. This is observed because torus network has more number of alternative paths than Benes network.

1 Introduction

The network performance of the data centers (DCs) primarily depends on the data center network (DCN) architecture. The data center networks are responsible for providing various Internet and cloud-based services. The data center networks need to be agile, dynamic, and scalable in nature with increasing traffic and application demands. There are four main categories of DCN topologies: direct network (e.g., torus [1], CamCube [2]), server centric (e.g., SprintNet [3, 4], BCube [5], DCell [6], and Benes [7]), tree-based topology [1] (e.g., Fat-Tree [8]), and hybrid network (e.g., Helios [9], c-Through [10]) [1]. Torus network topology belongs to the category of direct network topology in which servers are connected to other servers in

A. Sharma · R. G. Sangeetha (✉)
SENSE, VIT University, Chennai, India
e-mail: sangeetha.rg@vit.ac.in

G. Gnanagurunathan et al. (eds.), *Optical and Microwave Technologies*,
Lecture Notes in Electrical Engineering 468,
https://doi.org/10.1007/978-981-10-7293-2_8

a regular pattern. The servers are placed in close proximity to each other due to the regular torus topology which provides efficient communication in between them. The Benes network topology usually concentrates on the scalability of the network. The torus-based data center network is suitable candidate than Benes network because of its various unique advantages enlisted below. The traditional architectures are constructed with large number of links and switches, and the system reliability and network performance get affected greatly due to any faulty link or ToR (Top of the Rack) switch.

The torus topology incorporates high path diversity which provides better fault tolerance capability and good load balance ability even under permutation traffic to the data center network. As compared to traditional architectures, the torus topology exhibits better network reliability due to more number of alternative paths in between servers. This topology optimizes granularity of clusters and scalability of the topology because of its symmetric architecture. The DCN architectures based on torus topology can be scaled economically as compared to the traditional fat-tree architectures up to tens of thousands of servers. The torus topology is a suitable candidate to avoid network congestion because more number of equidistant paths which can reduce the queueing delays too. This architecture provides significant decrease in end-to-end delay and packet loss, with high throughput. This architecture has been implemented already as high performance computing network architectures and proved the high system performance in various dimensions. Some of the implemented architectures are 5D Torus as Blue Gene [11], 2D Torus as Cray SeaStar [12], 3D Torus as IBM's Blue Gene [13], and Cray Gemini [14]. So, torus topology can be implemented as data center architecture and yet analysis of reliability becomes a major concern for the evaluation of the whole network performance. The probabilistic description of any network architecture about its performance is given by the reliability analysis [15, 16]. In this paper, an analytical accurate method, reliability block diagram (RBD), has been used for the reliability analysis of torus and Benes network architectures. The rest of the paper is sectioned as follows: Sect. 2 presents an introduction to RBD method for analyzing reliability of complex systems. The evaluated expressions of network reliability for Benes and torus networks and their comparison are done in Sect. 3. Section 4 includes the conclusions drawn from the reliability analysis.

2 Reliability Block Diagram (RBD) Analytical Method

For the evaluation of network reliability, the network is assumed as a system and the nodes and links as the components of the system. So, the system reliability entirely depends on the components of the system and the relation in between the components. There are two conditions for each component, either it can be faulty or either in working condition. Basically, RBD is representing the network from reliability point of view as system and nodes as the components. In RBD, each component is connected in a series or parallel pattern to each other. The whole network is treated as a system of blocks in which a single block represents a single node and these locks are connected to each other in parallel, series, or series–parallel according to the network topology [17].

2.1 *Series RBD*

In the series system, the entire system fails, on the failure of single component. A basic representation of the series RBD with '*n*' components is shown in Fig. 1. Assuming that 'C_i' be the reliability if *i*th component, then the overall system reliability C_S is calculated as Eq. 1.

$$\begin{aligned} C_s &= C_1 \cap C_2 \cap C_3 \cap \ldots C_n \\ &= C_1(C_2|C_1)(C_3|C_1C_2)(C_4|C_1C_2C_3)\ldots(C_n|C_1C_2C_3\ldots C_{n-1}) \end{aligned} \tag{1}$$

If the components are independent components, then Eq. 1 is represented as Eq. 2.

$$C_S = \prod C_i, \quad i = 1 \text{ to } n \tag{2}$$

2.2 *Parallel RBD*

In the parallel system, there is no effect on the failure of a single component. The system fails if and only if all the components are faulty. A basic representation of the series RBD with '*n*' components is shown in Fig. 2.

Assuming that 'L_i' be the reliability if '*i*th' component, then the overall system reliability 'L_S' is calculated as Eq. 3.

$$\begin{aligned} L_S &= L_1 \cap L_2 \cap L_3 \cap \ldots L_n \\ &= L_1(L_2|L_1)(L_3|L_1L_2)\,(L_4|L_1L_2L_3)\ldots(L_n|L_1L_2L_3\ldots L_{n-1}) \end{aligned} \tag{3}$$

If the components are independent components, then Eq. 3 is represented as Eq. 4.

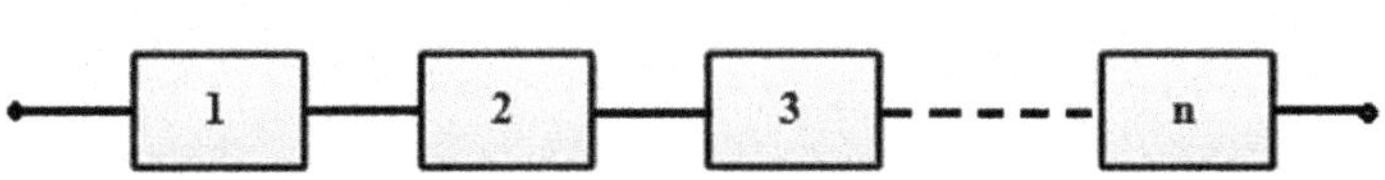

Fig. 1 RBD of series complexity

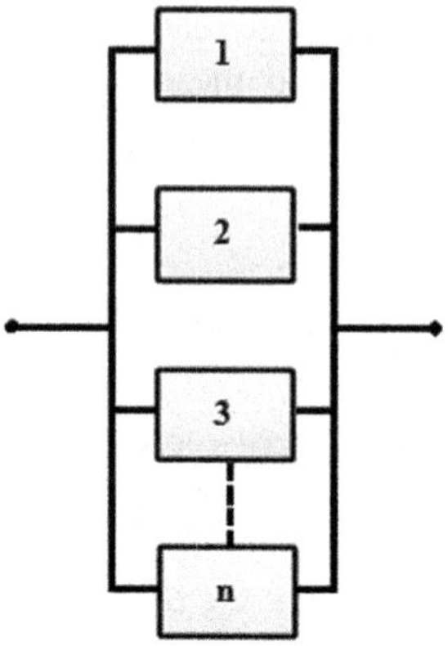

Fig. 2 RBD of parallel complexity

$$L_S = \prod L_i, \quad \text{where } i = 1 \text{ to } n \tag{4}$$

So, parallel reliability is calculated as Eq. 5.

$$\begin{aligned} C_S &= 1 - L_S \\ &= \prod L_i, \quad \text{where } i = 1 \text{ to } n \\ &= 1 - \prod (1 - C_i), \quad \text{where } i = 1 \text{ to } n \end{aligned} \tag{5}$$

2.3 *Series–Parallel RBD*

The complex systems incorporated with combination of components connected in series and parallel patterns come under this category. The reliability of such systems is calculated by breaking them into homogeneous units and calculating the reliability of those units independently. The final expression of total system reliability is obtained by replacing the homogeneous units back in the combination of series and parallel after obtaining their expressions as a single system. A graphical representation of such RBD is represented in Fig. 3.

The system reliability of this system is given by Eq. 6.

$$\begin{aligned} X_S = X_1[1 - (1 - X_2(1 - X_3))] \\ [1 - ((1 - (1 - (1 - X_4)(1 - X_5)))(1 - (1 - (1 - X_6)(1 - X_7))))] \end{aligned} \tag{6}$$

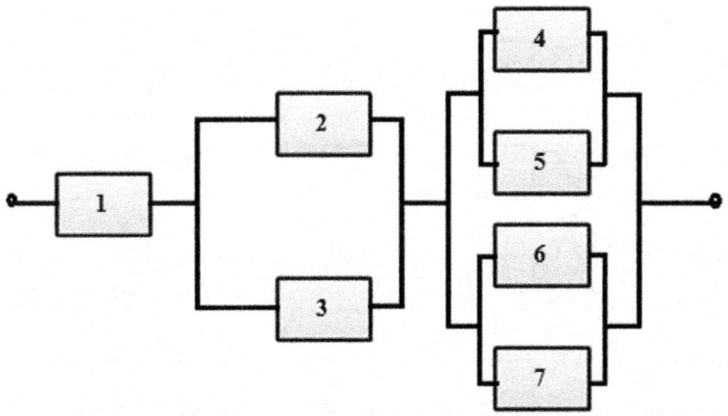

Fig. 3 RBD of series–parallel complexity

3 Network Reliability Analysis

In this section, the reliability of torus network using RBD is evaluated. The analysis is performed for the two-dimensional torus network and Benes network of size 8×8 and 16×16. For this analysis, the confidence levels from 0.99 (high switch reliability) to 0.90 (low switch reliability) have been considered. The overall system reliability has been evaluated considering both node and link failures. The results are validated and compared with existing Benes network using MATLAB. In this paper, each switching element is treated as a component and assumed to have a switch reliability equal to 'x', which is varying from high switch reliability to low switch reliability.

3.1 Network Reliability Analysis for Benes Network

Benes network is a type of non-blocking multistage interconnection. Benes network has 2(log2 P) − 1 stages with $P/2$ switching elements in each stage, with a network size of $P \times P$. The graphical representation of the network complexity of Benes architecture with network size of 8×8 is shown in Fig. 4. The network reliability RBD for Benes network of size 8×8 is shown in Fig. 5. Assuming that 'x' be the reliability of each switching component, then the overall network reliability for the network size 8×8 is calculated as Eq. 7.

$$X_8 = x^2[1-(1-(x^2(1-(1-x)^2)))^2] \tag{7}$$

Similarly, based on the RBD method the network reliability of Benes network of size 16×16 is given by Eq. 8.

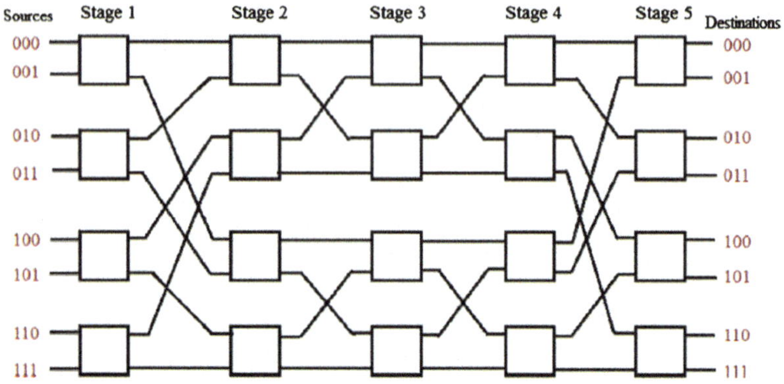

Fig. 4 Benes network architecture (network size 8 × 8)

Fig. 5 RBD for Benes network architecture (network size 8 × 8)

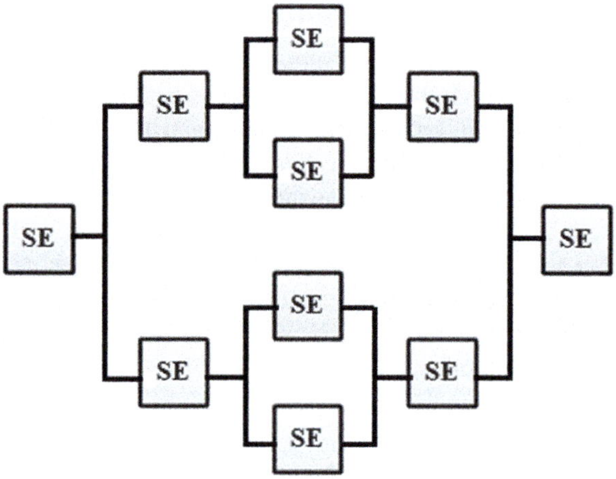

$$X_{16} = x^2\left[1 - (1 - X_8)^2\right] \tag{8}$$

3.2 Network Reliability Analysis for Torus Network

Torus network is a type of multidimensional interconnection network which is widely used in high performance computing and is a suitable candidate for data center networks. In this paper, for network reliability analysis two-dimensional torus network of size 8 × 8 and 16 × 16 is considered. The two-dimensional torus network of size $N \times N$ has a total of N_2 switching elements. The graphical representation of the network complexity of torus network architecture of size 4 × 4 is shown in Fig. 6. The network RBD for torus network of size 4 × 4 is shown in

Fig. 7. The torus network topology is symmetric topology, so the same pattern is obtained in the RBD with SE2 as obtained with SE1 as shown in Fig. 7. To reduce the complexity of the figure, it is shown by dashed line. Assuming that 'x' be the reliability of each switching component, then the overall network reliability is calculated as Eq. 9.

$$X_8 = x((x((1-x((1-x((1-x((1-x)^2))^2))^2)^2))^2)) \tag{9}$$

Similarly, based on the RBD method the network reliability of torus network of size 16 × 16 is given by Eq. 10.

$$P_{16} = x((1-x((1-x((x((1-x((1-x((1-x((1-x)^2))^2))^2)^2))^2)))^2))^2) \tag{10}$$

4 Comparison of Network Reliability Analysis

The analysis is performed for the two-dimensional torus network and Benes network of size 8 × 8 and 16 × 16. For this analysis, the confidence levels from 0.99 (high switch reliability) to 0.90 (low switch reliability) have been considered. The overall system reliability has been evaluated considering both node and link failures. The comparison of network reliability of torus and Benes architectures of network size 8 × 8 and 16 × 16 is represented in Figs. 8 and 9, respectively. This analysis shows that the reliability of torus network is better than Benes network. This is observed because torus network has more number of alternative paths than Benes network, due to which the probability of at least one non-faulty path under system failure is more than that of Benes network.

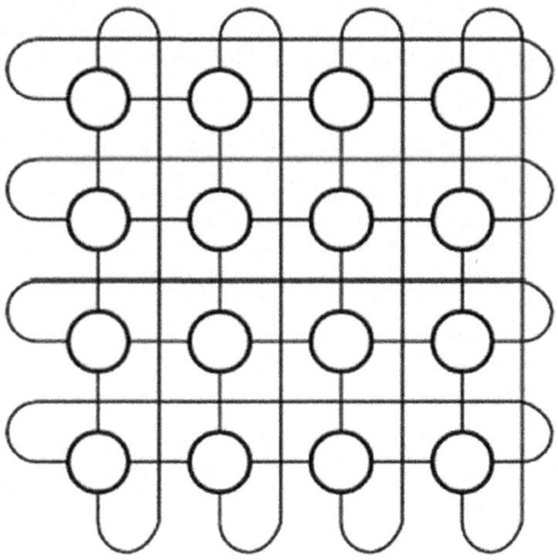

Fig. 6 Torus network architecture (network size 4 × 4)

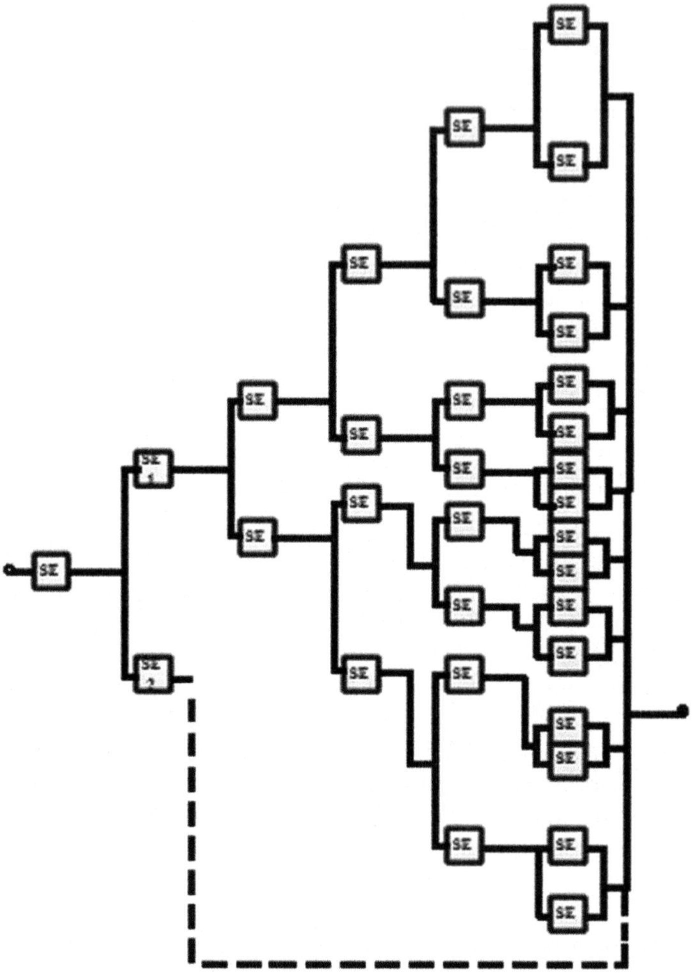

Fig. 7 RBD for torus network architecture (network size 8 × 8)

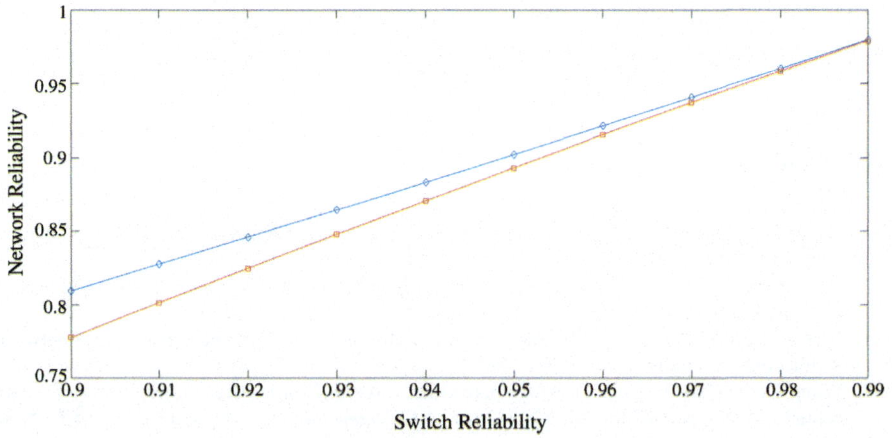

Fig. 8 Comparison of network reliability of Benes and torus network architectures (network size 8 × 8)

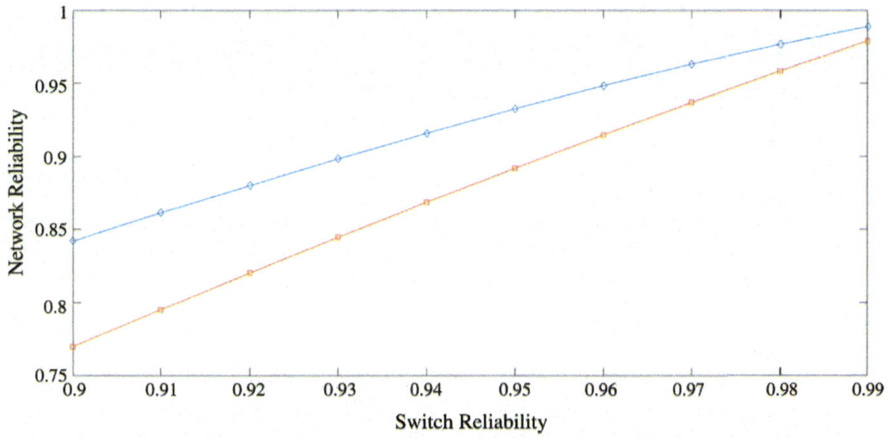

Fig. 9 Comparison of network reliability of Benes and torus network architectures (network size 16 × 16)

5 Conclusion

In this paper, the network reliability for torus and Benes networks is evaluated using an accurate analytical RBD method. For this analysis, the confidence levels from 0.99 (high switch reliability) to 0.90 (low switch reliability) have been considered. The overall system reliability has been evaluated considering both node and link failures. The results are validated and compared with existing Benes network using MATLAB. The analysis shows that the reliability of torus network is better than Benes network. This is observed because torus network has more number of alternative paths than Benes network.

References

1. Wang T et al (2016) Towards cost-effective and low latency data center network architecture. Comput Commun 82:1–12
2. Abu-Libdeh H et al (2010) Symbiotic routing in future data centers ACM SIGCOMM. Comput Commun Rev 40(4):51–62
3. Wang T et al (2015) Designing efficient high performance server-centric data center network architecture. Comput Netw 79:283–296
4. Wang T et al (2014) Sprintnet: a high performance server-centric network architecture for data centers. In: *Communications (ICC), 2014 IEEE International Conference on*, IEEE, New York
5. Guo C et al (2009) BCube: a high performance, server-centric network architecture for modular data centers. ACM SIGCOMM Comput Commun Rev 39(4):63–74
6. Guo C et al (2008) Dcell: a scalable and fault-tolerant network structure for data centers. ACM SIGCOMM Comput Commun Rev 38(4):75–86
7. Calabretta N et al (2013) On the performance of a large-scale optical packet switch under realistic data center traffic. J Opt Commun Netw 5(6):565–573
8. Al-Fares M, Loukissas A, Vahdat A (2008) A scalable, commodity data center network architecture. ACM SIGCOMM Comput Commun Rev. 38(4):63–74
9. Farrington N et al (2010) Helios: a hybrid electrical/optical switch architecture for modular data centers. ACM SIGCOMM Comput Commun Rev 40(4):339–350
10. Wang G et al (2010) c-Through: part-time optics in data centers. ACM SIGCOMM Comput Commun Rev 40(4):327–338
11. Chen D et al (2012) The IBM blue gene/Q interconnection fabric. IEEE Micro 32(1):32–43
12. Brightwell R, Pedretti K, Underwood KD (2005) Initial performance evaluation of the Cray SeaStar interconnect. In: Proceedings of the 13th symposium on High performance interconnects, 2005, IEEE, New York
13. Adiga NR et al (2005) Blue Gene/L torus interconnection network. IBM J Res Dev 49.2.3:265–276
14. Alverson R, Roweth D, Kaplan L (2010) The gemini system interconnect. IEEE 18th annual symposium on high performance interconnects (HOTI), IEEE, New York
15. Jahanshahi M, Bistouni F (2015) Improving the reliability of the Benes network for use in large-scale systems. Microelectron Reliab 55(3):679–695
16. Jahanshahi M, Bistouni F (2014) A new approach to improve reliability of the multistage interconnection networks. Comput Electr Eng 40(8):348–374
17. Birolini Alessandro (2007) Reliability engineering, vol 5. Springer, Heidelberg

Study of Microstrip Antenna Array with EBG Structure

R. M. Vani, K. Prahlada Rao and P. V. Hunagund

Abstract This paper discusses the performance enhancement of conventional two-element antenna array. The study is performed with electromagnetic band gap (EBG) structure loaded in the ground plane of the antenna array. The EBG structure employed consists of an array of swastika slots. The EBG structures possess the property of suppression and passage of electromagnetic waves in certain frequency bands. With the EBG structure, the mutual coupling of the conventional antenna array is reduced to −42.19 dB. Miniaturization of the antenna array is also achieved. An overall bandwidth of 79.72% is achieved with the EBG structure as compared to 3.562% of the conventional antenna array. With the introduction of EBG structure, radiation properties of the antenna array are better. The antenna array designed finds application in the L band of the microwave frequency region. The study has been performed using Mentor Graphics IE3D simulation software, and the measured results have been obtained using vector network analyzer.

1 Introduction

Antennas and antenna arrays play a very significant role in various fields of communications. There are various types of antennas like Yagi–Uda antenna, log periodic antenna, dipole antenna, microstrip antenna, etc. Each antenna possesses its own characteristics, advantages, disadvantages, and applications. Microstrip antennas are a class of antennas which are very versatile compared to other antennas, as they can be easily integrated with other active and passive devices. Additionally, they are compact, lightweight, and easy to fabricate. Microstrip

R. M. Vani (✉)
University Science Instrumentation Centre, Gulbarga University, Gulbarga, India
e-mail: vanirm12@rediffmail.com

K. Prahlada Rao (✉) · P. V. Hunagund
Department of PG Studies and Research in Applied Electronics, Gulbarga University, Gulbarga, India
e-mail: pra_kaluri@rediffmail.com

G. Gnanagurunathan et al. (eds.), *Optical and Microwave Technologies*,
Lecture Notes in Electrical Engineering 468,
https://doi.org/10.1007/978-981-10-7293-2_9

antennas are simple in construction. They are made of a radiating patch on top of the dielectric substrate and a ground plane at its bottom [1–4].

Microstrip antenna arrays have gained widespread attention among the researchers due to their large gain compared to single-element microstrip antennas. Miniaturization of antenna and reduction in mutual coupling are the two principal aspects that the antenna designers are focusing apart from other parameters toward the enhanced performance of microstrip antenna arrays. Antenna arrays suffer from the drawback of surface waves generated in the dielectric substrate. The surface waves pose a serious threat toward the antenna array performance. These surface waves are undesired as a part of the total radiated power is trapped in the dielectric substrate; thereby, the antenna array efficiency is reduced. Various technologies are available to overcome this serious disadvantage of microstrip antenna arrays [5–8].

Electromagnetic band gap (EBG) structures are suitable candidates in achieving this objective. Recent advancements and study of EBG structures have revealed their aid in improving the characteristics of antenna arrays. EBG structures are periodic arrangement of dielectric or metallic elements used to stop or pass electromagnetic waves in particular frequency bands. These stop or pass bands play a key role in forbidding or passing the surface waves. With the suppression of surface waves, EBG structures have made significant contribution in ameliorating the performance of antenna array in terms of reduction in back radiation, virtual size of antenna array, and mutual coupling between antenna elements [9–12].

2 Antenna Array Design Without EBG

The design frequency of the conventional two-element antenna array is 6 GHz consisting of two identical elements. The microstrip antenna array designed is a linear array where two elements are placed side by side in a straight line. The antenna array is fed by using corporate feeding technique. The distance between the two elements is 20.1 mm. The dielectric used in the design of the antennas is FR-4 glass epoxy with dielectric constant of 4.2. The height (*h*) and loss tangent of the dielectric are equal to 1.6 mm and 0.0245, respectively. The dimensions of the antenna array are calculated as per the equations mentioned in the literature. The length and width of the finite ground plane are equal to 115.8 and 62.7 mm, respectively. Figure 1 depicts the schematic of conventional two-element microstrip array antenna with finite ground plane.

Table 1 depicts the dimensions of two-element microstrip array antenna with finite ground plane.

By maintaining the same distance between the two antennas, the two antennas are fed independently. Such an arrangement is used to measure the value of mutual coupling between the two antennas under the assumption that the two antennas are fed with the same amount of power.

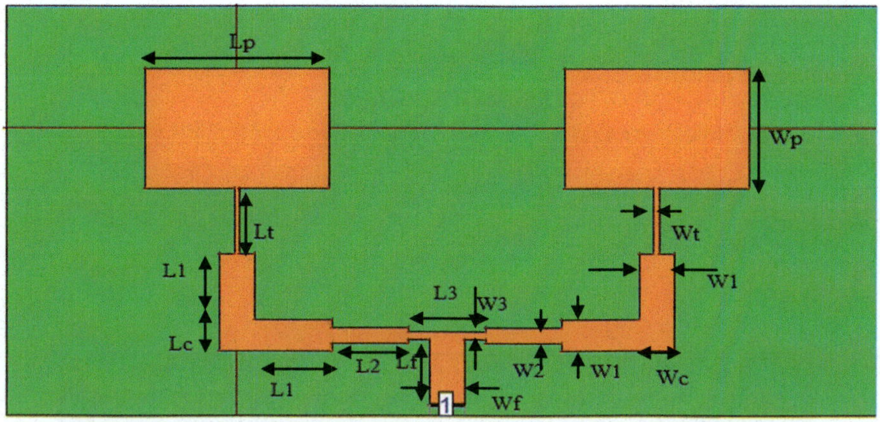

Fig. 1 Schematic of two-element antenna array

Table 1 Design parameters of two-element antenna array

Parameter	Value (mm)
Length of the patch (*Lp*)	15.73
Width of the patch (*Wp*)	11.76
Length of the quarter wave transformer (L_t)	6.47
Width of the quarter wave transformer (W_t)	0.47
Length of the 50 Ω line (L_1)	6.52
Width of the 50 Ω line (W_1)	3.05
Length of the coupler (L_c)	3.05
Width of the coupler (W_c)	3.05
Length of the 70 Ω line (L_2)	6.54
Width of the 70 Ω line (W_2)	1.62
Length of the 100 Ω line (L_3)	6.56
Width of the 100 Ω line (W_3)	0.70
Length of the feed line (L_f)	6.52
Width of the feed line (W_f)	3.05

3 Antenna Array Design with EBG Slots

The EBG structure used in the design of the array antenna consists of 4 × 7 matrix of swastika slots. Figure 2 shows the schematic of unit cell of the EBG structure employed and its dimensions.

These swastika slots are placed periodically in the finite ground of the conventional two-element antenna array. The distance between the two adjacent cells is 8.7 mm. Figure 3 depicts the schematic of antenna array with EBG.

To determine the mutual coupling of the array antenna with EBG, the EBG structure is loaded in the finite ground plane of array of two separately fed antennas as shown in Fig. 4.

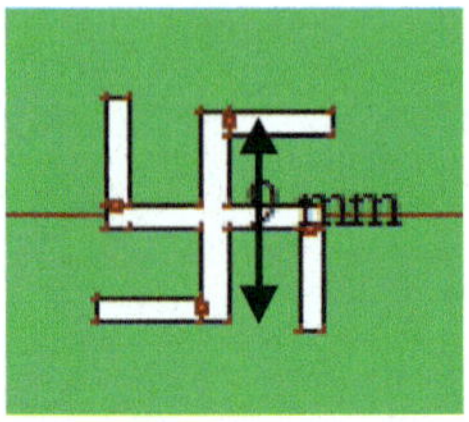

Fig. 2 Unit cell of EBG structure

Fig. 3 Schematic of two-element antenna array with EBG slot

Fig. 4 Schematic of two-element antenna array arrangement with EBG for mutual coupling measurement

4 Photographs of Fabricated Antennas

See Figs. 5, 6, 7, and 8.

(a) Front view (b) Back view

Fig. 5 Photograph of conventional two-element antenna array

(a) Front view (b) Back view

Fig. 6 Photograph of two-element antenna array arrangement for mutual coupling measurement

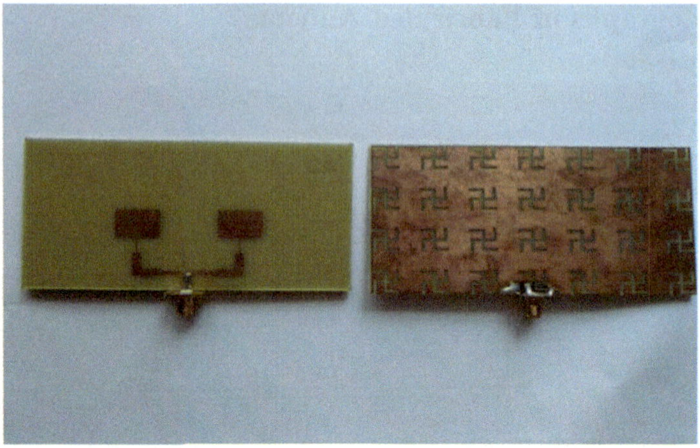

(a) Front view (b) Back view

Fig. 7 Photograph of two-element antenna array with EBG slot

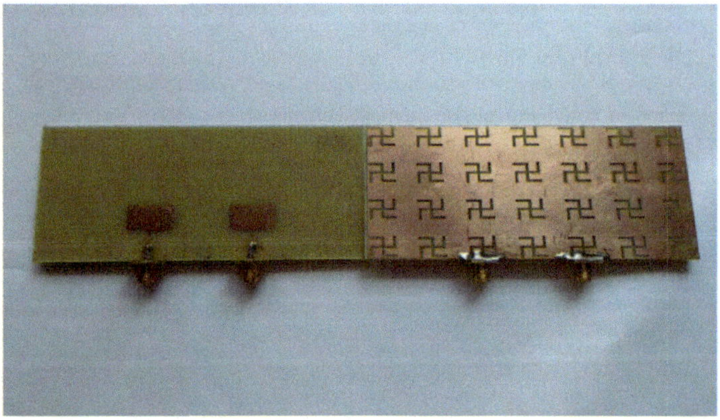

(a) Front view (b) Back view

Fig. 8 Photograph of two-element antenna array arrangement with EBG for mutual coupling measurement

5 Measured Results

The various characteristics of the antennas—resonant frequency, return loss, bandwidth, mutual coupling, size reduction, front-to-back ratio, and radiation pattern—have been measured. The return loss is designated by the *S*-parameter S_{11}, and mutual coupling is designated by the *S*-parameter S_{21}.

Figure 9 depicts the *S*-parameter characteristics of two-element antenna array without EBG. The *S*-parameters measured are return loss and mutual coupling, expressed in terms of decibels (dB).

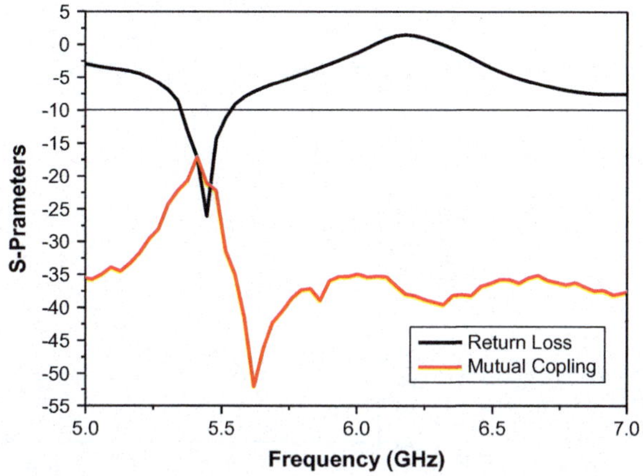

Fig. 9 Plot of return loss and mutual coupling versus frequency of two-element antenna array without EBG

The parameter return loss is defined as the amount power reflected back by an antenna. Resonant frequency is defined as the frequency where the return loss is minimum between the lower and upper frequencies. From Fig. 9, the value of return loss at the fundamental resonant frequency of 5.445 GHz is equal to −25.96. The return loss versus frequency plot also gives information about bandwidth and resonant frequency. Bandwidth is calculated by subtracting the lower frequency from upper frequency where the return loss is crossing the −10 dB line. From Fig. 9, the bandwidth of the conventional two-element antenna array is 194 MHz.

Mutual coupling determines the isolation between the two antennas. From Fig. 9, the mutual coupling measured at the fundamental resonant frequency of 5.445 GHz is −20.81 dB. This value of mutual coupling represents a strong value which needs to be decreased. The *S*-parameter characteristics of two-element antenna array with EBG are shown in Fig. 10.

The *S*-parameters measured are return loss and mutual coupling, expressed in terms of decibels (dB).

From Fig. 10, it is clear that the antenna array with swastika slot EBG structure is resonating at seven frequencies −1.469, 2.290, 2.85, 4.308, 7.105, 9.877, and 11.05 GHz, respectively. The return loss measured at the fundamental resonant frequency of 1.469 GHz is −11.12 dB. The lower value of resonant frequency of 1.469 GHz with EBG when compared to 5.445 GHz without EBG leads to a virtual size reduction in antenna array of 73.02%.

Figure 10 also depicts that the mutual coupling of the antenna array with EBG structure measured at the frequency of 5.445 GHz is reduced to −42.19 dB. The radiation patterns of antenna arrays without and with EBG are shown in Fig. 11.

Radiation pattern of an antenna provides the information about the distribution of power at different orientations of antenna. The main parts of a radiation pattern are major lobe and minor lobe. With the introduction of swastika slot EBG structure in the ground plane of the two-element antenna array, there is a significant decrease in the back lobe of total radiated power than that without EBG structure. The radiation plots of the antenna arrays without and with EBG structure are traced at

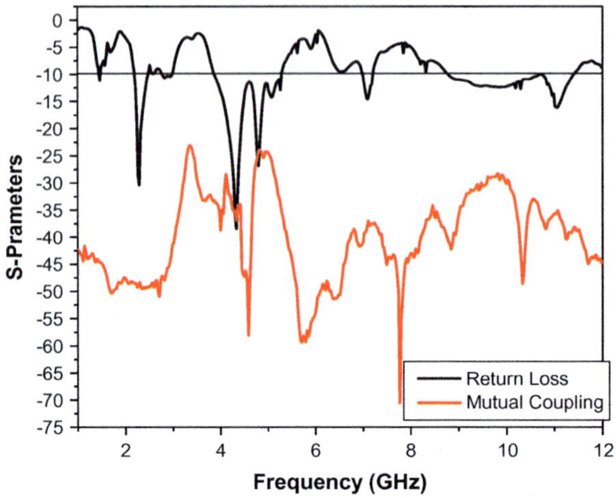

Fig. 10 Plot of return loss and mutual coupling versus frequency of two-element antenna array with EBG slot

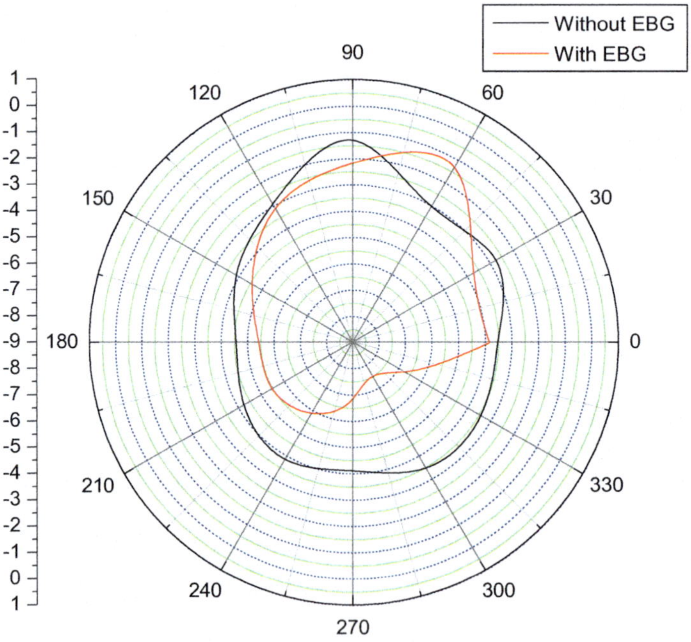

Fig. 11 Plot of radiation pattern of antenna arrays without and with EBG

5.445 and 1.469 GHz, respectively. The power radiated shown on *Y*-axis is expressed in terms of decibels (dB). From Fig. 11, the amount of backward power radiated without and with EBG is equal to −4 and −7.5 dB, respectively. With the introduction of EBG structure, the amount of back lobe is reduced by 3 dB.

Front-to-back ratio is defined as the ratio of power radiated in the desired direction to that in the backward direction. Front-to-back ratio is calculated by using the formula.

$$\begin{aligned}\text{Front to Back Ratio(FBR)} &= \text{Power radiated in the front direction(dB)} \\ &\quad - \text{Power radiated in the back direction(dB)}\end{aligned}$$

Power radiated in the front direction is the power measured at the angle of 90° and, in the back direction, is the power measured at the angle of 270°. Without EBG, antenna array has the front and back powers radiated are equal to −1.5 and −4 dB, respectively. With EBG, the corresponding values are equal to −2.5 and −7.5 dB, respectively. Hence, FBR calculated without and with EBG are equal to 2.5 and 5 dB, respectively. From the above discussion, FBR with EBG is greater than that without EBG.

The EBG slot structure etched in the ground plane of the microstrip antenna array disturbs its current distribution. This disturbance affects the transmission line characteristics such as line capacitance and inductance, which affect the band gaps of the EBG structure. Moreover, the introduction of EBG structure in the ground plane decreases the current density on the surface of the patch. This suppression of current density leads to reduction in mutual coupling and better isolation between antenna elements.

6 Summary of Measured Results

Table 2 depicts the summarized measured results.

Table 2 Summarized measured results

Type of antenna	Resonant frequency (GHz)	Return loss (dB)	Bandwidth (MHz)	Bandwidth (%)	Overall bandwidth (%)
Without EBG	5.445	−16.49	194	3.562	3.562
With swastika slot EBG	1.469	−11.12	44	2.99	79.72
	2.290	−28.66	334	14.58	
	2.850	−10.70	81	2.84	
	4.308	−38.17	1397	32.42	
	7.105	−14.46	267	3.75	
	9.877	−12.36	1951	19.75	
	11.05	−15.91	688	6.22	

7 Conclusion

In this paper, an attempt has been made to ameliorate the performance of conventional two-element antenna array with the swastika slot EBG structure loaded in the ground plane. The mutual coupling between the antenna elements has been significantly reduced with the introduction of EBG structure when compared to without EBG. The antenna array is also producing multiple resonances thus providing an overall bandwidth of 79.72% when compared to 3.562% without EBG. The antenna array is resonating at a lower fundamental resonant frequency with the EBG structure which results in a significant virtual size reduction in 73.02%. Moreover, reduction in back lobe and increase in front-to-back ratio have also been achieved. A considerable amount of return loss has also been achieved with the EBG structure. The results confirm that the antenna array with EBG structure has better performance compared to its counterpart.

References

1. James Scott; Lecture notes of EEET1071/1127 microwave and wireless passive circuit design
2. Gautam AK (2010) Antenna and wave propagation. S. K. Kataria & Sons
3. Constantine AB (1997) Antenna theory, analysis and design, 2nd edn, Wiley, London
4. Yang F, Rahmat-Sami Y (2009) Electromagnetic band gap structures in antenna engineering. The Cambridge RF and microwave engineering series
5. Kondori H, Mansouri-Birjandi MA, Tavakoli S (2012) Reducing mutual coupling in microstrip array antenna using metamaterial spiral resonator. Int J Comput Sci Issues 09 (3):51–56
6. Christodoulou CG, Parveen PF (2004) Fundamentals of antennas: concepts and applications. Prentice Hall of India, Wahid
7. Bahl IJ, Bhartia P (1980) Microstrip antennas. Artech House, USA
8. Bhavsar S, Singh B (2013) Electromagnetic band gap structures incorporated in antenna array: a review. Int J Comput Technol Electron Eng 03:22–28
9. Elsheakh DN, Iskander MF, Abdallah EA, Elsadek HA (2010) Microstrip array antenna with new 2D-electromagnetic band gap structure shapes to reduce harmonics and mutual coupling. Prog Electromagn Res C 12:203–213
10. Gupta S, Kumar D (2016) Compact patch antenna array using defective ground structure on feedline. Int J Comput Appl 149(09)
11. Abdulhamid M, Rahim MKA, Musa U (2015) Electromagnetic band gap structure for antenna design. IOSR J Electron Commun Eng 10(06):25–27
12. Mentor graphics IE3D user manual, April 2010

Gain Enhancement of Compact Multiband Antenna with Metamaterial Superstrate

Rengasamy Rajkumar and Kommuri Usha Kiran

Abstract In this paper, superstrate-based gain enhancement technique for a compact metamaterial multiband antenna is presented. The gain enhancement is achieved based on metamaterial-based superstrate layers. The proposed antenna consists of two superstrate layers which have 4 × 4 metamaterial unit cells and multiband antenna. Here, two superstrate layers are used to achieve the gain enhancement in the lower frequency bands of the proposed multiband antenna. The superstrate layer unit cells are analysed based on waveguide set-up method and etched in RT/duroid 5880. The air gap distance ($h1$ and $h2$) for both superstrate layers are fixed based on parametric study and placed over the multiband antenna. The proposed antenna has positive gain in all the desired frequencies.

1 Introduction

The modern communication needs compact antenna designs as well as multiple applications in a single device. In order to achieve that, multiband antenna technology is proposed. The compact broadband antennas cover various applications with compact size, but they have less gain [1] due to antenna miniaturization. In order to improve the gain and radiation characteristics, many techniques are proposed such as electromagnetic band gap techniques (EBG) [2, 3] and non-magnetic dielectric [4]. The above methods are effectively improved gain of the proposed antenna but they require thick dielectric layers, and it lead to increase the size or profile of the antenna. To realize compact designs, metamaterial-based designs are proposed. Metamaterials are artificial materials, and its properties are not found in nature. The metamaterials are very useful for enhancement of gain, radiation properties, broader bandwidth and

R. Rajkumar (✉) · K. Usha Kiran
School of Electronics Engineering, VIT University, Chennai Campus,
Chennai 600127, India
e-mail: rajkumar.r2013@vit.ac.in

K. Usha Kiran
e-mail: usha.kiran@vit.ac.in

G. Gnanagurunathan et al. (eds.), *Optical and Microwave Technologies*,
Lecture Notes in Electrical Engineering 468,
https://doi.org/10.1007/978-981-10-7293-2_10

agile performances [4–6]. Especially, metamaterials are used as gain enhancement structure in many designs based on various shaped SRR as [7–10]. The above metamaterial-based superstrate gain enhancement is about particular frequency region. Here in the proposed superstrate design, the gain is improved in the lower part region (application frequency region) while maintaining positive gain in the higher frequency region. The proposed technique is the effective method to increase the gain of the antenna. It occupies less space comparing other existing antenna designs. Its gain is improved in WLAN/WiMAX band applications.

2 Antenna Design

The proposed antenna is designed consists of two parts: first one is triangular metamaterial antenna and second is two superstrate layers. The triangular layer is designed as described in [1]. It covers three bands in which at lower frequency, antenna has less gain. In order to enhance antenna gain, two superstrate layers are designed on the RT/duroid 5880 with thickness of 0.5 mm. Two layers are placed above the antenna with air gap distance $h1$ of 4.45 mm and $h2$ of 5.5 mm as depicted in Fig. 3 (Table 1).

The $h1$ and $h2$ parameters are decided based on the parametric studies. In this, 4×4 square-shaped SRRs are present in the superstrate layer as illustrated in Fig. 1. Square-shaped SRRs are designed to resonate particular resonance frequency in order to have improved gain in the lower frequency (Fig. 2).

Table 1 Dimensions of the proposed antenna

Parameters	*L*	*W*	*L*1	*L*2	*L*3	*L*4	*L*5	*W*1	*W*2
Dimensions (mm)	25.2	23.7	10	14.42	9.42	3.9	2.5	10.9	7.7
Parameters	*G*	*G*1	*F*1	*F*2	*RL*1	*RL*2	*RG*	*WS*	*LS*
Dimensions (mm)	0.3	0.35	1.3	0.84	5.6	4.28	0.2	27.7	27

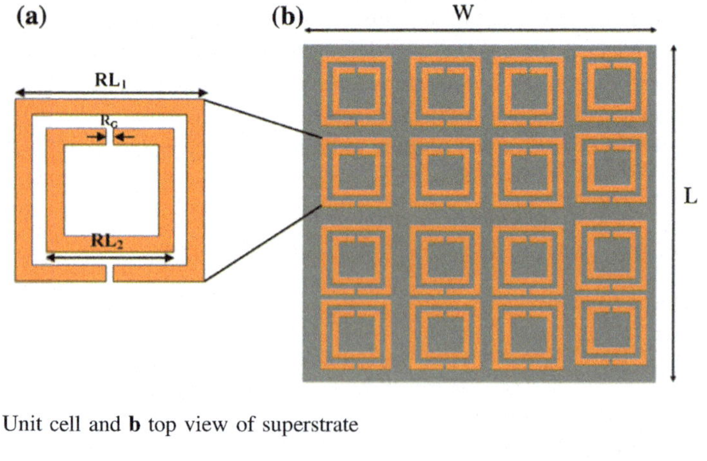

Fig. 1 **a** Unit cell and **b** top view of superstrate

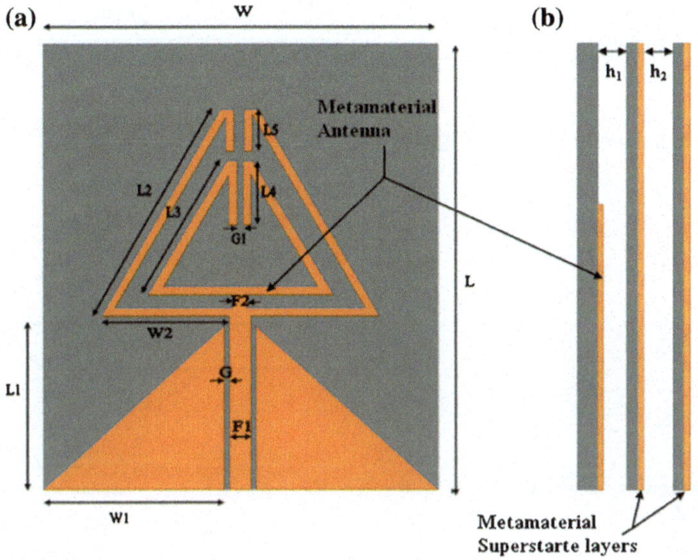

Fig. 2 Proposed antenna, **a** schematic diagram of multiband antenna and **b** perspective view

3 Result and Discussion

The proposed superstrate antenna covers multiple frequencies. The metamaterial antenna design covers desired applications. Metamaterial antenna (without superstrate) refection coefficient is shown in Fig. 7 (Fig. 3).

It covers WLAN/WMAX/ITU applications, and it has gain of −4.5, −2.5, 1.2, 1.8, 2.2 and 3.49 dB at 2.4/3.5/5.2/5.8/8.2 GHz in the lower frequency region. Due to more miniaturization nature, the antenna has gain low gain. To improve gain of the antenna, two metamaterial superstrate layers are used in above antenna with air gap. The square-shaped SRR was chosen owing to simple fabrication. To confirm resonance frequency of the proposed SRR structure, the waveguide set-up was followed as suggested [11]. Waveguide excitation is assigned as similar to [Elsevier] to the proposed design. The reflection and refraction ($S11$ and $S21$) are depicted in Fig. 4. From this, we can confirm the proposed SRR having resonance at 3.6 GHz. To improve gain in the lower band, 4 $\times$ 4 square-shaped SRRs on RT/duroid 5880 are etched. The first superstrate layer is placed with air gap $h1$ of 4.45 mm, and the second superstrate layer is placed with air gap of 5.5 mm. These two $h1$ and $h2$ parameters are concluded based on the parametric studies. The reflection coefficient of first layer ($h1$) and second layer ($h2$) is depicted in Figs. 5 and 6.

Comparison analysis with and without superstrate is shown in Fig. 7. We can infer that, after placing first superstrate, antennas reflection coefficient is shifted at higher frequencies (8.2 GHz) (Fig. 8).

Fig. 3 Waveguide set-up of the proposed antenna

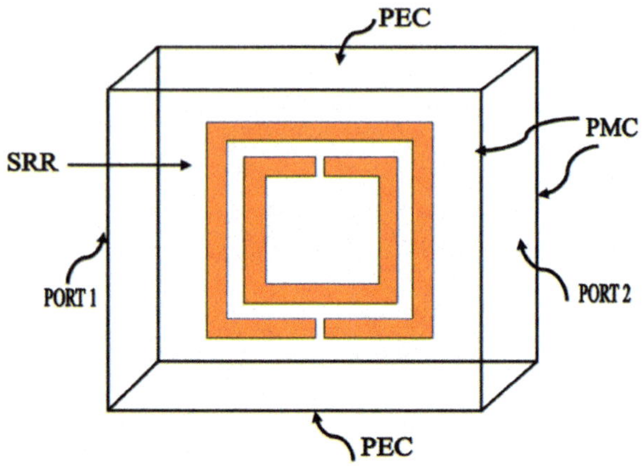

Fig. 4 Waveguide set-up of the proposed antenna

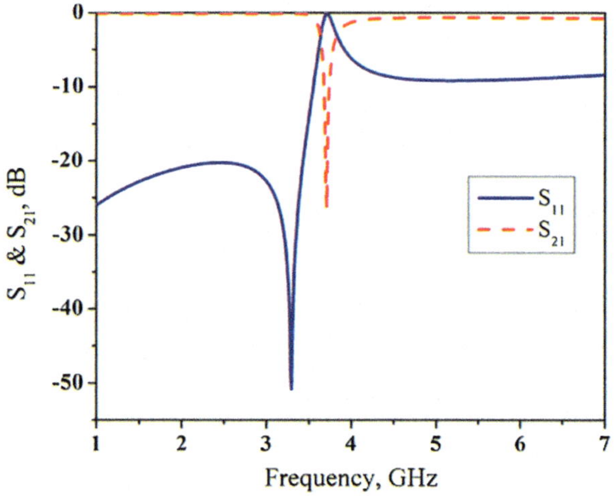

Fig. 5 Reflection coefficient of the proposed antenna with varying '*h*1'

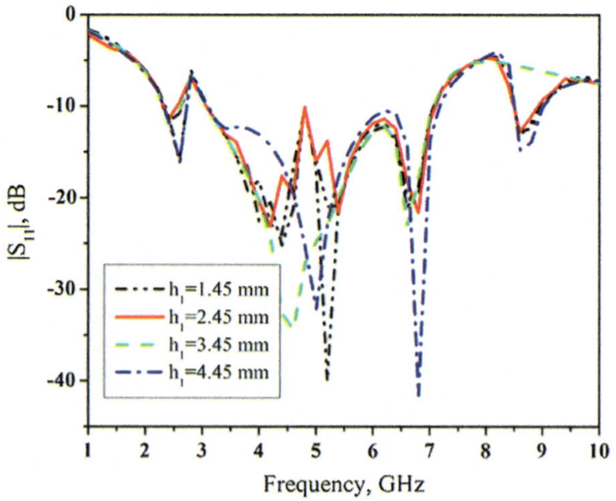

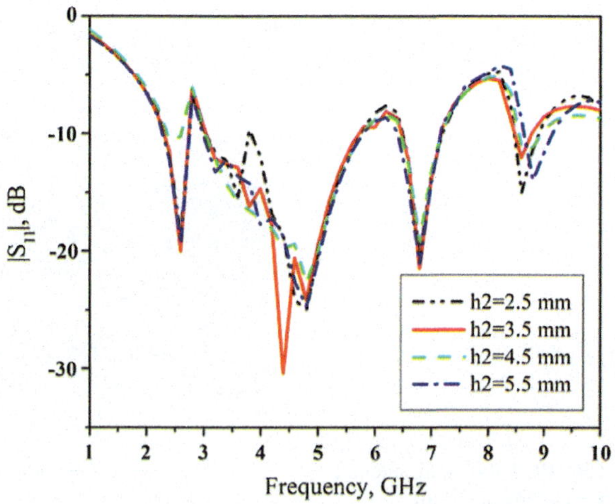

Fig. 6 Reflection coefficient of the proposed antenna with varying '*h*2'

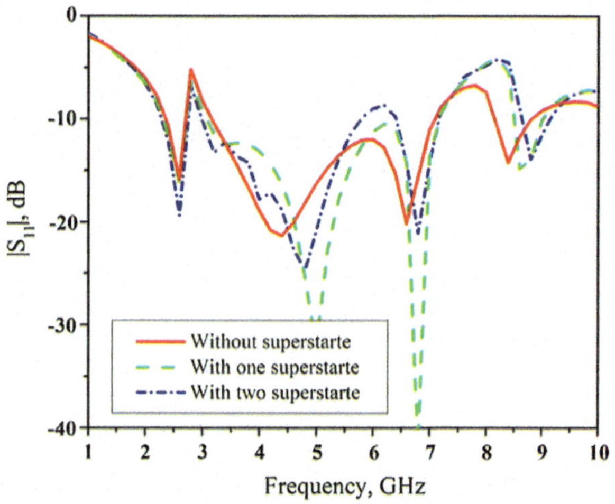

Fig. 7 Simulated reflection coefficient of various stages of the proposed antenna

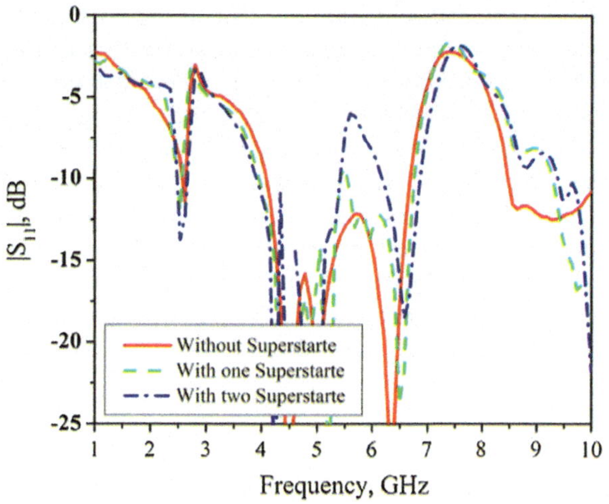

Fig. 8 Measured reflection coefficient of various stages of the proposed antenna

Fig. 9 Fabricated antenna measurement set-up

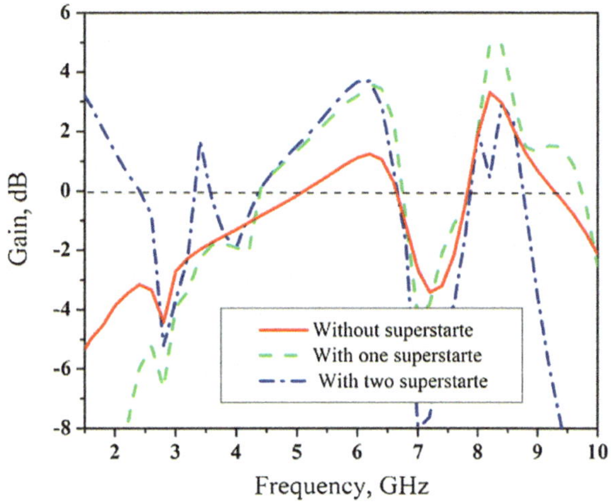

Fig. 10 Fabricated antenna measurement set-up

The measured reflection coefficients are similar to the simulated reflection coefficients. There are small variations that are present at higher frequencies due to fabrication and soldering loss. The proposed gain enhancement superstrate antenna and the measured antenna set-up are shown in Fig. 9. In this set-up, it has two superstrate layers placed over the metamaterial antenna with help spacers. The antenna characteristics are measured using Agilent vector network analyser N9918A. The proposed antenna E-plane and H-plane radiation characteristics are shown in Fig. 11. It shows that antenna has good radiation characteristics in all the desired frequencies. The gain of proposed antenna is improved after employing two metamaterial layers as shown in Fig. 10; all the energies radiated from printed antennas are converged by two substrates. Due to the no energy loss, the gain level is enhanced. It leads to the proposed antenna has gain of 0.40, 0.70, 2.64 and 1.7 dB at 2.4, 3.5, 5.5 and 8.3 GHz respectively (Fig. 11).

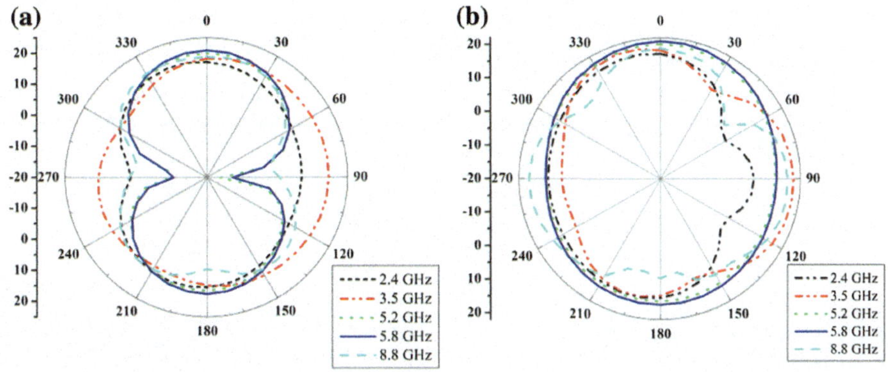

Fig. 11 Radiation pattern of the proposed antenna, **a** E-plane and **b** H-plane

4 Conclusion

The proposed superstrate antenna is designed and used for improving antenna gain in the lower frequency region. It used two 4 × 4 split-ring resonator superstrate to enhance gain of the antenna. The antenna has lower profile compared to other existing designs, and it enhanced gain level in whole lower working applications. The proposed antenna covers WLAN/WiMAX/ITU applications effectively.

References

1. Rajkumar R, Usha Kiran K (2016) A compact metamaterial multiband antenna for WLAN/WiMAX/ITU band applications. AEU Int J Electron Commun 70(5):599–604
2. Attia H, Ramahi OM (2008) EBG superstrate for gain and bandwidth enhancement of microstrip array antennas. In: Proceedings of IEEE AP-S International Symposium on Antennas Propagation, San Diego, CA, pp 1–4
3. Lee YJ, Yeo J, Mittra R, Park WS (2005) Application of electromagnetic bandgap (EBG) superstrates with controllable defects for a class of patch antennas as spatial angular filters. IEEE Trans Antennas Propag 53(1):224–235
4. Jacksonand DR, Alexopoulos NG (1985) Gain enhancement methods for printed circuit antennas. IEEE Trans Antennas Propag 33(9):976–987
5. Smith DR, Padilla WJ, Padilla WJ, Vier DC, Schultz S (2000) Composite medium with simultaneously negative permeability and permittivity. Phys Rev Lett 84(18):4184–4187
6. Caloz C, Itoh T (2006) Electromagnetic metamaterials: transmission line theory and microwave applications, 1st edn. Wiley-IEEE Press, Hoboken, NJ. ISBN-10: 0471669857
7. Attia H, Yousefi L, Ramahi OM (2011) Analytical model for calculating the radiation field of microstrip antennas with artificial magnetic superstrates: theory and experiment. IEEE Trans Antennas Propag 59:1438–1445
8. Bouzouad M, Chaker SM, Bensafielddine D, Laamari EM (2015) Gain enhancement with near-zero-index metamaterial superstrate. Appl Phys A 121:1075–1080
9. Tang H, Hou Q, Liu Y, Zhao X (2013) A high gain omnidirectional antenna using negative permeability metamaterial. Int J Antennas Propag 1–7

10. Liu Zhenzhe et al (2013) Enhancement of the gain for microstrip antennas using negative permeability metamaterial on low temperature co-fired ceramic (LTCC) substrate. IEEE Antennas Wirel Propag Lett 12:429–432
11. Chen H, Zhang J, Bai Y, Luo Y, Ran L, Jiang Q et al (2006) Experimental retrieval of the effective parameters of metamaterials based on a waveguide method. OptExpress 14 (26):12944–12949

Cooperative Communication for Resource Sharing in Cognitive Radio Networks

Lekha Abraham and R. G. Sangeetha

Abstract Increasing demand of wireless communication led to inefficiency in spectrum utilization. To solve this problem, cognitive radio networks came up as a key technology. In a Cognitive radio network, the secondary user uses the licensed spectrum of the primary user by monitoring the communication channel and instantly moves to the vacant channel when found, provided the transmission of the primary user is not interrupted. In a cooperative mode of communication, each entity is aware of each other and they cooperate to maximize the system performance. A cooperative spectrum sensing using energy detection with data fusion technique is implemented using MATLAB. The fusion center compares the energy of the received signal with the threshold value in order to detect the primary user presence in the channel. A two-stage dynamic spectrum sensing with adaptive thresholding using energy detection with data fusion technique in an QAM-16 platform is implemented using MATLAB, and the sensing efficiency is evaluated.

Keywords Cognitive radio networks · Adaptive thresholding · Spectrum sensing Cooperative communication

1 Introduction

In the recent decades, the cognitive radio networks (CRN) have been under research since it provides an efficient solution to the problem of resource scarcity. It enhances the radio frequency spectrum utilization by closely observing the channel and immediately using the vacant channel when found. The primary user (PU) or the licensed user do not use the allocated channel all the time, i.e., the spectrum wasted majority of the time, the cognitive radio network efficiently allots the unused

L. Abraham · R. G. Sangeetha (✉)
School of Electronics Engineering, VIT University, Chennai 600127, India
e-mail: sangeetha.rg@vit.ac.in

L. Abraham
e-mail: lekhaabh@gmail.com

G. Gnanagurunathan et al. (eds.), *Optical and Microwave Technologies*,
Lecture Notes in Electrical Engineering 468,
https://doi.org/10.1007/978-981-10-7293-2_11

channel to an unlicensed user or a secondary user (SU). The secondary user can use the channel till the primary user returns. The primary condition for allocating its spectrum to a secondary user for transmission is that the primary user should not be kept waiting. The secondary user should immediately vacate the channel when the primary user returns to use the channel. The presence of the primary user is determined by spectrum sensing techniques. Different sensing techniques like matched filter-based, cyclostationary featuring-based and energy-based sensing techniques are present. The non-requirement of a priori knowledge of primary user signal and its simplicity makes energy detection (ED) the most popular sensing technique in cooperative sensing. Energy detection method of sensing is a non-coherent method of detection, where the received signal energy determines the presence of the primary user.

2 Literature Survey

Soft combination and detection for cooperative spectrum sensing in cognitive radio networks introduce soft combinations decisions that are based on the energy detected by different secondary users. A two-bit hard combination requires lots of overhead for detection. The combination of both schemes where the area under observation is divided based on thresholds, and detection of the primary user is based on value of energy detected [1]. Primary-prioritized Markov approach for dynamic spectrum allocation proposes a continuous chain Markov approach used to model the interaction of primary and secondary user. An optimal access probability of secondary user is used to coordinate the spectrum access of the secondary user. A good trade-off between efficiency and fairness is achieved [2].

Optimal balancing between efficiency and fairness for resource management in cognitive radio network is given by a dynamic game theory approach. This gives the best trade-off between efficiency and fairness. It also proposes a joint resource allocation algorithm using dual decomposition algorithm [3]. Resource allocation and fairness in wireless powered cooperative cognitive networks, August, combine the advantages of the wireless powered communication network (WPCN) and cooperative cognitive radio network. It proposes sum-throughput optimal resource allocation (STORA) scheme that maximizes secondary user throughput. STORA increases the throughput of secondary user at a low fairness due to throughput-fairness trade-off [4].

Queue performance of cognitive radio networks with general primary user activity model proposes a 2D Markov chain which analyzes the queue performance of the SU. Effective capacity (EC) function is used to describe the data rate available, and effective bandwidth (EB) function is widely used to describe the traffic arrival for QoS [5].

Dynamic spectrum access and power allocation for cooperative cognitive radio networks, present a new cooperative framework for using underutilized resources. Cooperative relays are used for this purpose, where primary users rely on data for secondary users. It proposes an auction model where auctioneers cooperatively transmit power and allocate channel bands to secondary users. Secondary user takes assistance from primary user to purchase power and spectrum [6].

Cooperative-generalized sensing-based spectrum sharing approach for centralized cognitive radio networks, the secondary users are considered as mere receivers and secondary base stations (SBS) uses the power allocated to communicate. But no decision on the PU state is taken by SBS. Average transmit power shows improvement [7].

Coopetition methodology for resource sharing in distributed OFDM-based cognitive radio networks proposes coopetition technique that takes the best from cooperative and competitive spectrum sharing schemes. It also provides a balance between fairness and spectral efficiency for resource allocation [8]. Cooperative spectrum sharing with wireless energy harvesting in cognitive radio networks improves the EH (energy harvesting) efficiency. The secondary transmitter (ST) with the shortest distance from the PU is given the priority to transmit first. In this system, the energy is saved and used for data transmissions [9].

3 System Model

Spectrum sensing mechanism determines the presence of a PU. The following hypothesis is used to detect the presence of PU in the channel [10],

$$\begin{aligned} x(n) &= w(n),\ H_0; \\ x(n) &= s(n) * h(n) + w(n), H_1 \end{aligned}$$

where $x(n)$ is the received signal by each secondary user transmitter, $s(n)$ is the licensed signal of PU, and $w(n)$ is AWGN with 0 mean and σ_w^2 variance.

The system consists of seven primary users, whose licensed channel is to be shared between the secondary users. The Alamouti technique with two-transmit-antenna scheme is used as the transmit diversity in order to achieve the maximum transmit gain. The system is implemented in MATLAB R2013a and implemented energy-based spectrum sensing scheme using adaptive thresholding. Threshold setting is difficult due to noise uncertainty, and it often degrades the sensing performance of the cognitive radio networks. Detection between the noise and the primary signal is difficult, and the primary signal is often confused with the noise signal and vice versa (Fig. 1).

The region between the λ_1 and the λ_2 is the confusion region where the primary signal is often misinterpreted as the noise signal or vice versa.

The sensing performance of the cognitive radio is evaluated in terms of probability of detection (Pd), probability of false alarm (Pf), and probability of missed detection (Pmd). Pd is the probability that a primary signal is detected and is based on the threshold value (λ), which is dependent on the signal SNR [11].

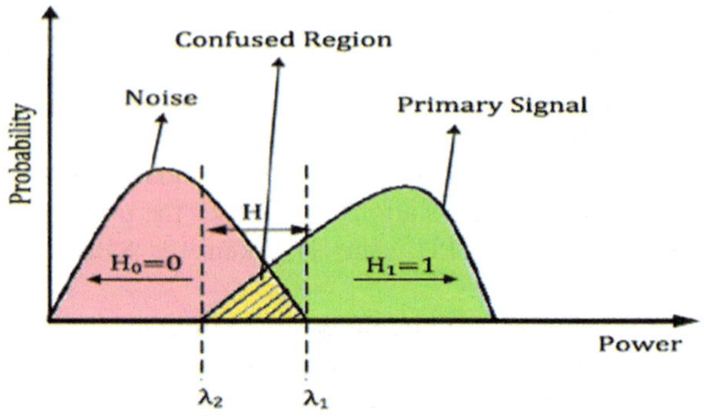

Fig. 1 Energy distribution of the primary signal [10]

If the received energy signal falls below λ, PU signal is not present and if it is above the particular λ value, the primary signal is present.

4 Results

The sensing performance of the cognitive radio network is evaluated by comparing the Pd versus Pf as in Fig. 2, the probability of detection increases as the probability of false alarm increases and eventually reaches a probability 1.

In Fig. 3, we can see that the probability of detection of the signal increases as the SNR increases.

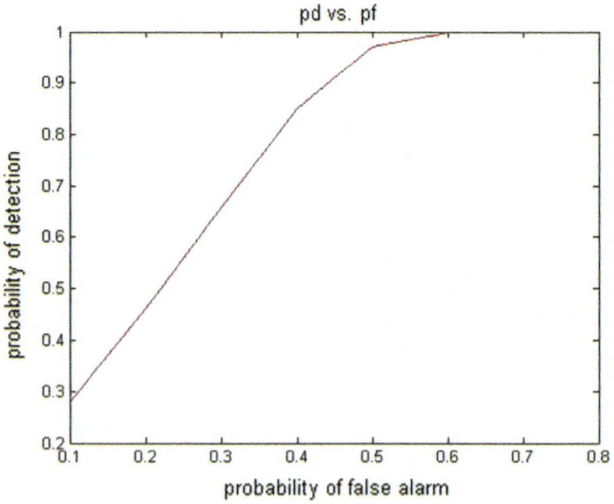

Fig. 2 Probability of detection (Pd) versus probability of false alarm (Pf)

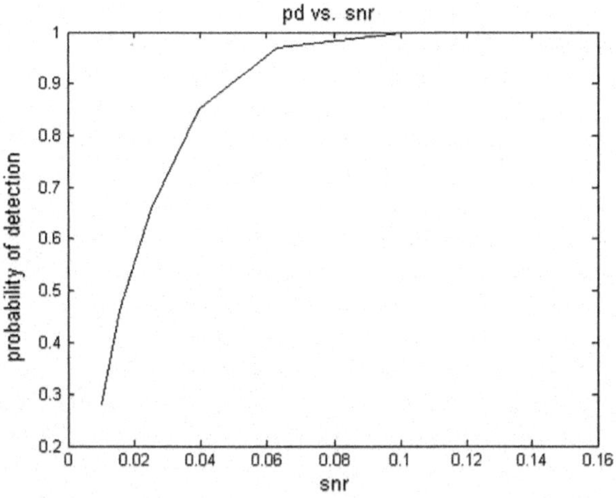

Fig. 3 Probability of detection (Pd) versus SNR

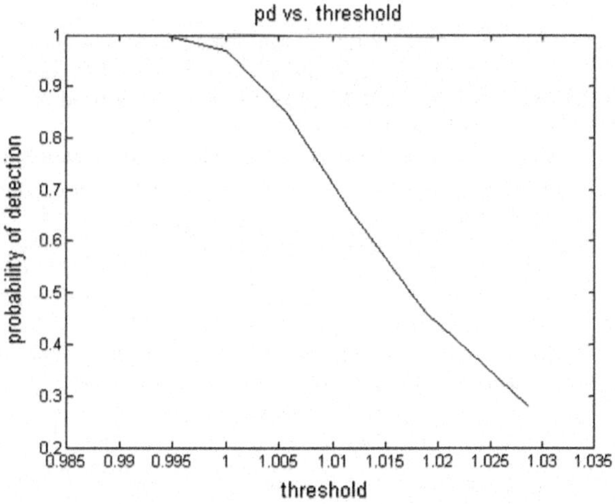

Fig. 4 Probability of detection (Pd) versus threshold

The probability of detection is high at lower threshold values and as the system has an adaptive threshold the probability of detection value decreases with increases in threshold as shown in Fig. 4. The probability of missed detection decreases with increase in SNR in Fig. 5.

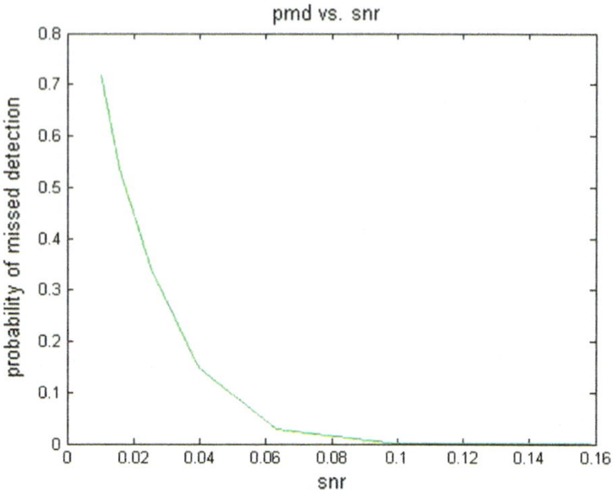

Fig. 5 Probability of missed detection (Pmd) versus SNR

References

1. Ma J, Zhao G, Li Y (2008) Soft combination and detection for cooperative spectrum sensing in cognitive radio networks. IEEE Trans Wirel Commun 7(11)
2. Wang B, Ji Z, Liu KJR (2009) Primary prioritized markov approach for dynamic spectrum allocation. IEEE Trans Wirel Commun 8(4)
3. Yang C, Li J, Chen D (2010) Optimal balancing between efficiency and fairness for resource management in cognitive radio network: a dynamic game-theory approach. In: IEEE CCNC proceedings
4. Kalamkar SS, Jeyaraj JP, Banerjee A (2016) Resource allocation and fairness in wireless powered cooperative cognitive networks. IEEE Trans Commun 64(8)
5. Jiao W, Sheng M, Lui KS, Liu G (2015) Queue performance of cognitive radio networks with general primary user activity model. IET J. ISSN 1751-8628
6. Zou J, Wu Q, Xiong H, Chen CW (2015) Dynamic spectrum access and power allocation foe cooperative cognitive radio networks. IEEE Trans Signal Process 63(21)
7. Chen Z, Gao F (2016) Cooperative generalized sensing based spectrum sharing approach for centralized cognitive radio networks. IEEE Trans Veh Technol 65(5)
8. Parzy M, Bogucka H (2014) Coopetition methodology for resource sharing in distributed OFDM-based cognitive radio networks. IEEE Trans Commun 62(5)
9. Zhai C, Liu J, Zheng L (2016) Cooperative spectrum sharing with wireless energy harvesting in cognitive radio networks. IEEE Trans Veh Technol 65(7)
10. Bagwari A, Tomar GS, Verma S (2013) Cooperative spectrum sensing based on two-stage detectors with multiple energy detectors and adaptive double threshold in cognitive radio networks. Can J Electr Comput Eng 36(4)
11. Yang G, Luo J, Wen OY (2016) Cooperative spectrum sensing in heterogeneous cognitive radio networks based on normalized energy detection. IEEE Trans Veh Technol 65(3)

Interdomain Traffic Engineering with BGP and MPLS VPN

M. Athira and R. G. Sangeetha

Abstract Flow of packets in network is done in a controlled manner by implementing set of techniques which confines within traffic engineering. Across border of domains, route advertisement packets of border gateway protocols are filtered to make the traffic engineering possible. Large organizations often have branches that are spread in different places. These distributed branches need to be connected with the similar level of security and privacy as in a LAN network. Virtual private networks (VPNs) were designed to deliver this need. A common VPN topology was designed of Multiprotocol extensions Border gateway protocol (MP-BGP) and Multiprotocol Label Switching (MPLS). And finally the performance analysis is done Switching (MPLS).

Keywords Traffic engineering · Interdomain · Internet protocol
Border gateway protocol · Route advertisements · Autonomous system
Interior border gateway protocol · Exterior border gateway protocol
Virtual route forwarding · Virtual private network · Multiprotocol label switching
Wide area networks · Multi-exit distinguisher · Internet service provider
Enhanced interior gateway protocol

1 Introduction

Larger enterprises have sites that are scattered over different locations, and these scattered sites need to be interconnected. BGP is the only protocol used in Internet. BGP is an exterior gateway classless path vector protocol. It is a routing protocol operated to exchange routing information among different autonomous systems

M. Athira · R. G. Sangeetha (✉)
School of Electronics Engineering, VIT University, Chennai 600127,
Tamil Nadu, India
e-mail: sangeetha.rg@vit.ac.in

M. Athira
e-mail: athira.m2015@vit.ac.in

G. Gnanagurunathan et al. (eds.), *Optical and Microwave Technologies*,
Lecture Notes in Electrical Engineering 468,
https://doi.org/10.1007/978-981-10-7293-2_12

over the Internet. An autonomous system is the collection of routers or networks that comes under single administrative domain. In an autonomous system, routers can use multiple interior gateway protocols for exchanging routing information inside an autonomous system, and it uses exterior gateway protocol to route packets between different autonomous systems.

Multiprotocol label switching (MPLS) is simple and flexible. In this case, label similar to an IP packet is used to forward the packet to the destination. The forwarding is based on the labels instead of the IP packet. All the packets with the same forward equivalence classes are forwarded along the same label-switched path. It protects from unauthorized access of network and denial of service which includes both control plane and data plane protection capabilities. Internet service providers and organizations can use these capabilities to implement robust and secure networks. It improves network reliability.

2 Exterior Gateway Protocol (EGP)

It is used to exchange routing information between different autonomous systems. It exchanges routing table between the hosts in an Internet.

2.1 Border Gateway Protocol (BGP)

Border gateway protocol is a path vector protocol. BGP does not work independently. Connectivity is required for establishing neighbourship. Thus, it requires an interior gateway protocol (OSPF, IGP and RIP) along with it [2, 3]. It exchanges routing and reachability information among different organizations on the Internet. This information contains the path in which a route takes to reach the destination. It allows loop-free interdomain routing between different autonomous systems. Inside an organization, routers use interior gateway protocols to exchange the routing information, and it uses exterior gateway protocols to route packets outside an autonomous system. A TCP connection exists between two BGP routers. Two BGP routers consisting of TCP connection between them to exchange routing information are known as peers or neighbours. Once TCP connection is established between the routers, they will send messages to exchange values and these values include autonomous system number, BGP version, the keepalive hold time and BGP router ID. When these values are confirmed and accepted, neighbourship is established between the routers. Otherwise, the routers cannot become neighbours and also it cannot exchange route updates. In BGP, neighbours are assigned manually. The BGP routers will exchange network reachability information. Border gateway protocol consists of iBGP and eBGP.

iBGP is used for peering between routers within an autonomous system. It is a mechanism to provide information to the internal routers. Internal BGP routers have to be configured as full mesh, where route transition takes place. Neighbourship is formed with all the routers mutually within an AS. Direct link connection is not required for iBGP connection. Only reachability is required to form iBGP peering.

eBGP is generally used for interconnecting networks in different autonomous systems. Direct link neighbourship is required for eBGP connection. eBGP is implemented only in the border router, which interconnects two or more autonomous systems. It requires an interior border gateway protocol like OSPF or IGP to transfer data from the external autonomous system to the internal autonomous system and vice versa. It serves as the primary protocol behind Internet connectivity.

Within an autonomous system, BGP information is redistributed to the interior gateway protocols (IGPs). A BGP table consists of lots of entries. A BGP router receives one route towards each destination. It will select best route among this set of routes using decision process.

3 Traffic Engineering

Traffic engineering is used to route the traffic according to our requirement by control flow of packets through the IP network [4–6].

BGP protocol has different attributes or network parameters. MED (multi-exit distinguisher) is a non-transitive attribute. MED is used to advertise the neighbours how they enter the autonomous system. It is propagated to all the routers within the neighbour autonomous system but not passed along any other autonomous system. MED is exchanged between autonomous systems, and it is used to let the other autonomous system know which path they should use to enter the autonomous system.

MED is an external metric of the route. The lowest MED value is preferred path. By default, the MED value is 90. Local preference is used for outbound traffic, and MED is used for inbound traffic.

4 MPLS VPN

MPLS VPN is a family of methods for using MPLS to create virtual private networks. A virtual private network (VPN) carries a private traffic over a public network. VPN can solve security problems associated with using the public Internet as a private wide area network (WAN) service. It provides an Internet-based WAN infrastructure to connect different branch offices, home offices, etc., securely through Internet without using expensive lease line connection. In Fig. 1, the customer edge routers are connected to the provider edge routers, which create a virtual private network in such a way that the traffic from one customer edge is

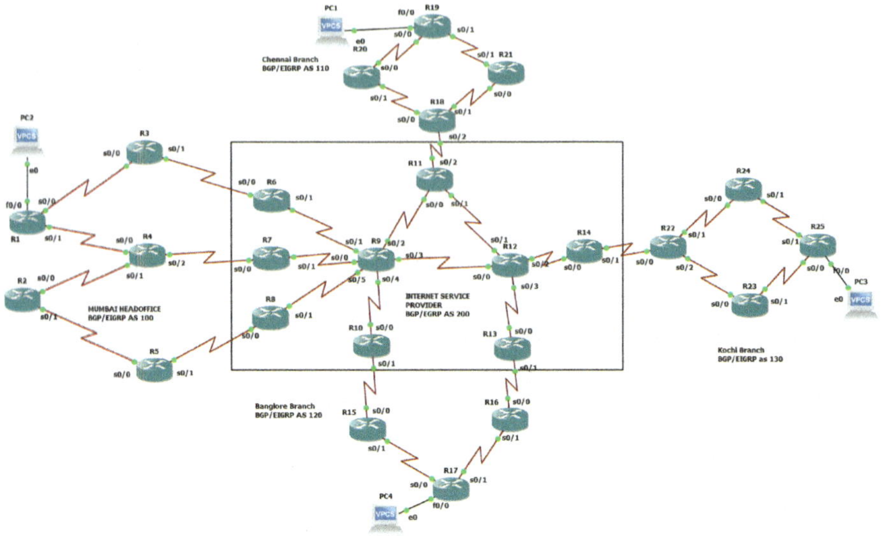

Fig. 1 Designed topology for analysis

encapsulated and sent to the other through provider edge router. The provider edge routers are at verge of the provider core network. All the routers in the Internet service provider consist of virtual route forwarding routing table along with virtual private network routing table. A link is formed between the customer edge routers and the provider edge routers.

In MPLS, when a packet reaches the ingress router, incoming packets are assigned with a label by the label edge routers. Adding labels to the packet is known as pushing. Labels are of fixed length and contain information. MPLS label is used to forward a packet. Label is different from an IP address. There is no format or structure for an MPLS label, and it differs at every hop in the network. MPLS header is 32 bit. It consists of 20-bit label value, three-bit experimental field, time to live (TTL) field of eight bits and one-bit stack. The labels help the label-switched router to process and forward the packet to the destination. Re-labelling a labelled packet is known as swapping. At each node, the current label is swapped with a new label when it goes to next node. Each label is unique. There is no need to do a lookup in routing table in this case. When the labels reach the pre-ultimate router, the labels are removed and the egress router will do an IP lookup in the routing table and forward the packet. The process of removing labels from the packet is referred as popping.

Multiprotocol label switching (MPLS) allows packets to be forwarded at switching level (layer 2) instead of passing it to network level (layer 3). It is sometimes referred as layer 2.5 protocol. Each packet is assigned a label when it enters the ingress router, and this label is swapped when it reaches the next router. The packets are forwarded based on the labels only, not on IP header. The labels can take much smaller range of value compared to IP address, and they can be easily looked up. But in case of IP addresses, they need to look up longest matching prefix.

Thus, MPLS improves the forwarding speed of the packets. The MPLS architecture is flexible and can be employed in any combination of layer 2 technologies.

5 Topology for Analysis

A topology consisting of four autonomous systems and an Internet service provider (ISP) is designed. Consider this topology as a bank network having head office at Mumbai and branches in Chennai, Bangalore and Kochi. Interior gateway protocol (IGP) is implemented within an autonomous system. Then, an exterior gateway protocol called border gateway protocol (BGP) is implemented between different autonomous systems [7–9]. The traffic engineering concept from head office to different branches has been implemented. Also, iBGP and eBGP peering and convergence to different routers has been checked. Performance parameters like fault tolerance due to link failure and delay are calculated and mentioned below.

6 Result

A. **Delay**
   Delay is the time taken to sent packets from the source to destination. Packets are sent from source to destination; the average time taken for transmission with and without MPLS VPN is calculated and is mentioned in Table 1.
B. **Fault tolerance due to link failure**
   When a path is failed, the router will take some time to update other routers about the failure and to calculate and find the next best path. Practically, this is known as fault tolerance.
   In order to calculate the fault tolerance, about 25 packets with timeout 2 s were sent from pc1 to pc3. Then, the path taken by the packets to reach the destination is traced. While sending a sequence of 25 packets we fail a link in that path and the packets starts to drop, this will continues till a new update is converged to all the routers. We find out the number of packets lost, and we

Table 1 Delay in BGP configuration and BGP with MPLS configuration

S. No.	Protocol	Average delay (ms)
1	BGP configuration	106.799
2	BGP with MPLS VPN configuration	84.37

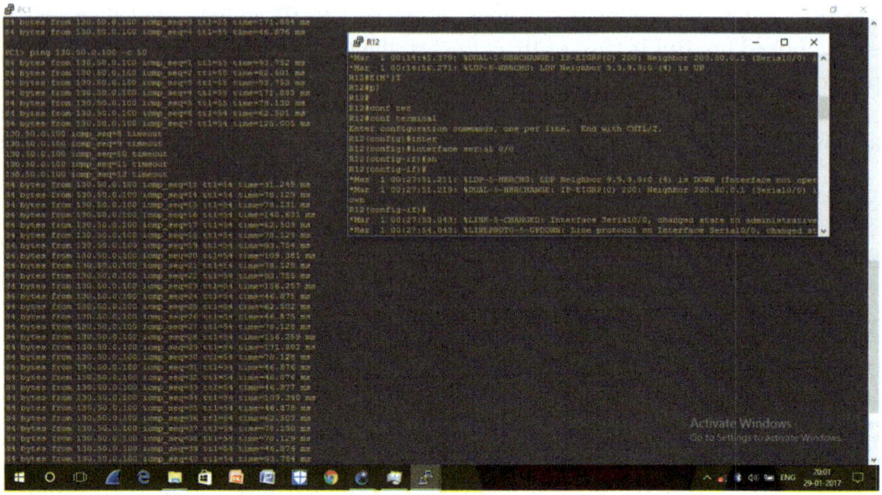

Fig. 2 Fault tolerance in MPLS VPN configuration when port 0/0 of r12 is blocked

know that each lost packet has a timeout of 2 s. The number of packets lost is multiplied with timeout to get the fault tolerance, and this process is continued five times and the average is taken. The same process is continued with and without MPLS VPN in the same topology (Figs. 2 and 3; Tables 2 and 3).

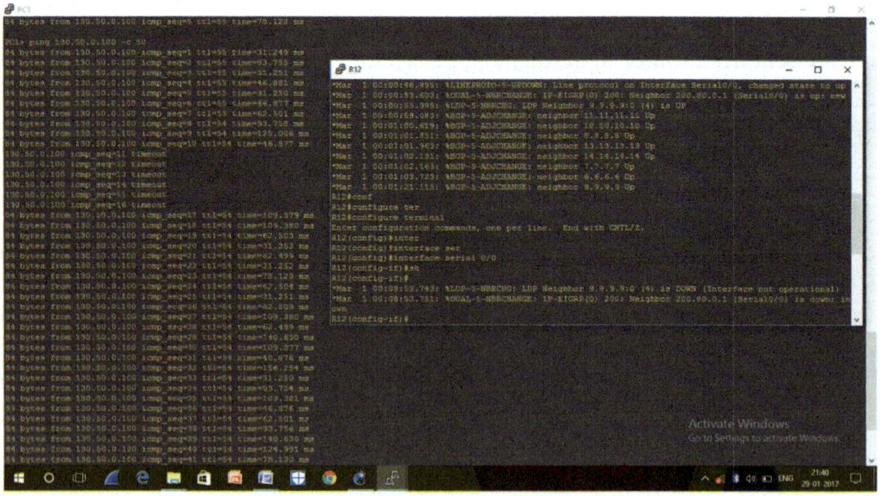

Fig. 3 Fault tolerance in BGP configuration when port 0/0 of r12 is blocked

Table 2 Fault tolerance in BGP configuration when port 0/0 of r12 is blocked

S. No.	Packets received	Packets lost	Fault tolerance	Average fault tolerance (s)
1	44	6	12	11.2
2	44	6	12	
3	46	4	8	
4	44	6	12	
5	44	6	12	

Table 3 Fault tolerance in MPLS VPN configuration when port 0/0 of r12 is blocked

S. No.	Packets received	Packets lost	Fault tolerance	Average fault tolerance (s)
1	49	1	2	7.68
2	45	5	10	
3	46	4	8	
4	45	5	10	
5	46	4	8	

7 Conclusion

In this paper, a topology consisting of four autonomous systems and an Internet service provider is designed, and it is implemented with two different configurations in the same topology. First, it is configured with BGP configuration and then with BGP and MPLS VPN. Then, the network parameters like delay and fault tolerance are calculated and mentioned in the above tables. Among the two configurations, the one with BGP and MPLS VPN configuration has better speed than the one with only BGP configuration. From the above calculations, it is clear that the MPLS VPN configuration has less delay and fault tolerance compared to BGP configuration. It has better performance than BGP configuration.

References

1. Podfrey P, Caesar M, Haken I, Stoica I, Shenker S, Singer Y (2015) Stabilizing route selection in BGP. IEEE/ACM Trans Network 23(1)
2. Francois F, Wang N, Moessner K, Georgoulas S, de Oliveira Schmidt R (2014) Leveraging MPLS backup paths for distributed energy aware traffic engineering. IEEE Trans Network Serv Manage 11(2)
3. Cohen R, Nakibly G (2010) Maximizing restorable throughput in MPLS networks. IEEE/ACM Trans Network 18(2)
4. Shao W, Iannone L, Rougier JL, Devienne F, Viste M (2013) Scalable BGP prefix selection for effective inter-domain traffic engineering. In: 2016 IEEE/IFIP network operations and management symposium. NOMS 2016

5. Mai J, Du J (2013) BGP performance analysis for large scale VPN. In: Third international conference on information science and technology 23–25 Mar 2013
6. Winter R (2011) The coming of age of MPLS. IEEE Commun Mag, April 2011
7. Cole J, Thain WE (2016) A small network for modeling MPLS. In: IEEE. 978-1-5090-2246-5/16/$31.00 ©2016
8. Minei I, Marques PR (2007) Scalabillity considerations in BGP MPLS IP VPNs. IEEE Commun Mag, April 2007
9. Labovitz C (2001) Delayed internet routing convergence. IEEE/ACM Trans Netw 9(3)

Analysis and Critical Parameter Extraction of an LED for Brain Implants

Rabinder Henry and Velmathi Guruviah

Abstract The applications of light-emitting diodes as light source have been extended to photodynamic therapy, nerve stimulation, and optical biosensors. This requires the diodes to be implanted inside specifically targeted organs for sensing and therapy using flexible substrates as a system which are biocompatible. The proposed work focuses on determining the parameters to be evaluated for designing of flexible bioimplant for optogenetics study.

Keywords Bioimplants · Optogenetics · Light-emitting diodes

1 Introduction

Biophotonics integrates photonics, microfabrication, nanotechnology, and biotechnology. It is the multidisciplinary domain defined as the science of generating and harnessing light to image, detect, and manipulate biological materials. Integrating inorganic light-emitting diodes (LEDs) with flexible substrates provides a flexible system suitable for curved, round, and soft morphology of the biological structures [1]. It allows for an internal light delivery system to manipulate damaged or tissues inside the body directly. LEDs built on stretchable substrates and optical microsystems for fiber coupling have been reported for neural implants [2], retinal implants [3, 4], heart [5], cutaneous implants for photodynamic therapy [1], and biotubes for artificial tissues [6]. The factors influencing such a flexible LED implant system are visualized using an Ishikawa diagram as shown in Fig. 1.

R. Henry (✉) · V. Guruviah
SENSE, VIT University, Chennai, India
e-mail: rabinder.henry2015@vit.ac.in

V. Guruviah
e-mail: Velmathi.g@vit.ac.in

G. Gnanagurunathan et al. (eds.), *Optical and Microwave Technologies*,
Lecture Notes in Electrical Engineering 468,
https://doi.org/10.1007/978-981-10-7293-2_13

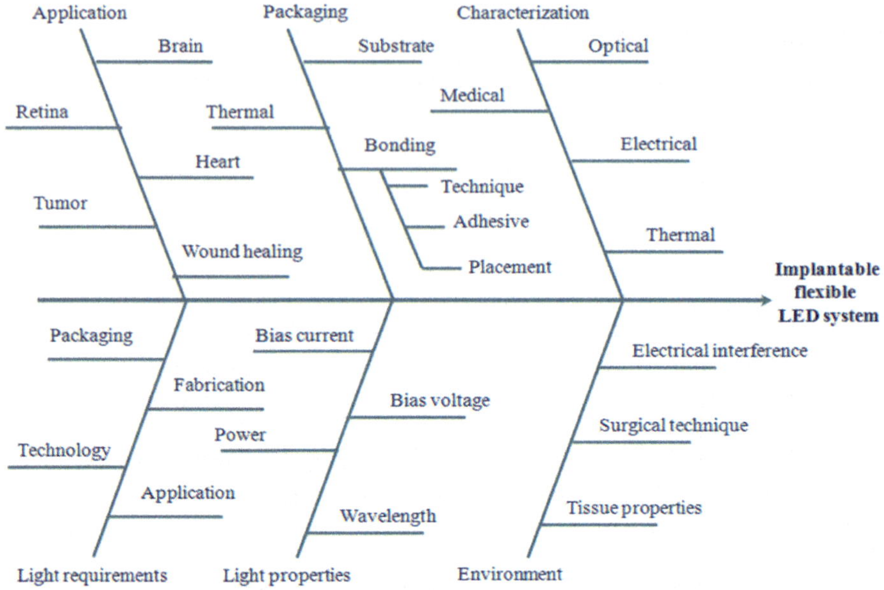

Fig. 1 Implantable LED system

The type of application or therapeutic nature determines the optical requirements from an LED. The properties of light generated depend on the operational characteristics of an LED in terms of applied voltage, current, power, and wavelength stability. The environment of operation influences the optical, electrical, and thermal characterization of the device. Similarly, implanted device influences the biological requirements in terms of safety. The implanted device is required not to cause any damage to the biological structure and its functionality. Neural implants for nerve stimulation or recording of neuronal signals especially in brain tissues are required to satisfy certain conditions. Structural biocompatibility of neural implant refers to mechanical interaction between the implantable microsystems and surrounding tissues. Detailed reviews of various techniques developed for neural implants to address mechanical effects are reported in [7]. The chemical and biological responses of nerve tissues to implants are defined as surface biocompatibility. The implantable microsystems must be biocompatible and should not cause any damage to the tissues. The test procedure and evaluation required to determine biocompatibility of neural implants are detailed in international standard ISO 10993 "Biological evaluation of medical devices" [8]. Similarly, the performance, integrity, and reliability of the implanted system are determined by biostability of the materials with respect to brain environment. Polymers have emerged as the most preferred substrate material along with metal wires and electrodes for neural implants. Such flexible system fulfills structural and surface biocompatibility and biostability requirement for bioimplants [9, 10].

2 LED for Direct Brian Implant

LED has emerged as a suitable light source for direct implantation in brain tissues owing to their smaller dimensions and low-power requirements. In contrast, lasers have limitation for direct implantation because of tissue damage through heating, retinal damage without heating (1064 nm), photochemical damage (446 nm), high-power output with respect to wavelength, narrowly focused with more hardware requirements for focusing and splitting and restriction in combining different wavelengths [11].

2.1 Selection of an LED

LEDs offer feasible solutions toward restrictions associated with lasers and other types of light sources for direct implantation in the brain. The advantages posed by LEDs for optogenetics, in vivo biosensors and cutaneous photodynamic therapy, can be summarized as high energy efficiency [12], high safety with respect to intensity [13], excellent portability for free moving animal studies [14], wide range of spectral characteristics (247–1300 nm) [12]; depending on application, no focusing is required except for single neuron excitation [15] and LED arrays provide precise surface area control for optogenetic experiments especially for multi-opsin studies and brain circuit mapping [15].

The selection of the LED is based on the spectral characteristic of the opsin required for the optogenetics experiments [9] and types of disorder with respect to photodynamic therapy [8]. The LED properties which influence the photostimulation of the nerve tissues are summarized in [10]. They are irradiation (wavelength), dose (fluence), intensity (power density or irradiance), irradiation time (treatment time), and continuous wave or pulsed mode.

2.2 Thermal Effects of an LED System

Integrating an LED onto a polymer substrate for neural implants poses new challenges and requirements with regard to managing device generated heat and light-induced heat [15]. The brain functionality is very sensitive to heat. The average brain temperature is 37 °C. Even a change of approximately one degree in temperature can have profound effects on single neuron or a network of neurons [16]. Tissue heating through temperature changes may lead to severe physiological [17] and behavioral side effects [18]. Figures 2 and 3 show simple model of LED on a flexible substrate for brain implant specifically for an optogenetic study. In Fig. 2, the LED is bonded on the bottom side of the substrate, so as to prevent any direct contact between brain tissues.

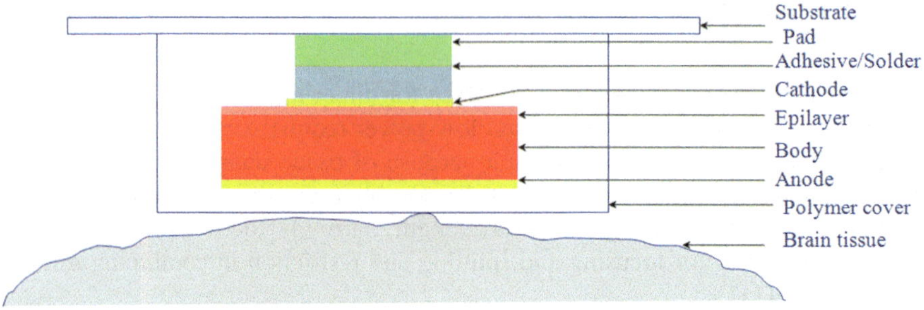

Fig. 2 Flexible LED implant for optogenetics

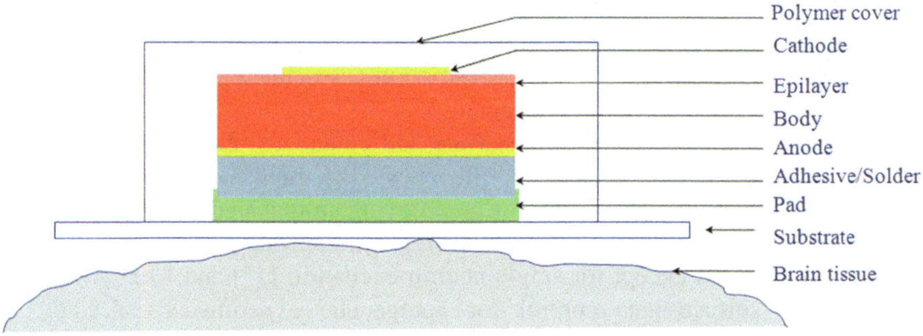

Fig. 3 LED for brain implant

In Fig. 3, the LED is bonded in such a way so as to face the brain tissues directly. In both cases, a polymer cover is required to prevent any contacts between LED and brain tissue. The heat generated in the LED and the heat generated by emitted light are the main sources for temperature change which may influence the functional temperature of the brain tissues.

The wire bonding between metal contacts and pads on the substrate is not been included in the thermal study. The dimension has been approximated based on the details available on the data sheet of EPIGAP Optronic GmbH LED [19]. The LED dimension and materials are shown in Table 1.

2.3 *Heat Generated by Light*

Constant light or pulsed activation over time may lead to temperature increment in the focused area of the brain tissue as shown in Fig. 4. Incident light on the inhomogeneous brain tissues is scattered and absorbed along the path of propagation. Depending on the intensity of the incident light, heat is generated at specific

Table 1 Dimension and Materials of LED

Dimensions of LED		
	Material	Dimension (μm)
Cathode	Gold (Au)	110 × 110 × 2
Anode	Gold (Au)	400 × 400 × 2
Body	GaN	360 × 360 × 70
Epilayer	InGaAs	360 × 360 × 6
Polymer cover	Polyimide	Variable
Solder		Thickness, 20
Pad	Copper	Thickness, 35

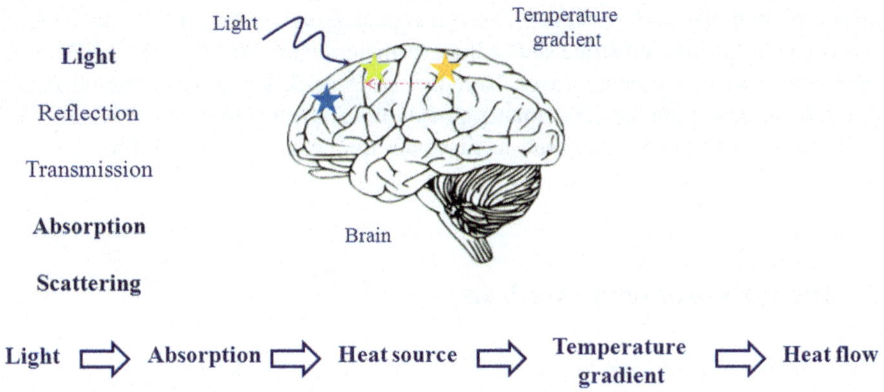

Fig. 4 Thermal effects of light

locations in proportion to the rate of absorption of photons by the tissue as shown in Fig. 4 [20].

The temperature gradient across the inhomogeneous tissues due to heat sources is counteracted by mass transfer of heat and also by metabolic heating due to excitation or inhibition. The amount of scattering and absorption of light depends on the wavelength (absorption of light is inversely proportional to wavelength) which results in a varying spatial structure of heat sources with respect to wavelength [20]. The heat generated by the incident light depends on total energy deposition which is the product of the local irradiation at the target and coefficient of absorption by brain tissue. Though many methods of estimation of the light-induced heating of tissues have been reported by [17] and [16] based on the total energy deposition, they are limited by the assumption of zero conductances at the heat source locations.

2.4 Heat Generated by Device

Direct placement of light sources inside the brain tissue using LEDs or other types of light sources may require additional thermal analysis and heat generation models to understand the effects of high-power light sources induced tissue heating as shown Fig. 4. When an LED is operated in a forward biased condition, heat is generated due to current flow across the diode. The p–n junction is the region in which most of the heat is dissipated.

The heat generated is transmitted across the diode structure anode, cathode, and epilayer, contact metals, through packaging structures and to the ambient. The heat transmission occurs through conduction and convection in the form of heat transfer in solids in diode and package. The basic packaging of bare LED die includes contact pads on the substrate and adhesive material for bonding the LED to the substrate. The heat transmission from the LED package to brain tissue occurs through bioheat transfer. The management of the heat generated from an LED package and technology required to dissipate the heat without affecting the brain tissues is the main focus of this work.

3 Design Parameter Analysis

The operational parameter voltage and current of the LED determine the electrical, optical, and thermal behavior. The device construction and packaging determine the thermal, optical, and electrical output in the operating environment. By optimizing the operational parameters of the LED's thermal behavior, output parameters may be constrained to operate within the brain conditions. The condition of the LED implant in the brain tissue is shown in Fig. 5.

The maximal junction temperature of an operating LED is approximately 125 °C [16]. The device generates huge amount of heat which has to dissipate through different layers of LED and the package. This dissipation may result in transfer of heat to the tissues. This necessitates proper heat sink structures and materials to protect the brain tissues. Similarly, the brain environment should not influence the operational characteristics of the LED as shown in Fig. 6.

The thermal analysis of the implantable system in the brain ambient includes failure modes of the operating LED and its effects on brain environment. To understand failure modes related to heat transfer in LED, package, and brain tissues, a detailed study of the various issues are evaluated using Ishikawa diagram and Pareto analysis. Figure 7 is the Ishikawa diagram showing the causes and factors to be considered in designing an implantable flexible LED system for brain tissues.

It can be observed from the diagram that the influencing causes are requirements and properties of light as described in Sect. 1 and operational parameters of LED. The possible effects such as thermal effect, light-induced effects, and mechanical effects on brain tissues are to be analyzed for such an implant. To understand the

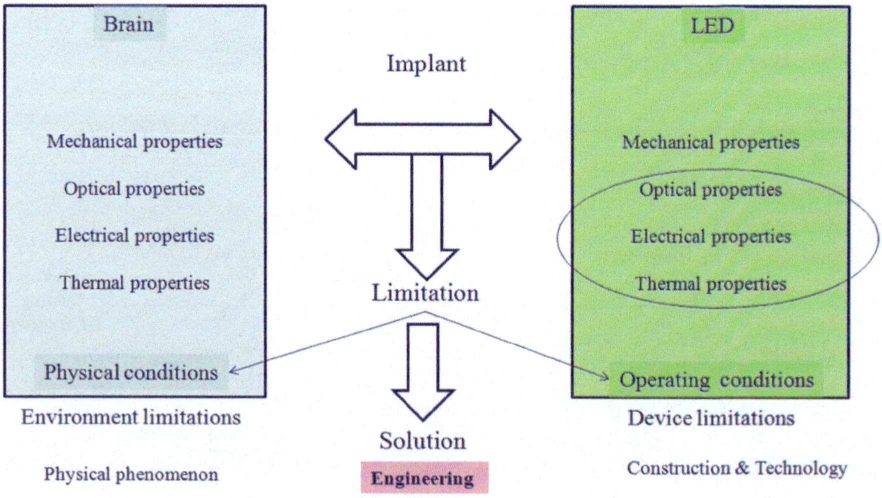

Fig. 5 Operating condition of an implanted LED

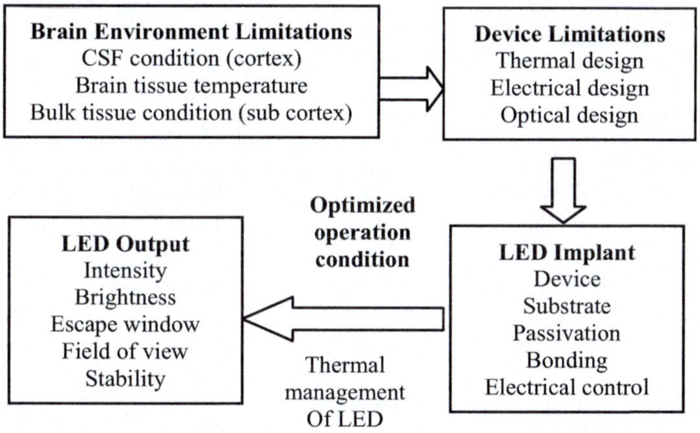

Fig. 6 Parameters influencing an LED

major factor, specifically influencing the operation of implantable LED system and its effect, a Pareto analysis is done based on the cause and effect Ishikawa diagram. The thermal, optical, mechanical, electrical, and biological factors affecting the implant and brain tissues are considered. The failures of the device with respect to these factors are summarized in Table 2.

The Pareto analysis for the given factors is shown in Fig. 8. It can be observed from the Pareto analysis that the most influencing factors are thermal and optical parameters. The total damages contribute to about 64% of total cumulative frequency.

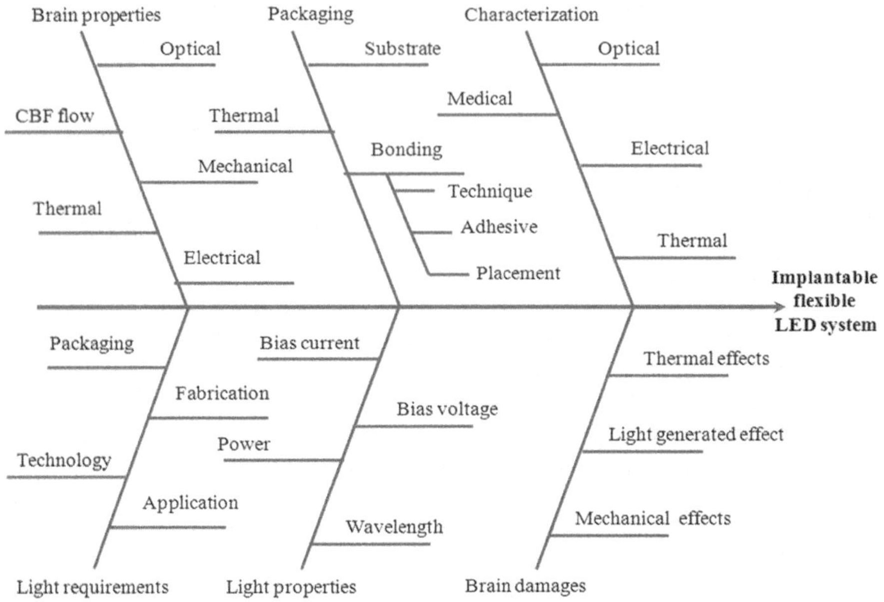

Fig. 7 Ishikawa diagram

Table 2 Effect and Failure Analysis of LED

Effects and failures				
Thermal	Electrical	Optical	Biological	Mechanical
Degradation of encapsulant	p–n junction failure	No light	Tissue bleeding	Tissue rupture
Tissue heating	Short circuit	Luminescence degradation	Inflammation	Structural damage of LED
Packaging failure	Open circuit	Non-uniform light emission	Tissue heating	Corrosion of LED
Abrupt LED failure	Tissue burning	Light-induced damages	Retinal damage	Blood vessel narrowing
Device failure			Tissue burning	Brian fluid

To further understand the influence of thermal factors, a subplot of thermal failures is shown in Fig. 9. The Pareto analysis of specific factors of thermal analysis shows that tissue heating is a major factor. The device failure and packaging failure also contribute to thermal effects which are to be addressed while designing the implant. The design of experiment is based on thermal factors responsible for device failures. These factors are deduced from Pareto analysis and its implications.

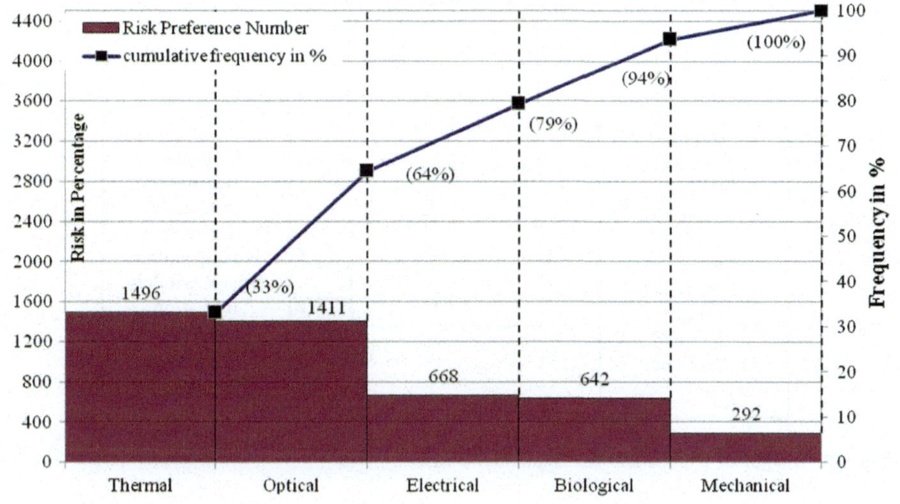

Fig. 8 Pareto analysis of influencing factors

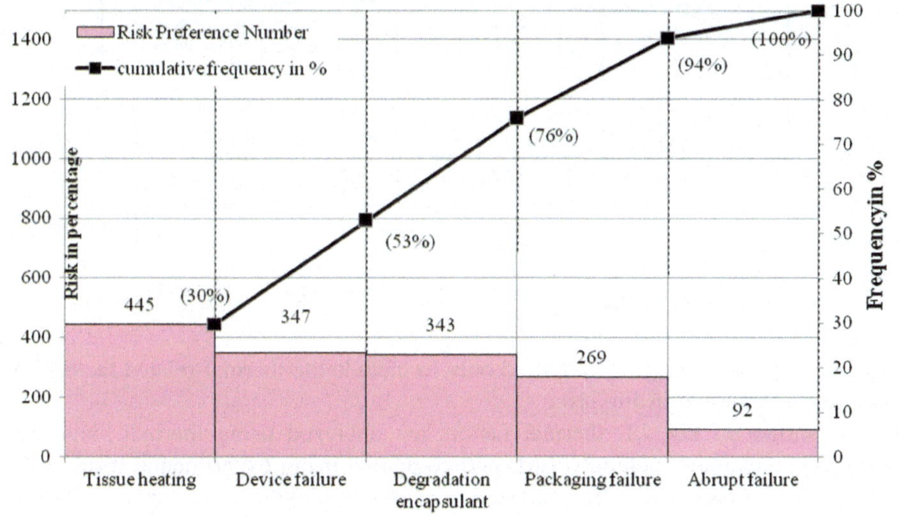

Fig. 9 Pareto analysis of thermal factors

The temperature change in the LED depends on the operational parameter voltage, current, and power. The influence of the thermal behavior of the LED on the optical output is also considered as shown in Fig. 2.

In normal atmospheric environment of temperature of 293 K and pressure of 101.325 kPa, thermal management of an LED is analyzed and designed for various applications [18]. Since the main influencing factors are thermal behavior of the

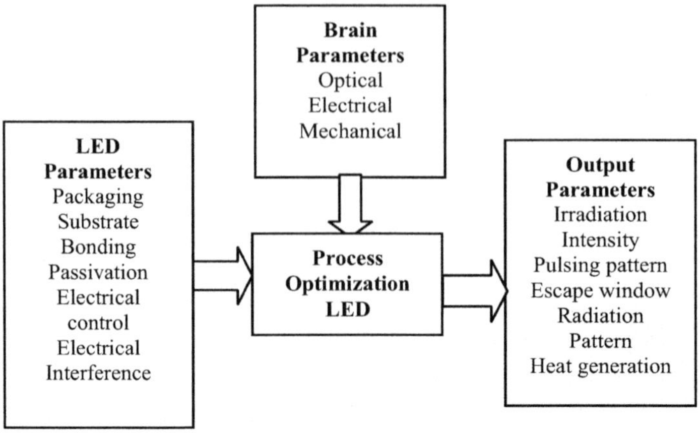

Fig. 10 Design of experiment for implantable LED system

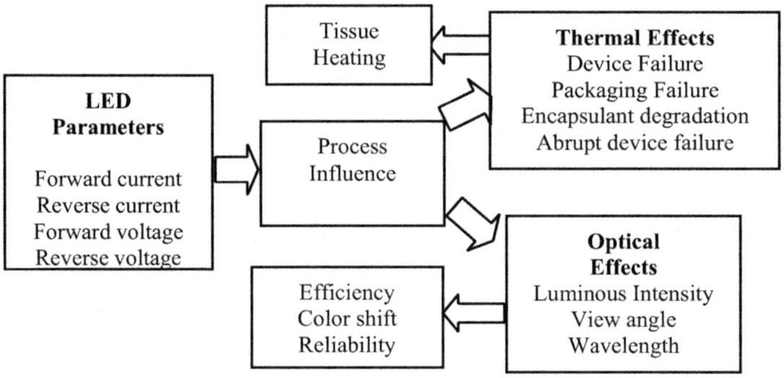

Fig. 11 Design of experiment of thermal behavior of implantable LED system

LED, experiment design is redrafted only to include the thermal-related factors and its effects as shown in Fig. 10.

As shown in Fig. 11, thermal factors are analyzed using thermal simulation using finite element method (FEM) and compared using experimental results. The optical output parameters are evaluated using gonio photometer using a LED board designed for testing the optical and thermal behavior of the implant system in brain environment.

The temperature change in LED as discussed earlier is the major factor to be considered as per the statistical analysis. The heat Generated by the LED's junction, its dissipation in the nearby environment are the main constraints. The detailed simulated thermal models and experimental analysis done to test the results are to be reported in a separate communication.

Acknowledgements The authors sincerely thank the guidance and support given by the IMOS team at Otto-Von-Guericke University, Magdeburg, Germany.

References

1. Nabutovsky Y, Pavek T, Turcott R (2012) Chronic performance of a subcutaneous hemodynamic sensor. Pacing Clin Electrophysiol 35(8):919–926
2. Polikov VS, Tresco PA, Reichert WM (2005) Response of brain tissue to chronically implanted neural electrodes. J Neurosci Methods 148(1):1–18
3. http://www.fda.gov/downloads/MedicalDevices/DeviceRegulationandGuidance/GuidanceDocuments/UCM348890.pdf. (2012)
4. Bruck SD (1991) Biostability of materials and implants. J Long Term Eff Med Implants 1 (1):89–106
5. Chen D et al (2012) Light-emitting diode-based illumination system for in vitro photodynamic therapy. Int J Photoenergy 1–6
6. Hassler C, Boretius T, Stieglitz T (2011) Polymers for neural implants. J Polym Sci Part B Polym Phys 49:18–33
7. Kim T et al (2013) Injectable cellular-scale optoelectronics with applications for wireless optogenetics. Science 340:211
8. Iwai Y, Honda S et al (2011) A simple head-mountable LED device for chronic stimulation of optogenetic molecules in freely moving mice. Neurosci Res 70(1):124–127
9. Yizhar O, Fenno LE et al (2011) Optogenetics in neural systems. Neuron 71(1):9–34
10. Madsen SJ (2013) Optical methods and instrumentation in brain imaging and therapy. Springer, New York. ISBN 978-1-4614-4977-5
11. Moser E, Mathiesen I, Andersen P (1993) Association between brain temperature and dentate field potentials in exploring and swimming rats. Science 259:1324–1326
12. Elwassif MM, Kong Q, Vazquez M, Bikson M (2006) Bio-heat transfer model of deep brain stimulation induced temperature changes. Conf Proc IEEE Eng Med Biol Soc 1:3580–3583
13. Long MA, Fee MS (2008) Using temperature to analyse temporal dynamics in the songbird motor pathway. Nature 456:189–194
14. Arik M, Becker C, Weaver S, Petroski J (2013) Thermal management of LEDs: package to system. In: Ferguson IT, Narendran N, DenBaars SP, Carrano JC (eds) Third international conference on solid state lighting, Proceedings of SPIE, vol 5187. EPIGAP Optronic GmbH data sheet EOLC-465-24
15. Childs C (2008) Human brain temperature: regulation, measurement and relationship with cerebral trauma, part 1. J Neurosurg 22:486–496
16. Bakk IP, Borsoi G, Favarolo PA (2012) Thermal management of LED systems. In: Elektrotechnik & Informationstechnik. Springer Verlag, Wien, Oct 2012
17. Aravanis AM, Wang LP, Zhang F, Meltzer LA, Mogri MZ, Schneider MB, Deisseroth K (2007) An optical neural interface: in vivo control of rodent motor cortex with integrated fiberoptic and optogenetic technology. J Neural Eng 4:S143–S156
18. Tay FR, Pashley DH, Loushine RJ, Weller RN, Monticelli F, Osorio R (2006) Self-etching adhesives increase collagenolytic activity in radicular dentin. J Endod 32:862–868
19. EPIGAP Optronic GmbH, data sheet EOLC (2011) pp 465–24
20. Yaroslavsky AN, Schulze PC, Yaroslavsky IV, Schober R, Ulrich F, Schwarzmaier HJ (2002) Optical properties of selected native and coagulated human brain tissues in vitro in the visible and near infrared spectral range. Phys Med Biol 47:2059–2073

Review of Thermal Management of an LED for Brain Implants

Rabinder Henry and Velmathi Guruviah

Abstract The brain implant based on implantable microdevices is restricted by the temperature changes which may damage the brain tissues. This requires to study of the thermal behavior of the implanted devices. In this case, a light-emitting diode is being implanted for optogenetics study. The paper describes theoretical thermal models to study the heat flow across the diode and biological tissues.

Keywords Bioimplant · Light-emitting diodes · Thermal management
Optrodes

1 Introduction

Bioimplants are medical devices designed to replace missing biological structure, replace a damaged biological tissue or structure, improve the functionality of existing organ or tissue, and interface to interact with biological systems [1]. The biomedical implants with increased complexity, functionality, and microfabrication have resulted in new engineering challenges [2]. Implants specifically designed for neural interfaces which are two-way exchange of information between neural structures and digital analytical systems are offering more complex requirements and design constraints. Such neural interfaces happen at multiple levels including peripheral nervous system, spinal cord, and brain tissues. The challenges with regard to neural interfaces include selectivity, stability, resolution versus invasiveness, implant-induced injury, and the host-interface response [3]. Traditionally, neural interfaces are limited to electrical stimulation using electrodes and also for recording neuronal responses as patterns of action potentials. These potentials per unit time basically code the functionality of the nervous system. The neu-

R. Henry (✉) · V. Guruviah
SENSE, VIT University, Chennai, India
e-mail: rabinder.henry2015@vit.ac.in

V. Guruviah
e-mail: Velmathi.g@vit.ac.in

G. Gnanagurunathan et al. (eds.), *Optical and Microwave Technologies*,
Lecture Notes in Electrical Engineering 468,
https://doi.org/10.1007/978-981-10-7293-2_14

roanatomical, neurochemical hierarchy and architecture of the nervous system are functionalized based on spatiotemporal patterns of action potentials [3]. In the last decade, use of light as a therapeutic agent for treating skin diseases has been extended for novel optical stimulation of various cells and tissues. This has led to development of implantable light-emitting structures inside the body for treating or addressing different types of disorders and diseases [4]. Such implantable system includes small dimensional light-emitting diodes (LED) as light sources for nerve stimulation in tandem with electrodes inside the brain. Cutaneous photodynamic therapy and in vivo biosensors are a couple of early technologies based on implantable light sources. These implantable systems include LEDs, electronic control devices, and wireless systems to transfer data. These implantable devices generate huge amount of heat which may limit their functionality and characteristic behavior [5].

Heat generated, if not properly engineered to dissipate through design, may cause serious damages to biological systems [6]. Thermal management of bioimplants in the operating biological environment is a major concern and ever increasing design constraint. There are numerous factors and parameters which influence the thermal behaviors of implanted electronic and wireless devices [5]. Figure 1 shows potential causes for temperature increase in brain tissues. The power dissipated by operating devices like LED, electrode, and wireless device, and heat induced by incident electromagnetic wave like light or microwave lead to heating of surrounding tissues through bioheat transfer [5].

Such domain is optogenetics wherein heat plays a major role. Optogenetics is basically a technique which combines optics and genetics to understand the behavior of neuronal tissues. Optrode is a device which is implanted on the surface

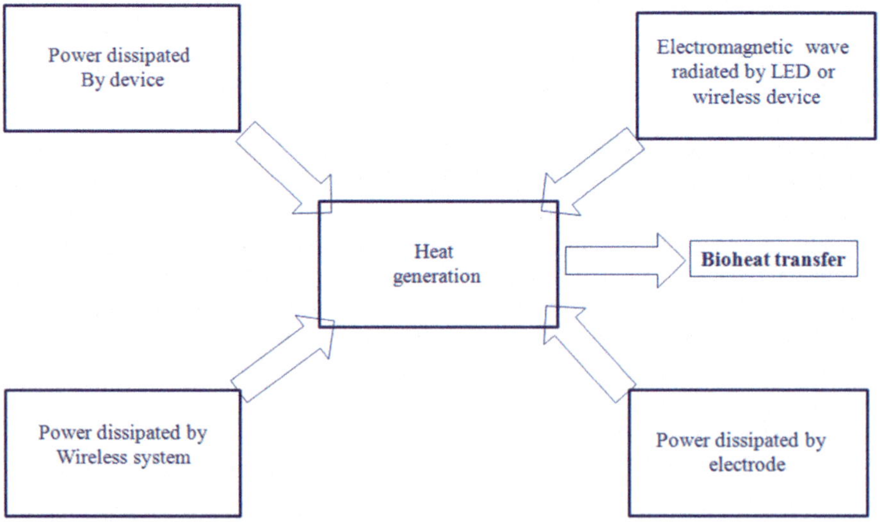

Fig. 1 Overview of heat generated with brain implants

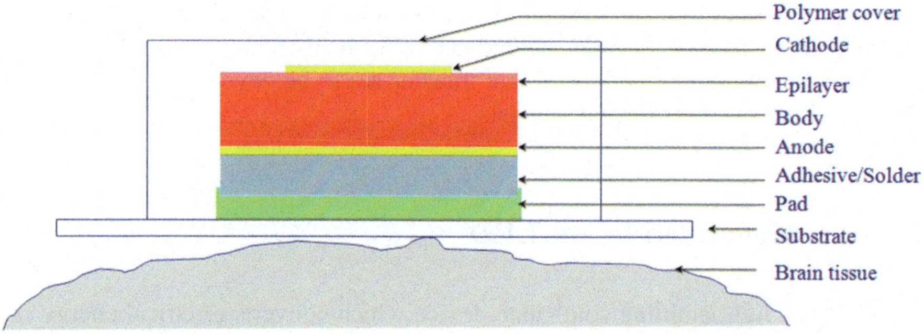

Fig. 2 LED for brain implant

cortex of mammals. The implantable device includes electrodes and light sources like LED or optical fiber which carry light to the tissues. The neuronal cells are stimulated by light, and the responses are recorded using electrode. Such a device is defined as optrode [1, 7]. A detailed review of optogenetics as neuromodulation technique has been done in [7], and why analyzing heat generation and methods to dissipate is important for a brain implant has been presented in the reference [8].

Figures 2 and 3 show simple model of LED on a flexible substrate for brain implant specifically for an optogenetic study.

In Fig. 2, the LED is bonded on the bottom side of the substrate, so as to prevent any direct contact between two brain tissues.

In Fig. 3, the LED is bonded in such a way so as to face the brain tissues directly. In both cases, a polymer cover is required to prevent any contacts between LED and brain tissue. The heat generated in the LED and the heat generated by the emitted light are the main sources for temperature change which may influence the functional temperature of the brain tissues.

In order to understand and analyze the implication of heat generated in an LED, it is required to understand and analyze the heat generation mechanism in an LED and how the heat generated in the LED junction spreads across the implanted system and affects neuronal tissue. The major aspect would be to prevent such an

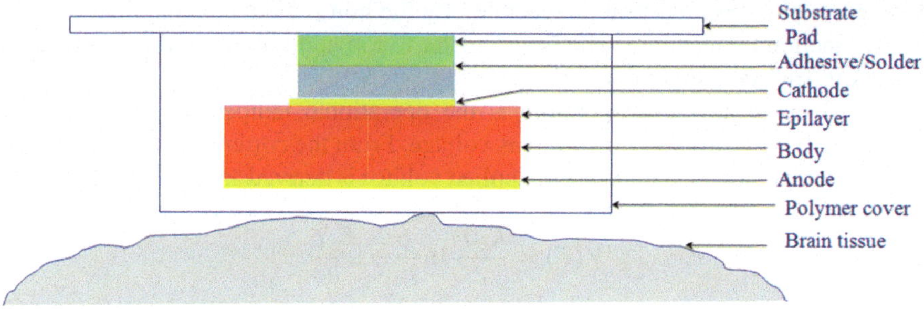

Fig. 3 LED for optrodes

incident through design and development of structures which dissipate heat without affecting brain tissues. In order to understand heat dissipation in an LED and to brain tissue, simulation models of heat transfer are to be developed before starting with any kind of experimental analysis.

2 Heat Generated by an LED

An LED is a light-emitting solid-state device which converts electrical energy into light energy. It is a p–n junction diode made from direct bandgap semiconductor materials which emits incoherent narrow spectrum of light when biased in the forward direction of the p–n junction. The phenomenon by which material emits light in response to an electric current passed through it or in the presence of a strong electric field is defined as electroluminescence discovered in the year 1907 by Henry Joseph Round [9]. The wavelength of the emitted light is dependent on the energy gap E_g of direct bandgap material as given in Eq. 1

$$\lambda = \frac{hc}{\vec{E}} \tag{1}$$

where h is Planck's constant, c velocity of light, and E the bandgap energy in eV.

2.1 Thermal Properties of an LED

When an LED is operated in forward bias condition, heat is generated by non-radiative recombination and photon absorption in the material. The induced heat by external voltage increases the junction temperature until the heat is transferred and dissipated along the structure [10].

The LED reaches its operational temperature at equilibrium between heat generation and dissipation. The heat transfer across the LED results in three different temperatures as shown in Fig. 4. Some of the applied power at the p–n junction is emitted as photons, and heat is generated by the non-radiative recombination. This is the junction temperature. The heat transfer to top and bottom of the p–n junction results in surface temperature.

The average device temperature is the operational temperature which can be measured by determining forward bias voltage. From the current voltage equation of a diode, temperature dependence of applied voltage is derived as shown in Eq. 2 [10].

$$V(T) = \frac{K_B T}{e} \ln\frac{I}{I_0} + \frac{E_g T}{e} \tag{2}$$

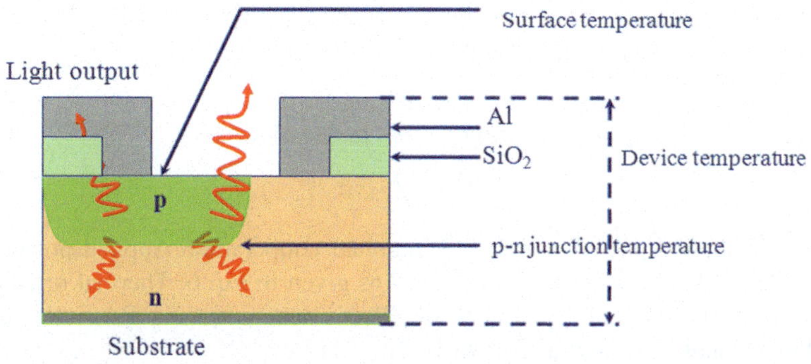

Fig. 4 Temperatures in an LED

where K_B is the Boltzmann constant, T is temperature, I is current, I_0 is saturation current, e is electron charge, and E_g is the band gap energy which changes with temperature.

The wavelength of photon emitted depends on bandgap energy as shown in Eq. 3

$$\begin{gathered} h\upsilon = E_g + \frac{1}{2}kT \approx E_g \\ V_{th} \approx V_0 \approx E_g/e \end{gathered} \tag{3}$$

The bandgap energy is dependent on material composition and temperature. This is derived empirically by Eq. 4.

$$E_g(T) = E_g(T = 0\ \text{K}) - \frac{\alpha T^2}{T + \beta} \tag{4}$$

where α and β are empirical material constants determined experimentally [11]. From Eq. 4, it is clear that increasing temperature reduces bandgap energy. This results in shifting of wavelength of the light emitted by LED with temperature. The forward voltage decreases with constant current for increasing temperature as deduced from Eq. 3. When an LED is forward-biased with constant current, the junction temperature T_J raises and stabilizes at a certain point and is given by Eq. 5 [12].

$$T_J = R_{th}P + T_{amb} \tag{5}$$

where R_{th} is thermal resistance, P is applied power, and T_{amb} is ambient temperature. Thermal resistance is the measure of resistance offered by a material (specific thermal resistance or resistivity mK/W) or a component (absolute thermal resistance K/W) to the flow of heat. Thermal conductivity is the material property which describes the flow of heat in a homogeneous medium. Thermal resistance can be calculated using geometry of a component or material and its thermal conductivity.

For complex structures, FEM modeling is used to simulate the thermal resistances. The thermal resistance is calculated using Eq. 6.

$$R_{\mathrm{th}} = \frac{T_{\mathrm{J}} - T_{\mathrm{amb}}}{P} \quad P = VI \tag{6}$$

where P is applied power, and T_{amb} is ambient temperature. Applied power is the product of applied voltage V and current I as given in Eq. 6. Thermal resistance is evaluated using thermal equivalent of Ohm's law which is Fourier's law for one-dimensional heat flow as shown in Eq. 7 [12].

$$I = \frac{1}{\rho}\frac{A}{L}\Delta V \quad q = K\frac{A}{L}\Delta T \tag{7}$$

where A is the area, L is length, is resistivity, and V is change in applied voltage. The equivalent of current is heat flux q, T is change in temperature analogous to voltage, and K is thermal conductivity in Fourier's law. Thermal resistance at junction with respect to ambient consists of two components [13]. The device internal thermal resistance R_{thJS} is given between junctions with respect to outside surface that is up to solder joint. The value of thermal resistance is given by geometry, material, and chip size. The external level resistance R_{thSA} is the application specific resistance to heat flow from surface to ambient. Figure 5 shows thermal resistance network for an LED chip.

Thermal resistance is also defined for optical power of an LED. The thermal resistance in Eq. 6 is defined as electrical thermal resistance. The real thermal resistance is then given by Eq. 8.

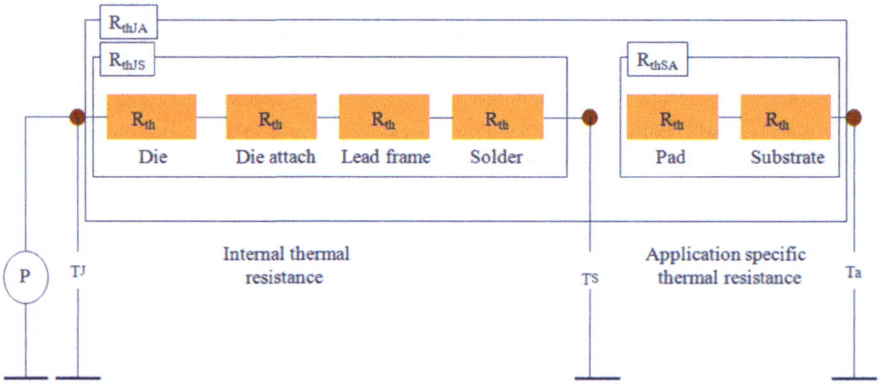

Fig. 5 Thermal resistance network of an LED

$$R_{\text{threal}} = \frac{T_{\text{J}} - T_{\text{amb}}}{P - P_{\text{optical}}} \tag{8}$$

where P_{optical} is the optical power. Since most of the power is dissipated at the junction, heat is transferred to the top and bottom of the LED. For steady-state analysis, the differential form of Fourier's law of conduction is used as given by Eq. 10.

$$\vec{q} = -k(T)\nabla T \tag{9}$$

where q is the product of thermal conductivity k (T) and the negative local temperature gradient $\tilde{N}T$. The heat flux is the amount of heat flowing through unit area in unit time.

2.2 *Junction Temperature Measurement*

The junction temperature influences the performance of an LED in terms of electrical and optical power output. There are many methods to measure junction temperature of semiconductors. The LED can be modeled using 3D full-wave software solver tools like ANSYS or COMSOL Multiphysics. In this thesis, COMSOL Multiphysics is used to model the LED and solve thermal management of an LED. The COMSOL includes heat transfer module for simulating heat transfer in solids using Fourier's conduction law. AC/DC module is used for DC current analysis of the LED. Since COMSOL allows coupling of modules, the simulation is done for current generated heat transfer in the system. The bioheat transfer module is used to analyze the heat spreading to brain tissues [14]. The model includes the necessary boundary conditions to simulate the brain properties. Full-wave model allows both static and transient time analysis of heat transfer in LED, package and brain tissue. The heat analysis includes temperature simulation for parametric analysis. By varying the diode operational parameters, dimensional parameters, and types of substrates, changes in temperature are computed to give complete thermal analysis of heat generation by the device and methods to prevent the heat dissipation to the brain tissues. Using the simulated temperatures for different layers of the implantable flexible LED system, the junction temperatures for the LED are calculated. The influence of LED parameters including junction temperature on optical output of the LED is evaluated as described in Eqs. 2 and 3 and shown in Fig. 6 [10].

In direct physical contact measurement, temperature-sensing device is directly placed on anode or near the junction to measure changes. Thermocouples, scanning thermal probes, liquid crystals, or thermographic phosphorous are some of the devices used for junction temperature measurement [15]. This method provides good spatial resolution. The spatial resolution depends on the contact surface

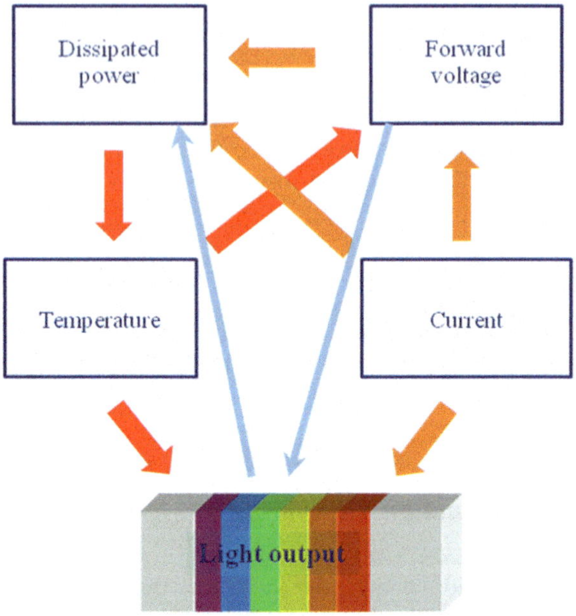

Fig. 6 LED parameters

available for measuring, and time response depends on the thermal response of the measuring probe. In the experimental verification of the simulated data thesis, a thermocouple is used for measuring junction temperature.

In electrical junction temperature measurement method, the junction temperature is utilized as a sensor. The forward voltage drop across the p–n junction is used as a temperature-sensitive parameter. From Eq. 2, the relationship between forward voltage and junction temperature is linear for constant current. This current is defined as sense current [16]. By including series resistance in diode current equation, the sensing current of the LED is given by Eq. 10

$$I_{\mathrm{f}} = I_0\left[\exp\left(\frac{e(V_{\mathrm{f}} - VR_{\mathrm{S}})}{\eta K_{\mathrm{B}} T}\right) - 1\right] \tag{10}$$

where V_{f} - applied forward voltage, R_{s} - series resistance, I_0 - reverse saturation current, e - is charge of electrons, and K_{B} - Boltzmann constant. The forward voltage with respect to junction temperature is derived using Eq. 11 and Eq. 2 as shown in 11 [16].

$$V_f = \frac{K_{\mathrm{B}} T}{e}\ln\frac{I_{\mathrm{f}}}{CT^r} + \frac{E_{\mathrm{g}} T}{e} + nV_{\mathrm{g}0} + I_{\mathrm{f}} DT^{-3/4}\exp\left(\frac{-E_{\mathrm{a}}}{kT}\right) \tag{11}$$

where E_{a} is acceptor activation energy, $V_{\mathrm{g}0}$ is potential difference at the conduction-band and valence-band edges at temperature of absolute zero, r is decided by the temperature dependence of the minority carrier mobility, and C and

D are the constant decided by the junction area and dopants with concentration. From above equation, the junction temperature is estimated using Eq. 13 [13].

$$T_J = \frac{\Delta VT}{K} + T_{amb} \tag{12}$$

where V_T is temperature-sensitive parameter, and K is calibration constant given by Eq. 13 [16].

$$K = \left| \frac{V_H - V_L}{T_H - T_L} \right| \tag{13}$$

where V_H and V_L are high and low applied voltage, and T_H and T_L are corresponding junction temperatures.

2.3 Thermal Analysis

The thermal properties of materials used in diode and package determine the thermal behavior of the system. In steady-state analysis, the past temperature of the devices is not considered. Generally, instantaneous power is supplied to the LED and kept constant over time. During this time, varying heat dissipation results in change of temperature. If previous temperature of the LED is not influencing, the device reaches its thermal equilibrium [17]. Though it is required to perform analysis for infinite amount of time in an ideal case, practically the simulation is done for a fixed time. In this analysis, the ambient temperature and cooling conditions are kept constant. The analysis also includes coupling of Joule heating and AC/DC module to the heat transfer analysis. In these analyses, the applied voltage is kept constant and the LEDs thermal equilibrium is not reached instantly. This is due to the thermal capacitances of materials. The thermal capacitances take time to get charged completely. Since the LEDs are made of different material layers, a system reaches its thermal equilibrium only when capacitances of all materials are completely charged. The thermal resistance in case of steady-state analysis is done using Eq. 8. The thermal behavior of LED for different power supplies can be predicted using this thermal analysis from transient to steady state and is then thermally characterized [14]. The thermal behavior of the LED under time-dependent analysis is simulated using short pulse voltages. This gives the LED transient thermal properties. The diode is allowed to cool down after application of short pulse to reach the ambient temperature. The thermal time constant is given by Eq. 14

$$T_J = PR_{th}\left[\frac{4}{\pi^{\frac{3}{4}}}\right]\left[\frac{t}{\tau}\right]^{\frac{1}{2}} \quad \text{for } t < \tau \tag{14}$$

where T is total time and τ is thermal constant. The time constant can also be calculated with Eq. 15.

$$\tau = \left[\frac{2F}{\pi}\right]^2 \left[\frac{\rho C_\mathrm{P}}{k}\right] \tag{15}$$

where F is the thickness of die, ρ is density of the material, Cp is heat capacity, and k is thermal conductivity [15].

3 Bioheat Transfer

The main focus of this work is to analyze possible conditions under which the heat dissipating from LED may be transferred to the brain tissues and to provide solutions based on packaging and design to prevent tissue heating. The heat transfer in tissues occurs through conduction, convection, evaporation, and radiation. Heat is conducted in tissues through energy transfer between high energetic particles and less energetic particle and is modeled using Fourier's conduction law as given by Eq. 16.

$$\rho C_\mathrm{P} \frac{\partial T}{\partial t} = \nabla (k \nabla T) \tag{16}$$

where ρ is the density, C_p is specific heat capacity, k is thermal conductivity, and T (K) is tissue temperature [18]. Thermal properties of tissues depend on type of tissue and temperature. Tissues are composed of combination of water, fat, and proteins. Thermal behavior of tissues differs based on the amount of water content, since it is a good conductor of heat. Convection is the process of heat transfer between solid body and fluid medium. Tissues are surrounded by body fluids and blood vessels. The convective heat transfer in tissues and fluids is modeled using Newton's law of cooling as given in Eq. 17.

$$q = h(T - T_1) \tag{17}$$

where h is local heat convection coefficient, T is surface temperature of bulk tissue, and T_1 is temperature of the fluid [18]. In most of the tissue surfaces, there are no fluid motion therefore conduction occurs. Hence, convection analysis is not included. Thus, convection analysis is not included. Tissue is not a perfect black body, and its emissivity is less than one. So the radiative heat transfer of internal tissues is neglected since their contribution is negligible [18].

4 Summary

The COMSOL Multiphysics software is used to model the implantable flexible LED system to simulate thermal characteristics. With respect to thermal analysis, the tool includes heat transfer in solids module, Joule heating module, AC/DC module, and bioheat transfer module. The module coupling allows simulating current induced heating in solids and biological tissues. The joule heating module is a coupled module which includes both current and heat transfer modules. The LED system is built in a 3D geometry. The associated material properties are assigned to different layers of the LED. The boundary conditions are specified for different physics models. The different scenarios of heating and cooling mechanisms are given as input depending on simulation and model requirements. Figure 7 shows the insulation model using thermal grease as part of the analysis to insulate the brain tissues.

To measure the device temperature at room temperature, initially a box made of black interior is designed as shown in Fig. 8. This insulates the LED from environmental disturbances and reflection. This is used only to measure the device thermal characteristics in the ambient temperature without vacuum chamber as shown in Fig. 7. This setup includes a PCB holder on the bottom plate. The air inside the hot chamber is heated to a constant temperature. Since the PCB is not placed directly on the hot plate, only the environment temperature is controlled.

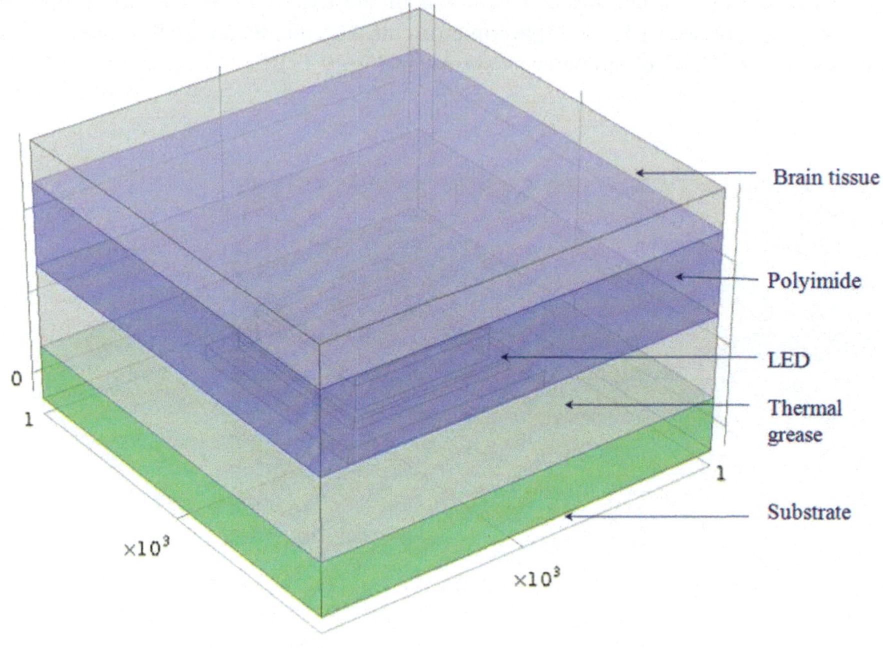

Fig. 7 Thermal insulation using thermal grease and polyimide layers

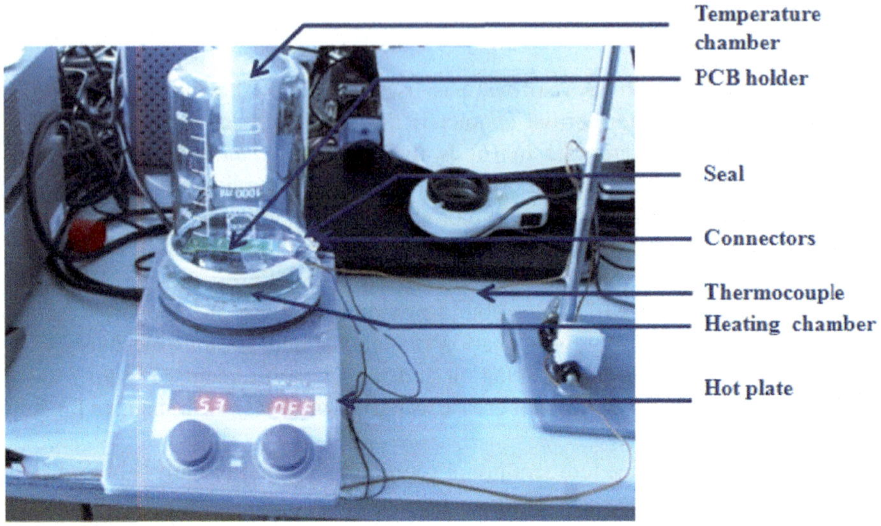

Fig. 8 Constant heat chamber

This hot chamber is covered with black surface to minimize interference or reflection from the ambient light.

To understand the optical characteristics of the LED which are not given in the datasheet and to correlate thermal behavior of the LED with respect to its optical performance, radiometric and photometric measurements are determined using LEDGON 100 Goniophotometer as shown in Fig. 9 [7].

Fig. 9 LEDGON 100 goniophotometer

5 Conclusion

The mathematical modeling using COMSOL and verification of simulated results with experimental results along optical parameter characterizations have been communicated in separate paper. This paper only focuses on the review theoretical background required to do such an analysis before going in for clinical trials.

Acknowledgements The authors sincerely thank IMOS team at Otto von Guericke University, Magdeburg.

References

1. Fries RE (2000) Handbook of medical device design. CRC Press, 14 Sept 2000. ISBN-0-8247-0399
2. Greenbaum E, Zhou D (2009) Implantable neural prostheses 1, devices and applications. Springer, Dordrecht Heidelberg. ISBN: 978-0-387-77260-8
3. Grill WM, Norman SE, Bellamkonda RV (2009) Implanted neural interfaces: biochallenges and engineered solutions. Annu Rev Biomed Eng 11:1–24
4. Koo M, Park SY, Lee KJ (2012) Biointegrated flexible inorganic light emitting diodes. Nanobiosens Dis Diagn 1:5–15
5. Lazzi G (2005) Thermal effects of bioimplants. IEEE Eng Med Biol Mag
6. Mrozek S, Vardon F, Geeraerts T (2012) Brain temperature: physiology and pathophysiology after brain injury. Anesthesiol Res Pract 1
7. Instrument Systems GmbH, LEDGON Goniometer, Benutzerhandbuch, Version 1.2 (2008)
8. Henry R, Guruviah V (2016) Review of neuromodulation techniques and technological limitations. IETE Tech Rev 33:368–377 (Taylor & Francis)
9. Schubert EF (2006) Light-emitting diodes, 2nd edn. Cambridge University Press, Cambridge. ISBN: 9780521865388
10. Gu Y, Narendran N (2003) A noncontact method for determining junction temperature of phosphor-converted white LEDs. Proc SPIE 5187:107–114
11. Varshni YP (1967) Temperature dependence of the energy gap in semiconductor. Physica 34 (1):149–154
12. Bakk IP, Borsoi G, Favarolo PA (2012) Thermal management of LED systems. In: Elektrotechnik & Informationstechnik. Springer Verlag, Wien, Oct 2012
13. OSRAM opto semiconductors, LED fundamentals series, internal thermal resistance of LEDs (2011)
14. Technologies of high frequency transistor primer, part iii thermal properties. Agilent Technologies (2012)
15. Huber R (2004) Temperature measurement with thermocouples, Application notes Osram opto semiconductors GmbH, May 2004
16. Chen Q et al (2011) Dynamic junction temperature measurement for high power light emitting diodes. Rev Sci Instrum 82:084904
17. Kisin MV, Brown RGW, El-Ghoroury HS (2009) Software package for modeling III-nitride QW laser diodes and light emitting devices. In: Proceedings of the COMSOL Conference, Boston
18. Xu F, Lu T (2011) Introduction to skin biothermomechanics and thermal pain. Springer, Berlin, pp 24–54

Band Gap Analysis in Defectless Photonic Crystals

T. Sridarshini and S. Indira Gandhi

Abstract Recent advancement in the field of optical communication has made photonic crystals to emerge as a most interesting area for research. Photonic crystal is periodic arrangement of two materials of different dielectric materials that bring about photonic band gap effect. This band gap effect could be understood by analyzing the optical properties of photonic crystals. In this paper, photonic band gap calculations are performed for two-dimensional photonic crystals using MATLAB. The band structure is obtained for different radius to lattice constant ratios and refractive index values.

Keywords Photonic crystals · Photonic band gap · Maxwell's equations
Refractive index

1 Introduction

Arrangement of ordered microscopic structure extending in all directions constituting of atoms, molecules, or ions is termed as crystals or crystalline solids. Analogy to these crystals, photonic crystals is those with periodic contrast in dielectric constant. The similarities between photonic crystals and solid-state physics are that: periodic potential due to a lattice of periodic dielectric contrast resembles that created by atomic lattice in solid-state crystals. That is the behavior of photons inside the photonic crystals is optical analogy of behavior of electrons in semiconductor crystals. Hence, it also possesses band gap called photonic band gap (PBG) or forbidden band gap [1], where the photon density is zero [2].

Band structure relates the light wave properties to the medium in which the wave is propagating [3]. This band structure provides the basic and general information of

T. Sridarshini (✉) · S. Indira Gandhi
Department of Electronics Engineering, Madras Institute of Technology,
Anna University, Chennai 600044, India
e-mail: sridarshiniv@gmail.com

G. Gnanagurunathan et al. (eds.), *Optical and Microwave Technologies*,
Lecture Notes in Electrical Engineering 468,
https://doi.org/10.1007/978-981-10-7293-2_15

the photonic crystal, which is necessary to design the structures for various applications.

Photonic crystals find its application extensively in the field of optical communication and networks. Starting from the basic works in this field carried out by Eli Yablonovitch, who used 3D photonic crystals to avoid spontaneous emission in lasers [4] and Sajeev John founded that localization of light is possible in disordered periodic structure [5] both in 1987, researchers are working on these crystals to realize all-optical networks. Work has been carried out in designing multiplexer, demultiplexer, filters such as band stop, band pass, channel add–drop filters, narrow band filters, wavelength filters [6], couplers, splitters, and circulators [7]. These structures are obtained by introducing defects in the ideal crystal lattice such as point defect and line defects.

Band gap computation in 2D photonic crystals for different lattice using finite difference time domain (FDTD) has been done by Kaur [8]. Photonic band gap for 1D photonic crystals has been done by Zhao et al. [9]. In this work, band gap of defectless structure is analyzed.

2 Mathematical Formulation for Band Gap

Light wave propagation is always governed by Maxwell's equation. Photonic crystals being a non-conductive with no free charges and no electric current density, the parameters for the same ρ and J are taken to be zero [10]. And also considering the medium to be non-magnetic, permeability, μ is taken as 1. After simplification and eliminating electric field, one can obtain the master equation representing photonic crystals as,

$$\nabla \times \frac{1}{\in (r)} \nabla \times H(r) - \left(\frac{\omega}{c}\right)^2 H(r) = 0 \tag{1}$$

where $(\frac{\omega}{c})^2 H(r)$ is the Hermitian, and therefore, the values $(\frac{\omega}{c})^2$ indicate the frequencies of the allowed modes for $H(r)$. These frequencies are termed as Eigen frequencies.

The master equation is considered as Eigen value problem and is then solved for obtaining band structure of the photonic crystals. Band structure computation involves relating the Eigen frequencies to wave vector, k. It is given by

$$\omega^2 \varepsilon - k^2 = 0 \tag{2}$$

is obtained from solving the Eigen problem of the photonic crystals using Bloch's theorem [3]. Various methods can be used to obtain the band structure, whereas in this paper, plane wave expansion method is used.

3 Photonic Crystal Design

A 2D photonic crystal of infinite length is considered. The background material is taken as air with dielectric constant, ε = 1, and the circular rods which are arranged periodically in the air medium have the dielectric constant, ε = 12.11. Refractive index (RI) of the materials can be calculated from its dielectric constants using the formula, RI = $\sqrt{\varepsilon}$. Hence, the RI of the circular rods corresponds to silicon. Refractive index for silicon is selected in the III window wavelength range of 1.55 nm. Circular rods are arranged in square lattice with lattice constant '*a*' and radius '*r*' as shown in Fig. 1a. Band gap calculation is done for the smallest zone termed as Brillouin zone.

This Brillouin zone can be used to explain light propagation [2]. In Fig. 1b, k_x and k_y represent the wave vector, and Γ, M, and X represent three symmetric points in the zone with coordinate axis Γ(0, 0), M(π/a, π/a), and X(π/a, 0) [3]. Band structure is computed in MATLAB using plane wave expansion method. Various iterations are carried out in order to analyze the effect of ratio of radius to lattice constant and effective refractive index on photonic band gap.

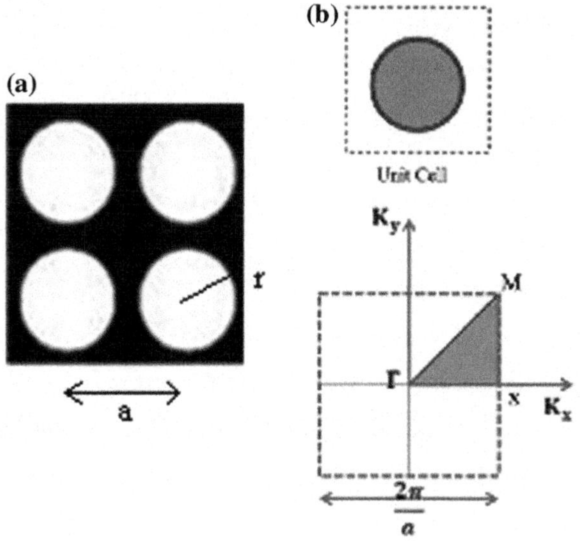

Fig. 1 **a** 2D photonic crystals in a square lattice arrangement with radius of circular rod '*r*' and the lattice constant '*a*', **b** unit cell definition and Brillouin zone of square lattice structure

4 Results

4.1 Influence of Refractive Index

To analyze the influence of refractive index (dielectric constant) on photonic band gap, simulation is done for various values of dielectric constant of the circular rod having lattice, $a = 1$ µm and radius as the rod, $r = 0.45$ µm. The photonic band gap obtained is shown in Fig. 2a, d. From these figures, it is observed that for $n_2 = 2.0$, PBG appears in between band 4 and band 5. This frequency range of light waves is forbidden inside the crystal or in other words, it is completely reflected out of the crystal. As RI of circular rod increases the PBG bands increase and also the forbidden gap shifts toward lower frequency range, as shown for RI = 2.8, 3.4 and 4.0 in Fig. 2b–d.

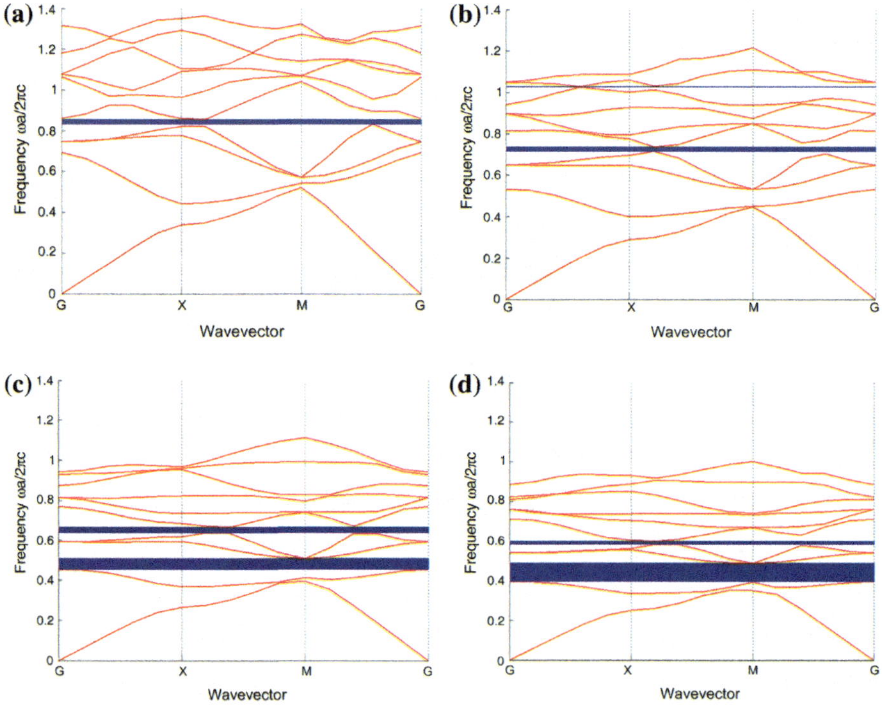

Fig. 2 **a** PBG of 2D photonic crystal with $n_2 = 2.0$, **b** PBG of 2D photonic crystal with $n_2 = 2.8$, **c** PBG of 2D photonic crystal with $n_2 = 3.4$, **d** PBG of 2D photonic crystal with $n_2 = 4.0$

4.2 Influence of Ratio of Radius to Lattice Constant

The effect of ratio of radius of the circular rod to the lattice constant (*r*/*a*) is also been analyzed for the same parameter of the 2D photonic crystal layout. For various values of RI of the circular rod, the parameter *r*/*a* is varied and PBG of the structure is obtained. Some of the results are shown in Fig. 3a, b.

From the results, it is clear that as the ratio *r*/*a* increases the PBG appears in the band structure. For the refractive index of 3.42 for the circular rod, the PBG appears in between band 4 and band 5, and for refractive index of 3.48, the PBG appears in between the same bands but with increased band gap.

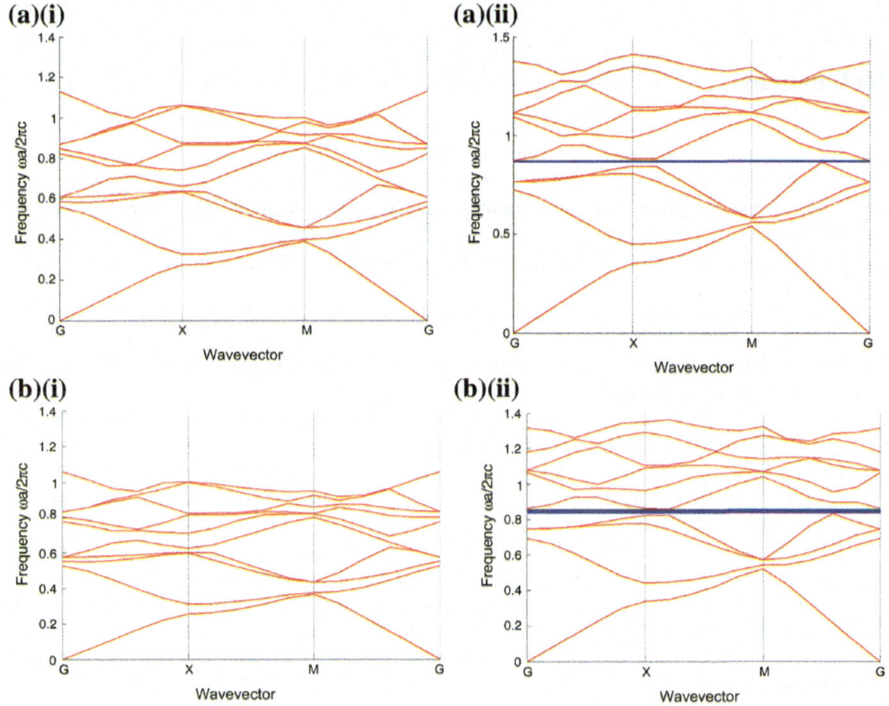

Fig. 3 **a** (i) PBG of 2D photonic crystal with $n_2 = 3.42$ and $r/a = 0.25$, (ii) PBG of 2D photonic crystal with $n_2 = 3.42$ and $r/a = 0.45$, **b** (i) PBG of 2D photonic crystal with $n_2 = 3.48$ and $r/a = 0.25$, (ii) PBG of 2D photonic crystal with $n_2 = 3.48$ and $r/a = 0.45$

5 Conclusion

Thus, the effect of the parameters such as lattice constant '*a*', radius of the rod '*r*', and refractive index of the material is used as rods on PBG for defectless 2D photonic crystal structure has been reported. This approach can be used for designing various photonic crystal-based optical components. Similar work computing the effect of size of the rod in 2D photonic crystals has been investigated [11]. Photonic band gap for required frequency range can be obtained by suitable choice of refractive index and other parameters.

Acknowledgements The authors would like to thank 'Visvesvaraya Ph.D. Scheme for Electronics and IT'—DeitY, for providing the financial support to carry out this research work.

References

1. Joannopoulos JD, Johnson SG, Winn JN, Meade RD (2008) Photonic crystals molding the flow of light. Princeton University Press
2. Gong Q, Hu X (2013) Photonic crystals principles and applications. DRC Press, Taylor and Francis Group, LLC
3. Sukhoivanov IA, Guryev IV (2009) Physics and practical modeling. Springer series in optical sciences
4. Eli Y (1987) Inhibited spontaneous emission in Solid-state physics and electronics. Phys Rev Lett 58(20):2059–2062
5. John S (1987) Strong localization of photons in certain disorder dielectric superlattices. Phys Rev Lett 58(23):2486–2489
6. Sathyadevaki R, Shanmuga sundar D, Raja AS (2016) Design of dual ring wavelength filters for WDM applications. Opt Commun 380:409
7. Umamaheswari C, Shanmuga Sundar D, Sivanantha Raja A (2016) Exploration of photonic crystal circulator based on gyromagnetic properties and scaling of ferrite materials. Opt Commun 380:186–195
8. Kaur A, Sappal AS (2014) To calculate band gap in photonic crystal in 2D using finite difference time domain. Int J Innovative Res Sci Technol 1(3)58–60
9. Zhao J, Li X, Zong L, Chen G (2009) Calculation of photonic bandgap of one dimensional photonic crystal. J Phy Conf Ser 183
10. Sakoda K (2005) Optical properties of photonic crystals. Springer series in optical sciences, 2nd edn
11. Anthrayose M, Sridarshini T, Indira Gandhi S (Unpublished) Performance analysis and characterization of 2D photonic crystals—an analytical approach

Multiband High-Gain Antenna with CPW Feed for Wi-Fi, WI-MAX and X Band Application

Anish Mukherjee, Abhishek Kanaujia and Ravi Prakash Dwivedi

Abstract Due to the recent developments in the domain of wireless network, the demand of multi frequency antenna has increased significantly. Wi-Fi has become a very popular choice available to the end users, providing them broadband connectivity and an alternative to cable and digital subscriber line, while X band finds extensive use in RADAR and satellite communication systems, and the proposed design can be put to use where there is a need for integration of both technologies. In this study, the authors have developed a novel U-shaped slot antenna for Wi-Fi and X band applications which gives a significantly higher gain and better performance than the existing antennas aimed for the same application to validate the model the structure was simulated in HFSS and operating frequencies for the proposed antenna were in compliance with the IEEE 802.11 WLAN protocol and the X band range. The structure has a fairly simple design and is having a fairly compact dimension of 40 mm × 40 mm × 1.6 mm. Moreover, the obtained radiation pattern demonstrates that proposed antenna has a significantly unidirectional pattern in both E-plane and H-plane. In addition to the unidirectional radiation pattern, a significant reduction in conduction loss was obtained.

Keywords CPW · Wi-Fi · Wi-MAX · X band · Multiband

A. Mukherjee · A. Kanaujia · R. P. Dwivedi (✉)
School of Electronics Engineering, VIT University, Chennai, Tamil Nadu, India
e-mail: raviprakash.dwivedi@vit.ac.in

A. Mukherjee
e-mail: anish.mukherjee2014@vit.ac.in

A. Kanaujia
e-mail: abhishek.kanaujia2014@vit.ac.in

G. Gnanagurunathan et al. (eds.), *Optical and Microwave Technologies*,
Lecture Notes in Electrical Engineering 468,
https://doi.org/10.1007/978-981-10-7293-2_16

1 Introduction

In the present time, there has been unprecedented development in the domain of wireless communication due to the increase in demand by the ever-increasing number of the end user of this technology. The low cost, the high reliability and the longevity of the devices have made this technology available to multitude of people. To support the growth in the wireless technology, there is a continuous want for the development of the antenna technology without which the whole wireless communication industry will come to a standstill. Wireless communication has presented so many applications like WLAN, GPS, mobile telephony, remote sensing and RADAR [1]. These applications find their use in all the domains, be it the customer-oriented application, medical application or military applications. One of the reasons of the rapid development of the wireless communication systems is the advent of the microstrip patch antenna which is lightweight, cheap and the ease of fabrication of the antenna. The simple microstrip patch antenna although is cheap and easy to fabricate can provide very low gain and requires an amplifier to be attached to the circuitry before the signal can be used [2]. Fringing fields that are present at the edge of the substrate are yet another source for radiation loss. Co-planar waveguide provides a higher gain, and the loss is also significantly reduced without increasing the cost and reducing the ease of fabrication, and a CPW feed is easy to integrate with the microwave integrated circuits [3].

Multiband antennas are becoming more and more popular as the need for integration of the different technologies is becoming imminent [4]. Multiband antennas are becoming more and more imminent as the need for integration of different technologies is increasing. WLAN has become a viable option for mobile communication and inter communication between wireless sensor nodes [5–9]. C band frequency range has found extensive use in satellite communication and in television broadcast as well. Ku band is used in RADAR applications mostly. The biggest advantage with C band and Ku band is that parabolic reflectors become more and more efficient with the increase in frequency, and therefore, these frequency bands are preferred in long-distance communication usage. Another advantage is the reduction of antenna size with the increase in frequency, enabling us to have even more compact communication devices. Ku band additionally comes with another advantage that it is generally cheaper than the other band, and extensive utilization of this band would be beneficial economically. The proposed structure can be deployed where there is a need for the integration of the aforementioned technologies such as military application where there is a need of RADAR and the information obtained by RADAR has to be relayed to other mobile devices.

The antenna design is discussed in the next section followed by the results. The last section of the paper summarizes and concludes the entire work.

2 Antenna Design and Simulation

The antenna is designed and simulated using the software HFSS 13.0. The antenna is fabricated using FR4 epoxy as the substrate having dielectric constant of 4.4. Initially, a microstrip patch antenna is designed for a single resonant frequency. The length L and the width W of the microstrip antenna are calculated using the formula derived from the transmission line model. The formula is mentioned below:

$$W = \frac{c}{2f_0}\sqrt{\frac{\varepsilon_r + 1}{2}} \tag{1}$$

$$L = \frac{c}{2f_0\sqrt{\varepsilon_e}} - 2\,\Delta L \tag{2}$$

where c denotes the speed of light in vacuum, ε_r is the relative dielectric constant, f_0 is the center frequency, and ε_e is known as the effective dielectric constant which can be calculated using the formula,

$$\varepsilon_r = \frac{\varepsilon_r + 1}{2} + \frac{\varepsilon_r - 1}{2}\sqrt{1 + \frac{10h}{W}} \tag{3}$$

where h is the height of the substrate, and ε_r is the relative dielectric constant of the substrate. Fringing fields introduce a term ΔL which can be calculated by the Hammerstad formula:

$$\Delta L = 0.412h\frac{(\varepsilon_r + 0.3)\left(\frac{W}{h} + 0.264\right)}{(\varepsilon_r - 0.258)\left(\frac{W}{h} + 0.8\right)} \tag{4}$$

Figure 1a shows the CPW patch antenna. Figure 1b and c shows the return loss and the current distribution of the CPW patch antenna.

A U-shaped slot is made in the patch antenna to increase the current distribution in the center. Instead of the microstrip feed, the patch is fed using a CPW feed which reduces the loss due to the fringing fields. The slots made in the patch are optimized to obtain three resonant frequencies 3.3, 6.6 and 12.2 GHz which are in compliance with WLAN band, C band and Ku band. The co-planar ground of the CPW feed is extended to further reduce the fringing fields and to further increase the gain. Slots are made in the ground plane to optimize the gain obtained from the antenna. The size of the antenna is 40 mm × 40 mm made on FR4 substrate having a thickness of 1.6 mm and the dielectric constant 4.4. The geometrical layout of the proposed antenna is shown in Fig. 2, and the dimensions are shown in Table 1.

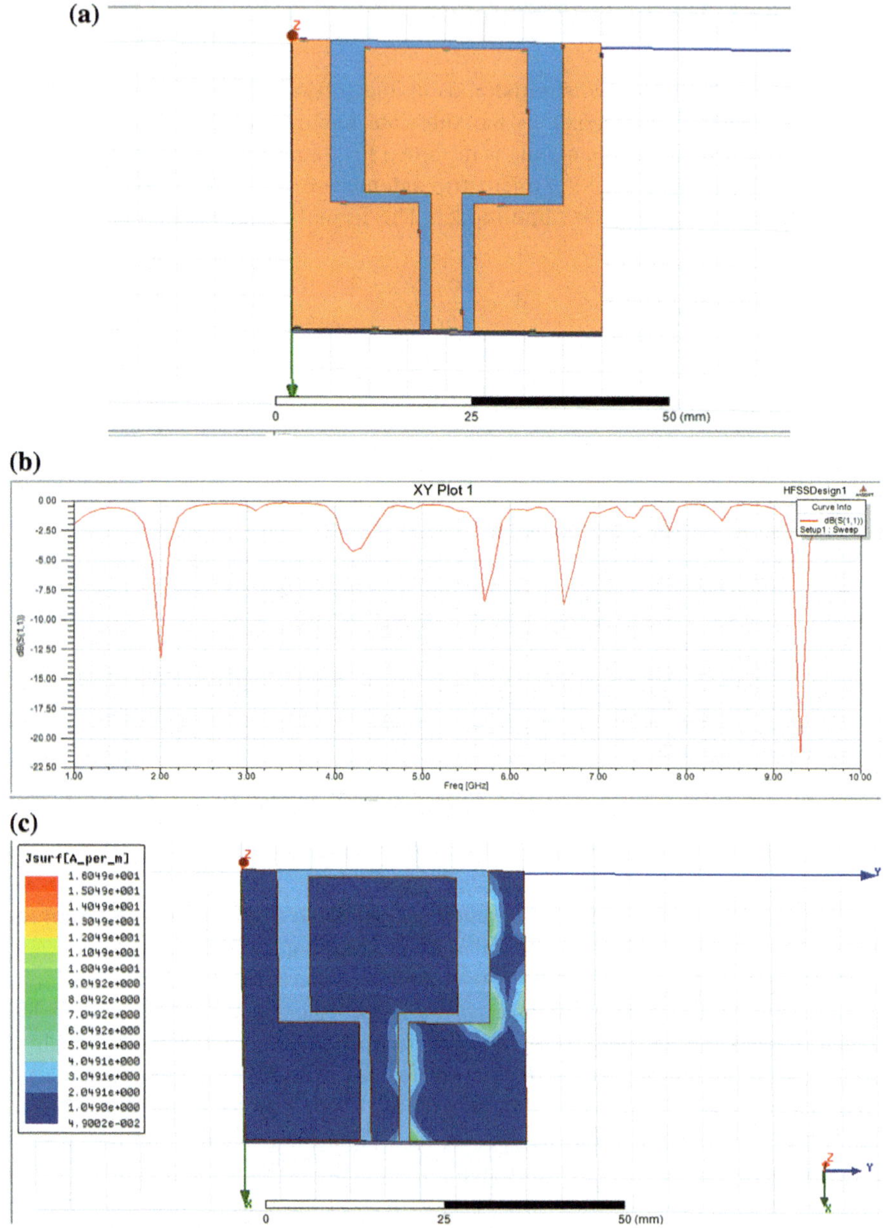

Fig. 1 **a** CPW patch antenna, **b** S_{11} parameter of the proposed antenna, **c** current distribution

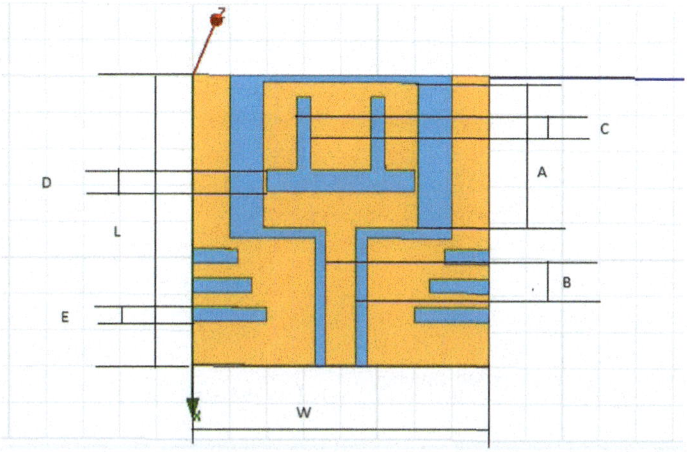

Fig. 2 Dimension of the proposed antenna

Table 1 Dimension (All dimensions are in mm)

L	40
W	40
A	20
B	4
C	2
D	3
E	2

3 Results and Discussion

The return loss of the proposed design is shown in Fig. 3. The radiation pattern for the frequency band of operation is shown in Figs. 4 and 5. The obtained gain is about 2.42 dB in the 12.2 GHz band and 0.6 dB in 6.6 GHz band. The radiation pattern is highly unidirectional.

The current density distribution of the proposed design for the different operating frequencies is shown in Figs. 6 and 7. The VSWR plot is for the proposed structure, and it is shown in Fig. 8. It is well below standard VSWR < 2 for desired frequency of operation.

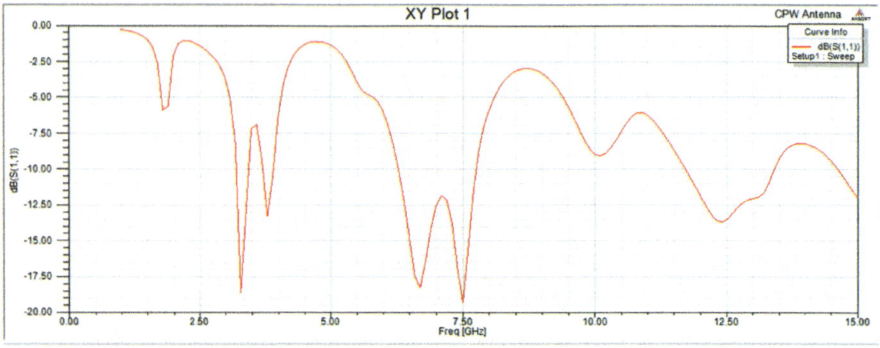

Fig. 3 Return loss of the proposed structure

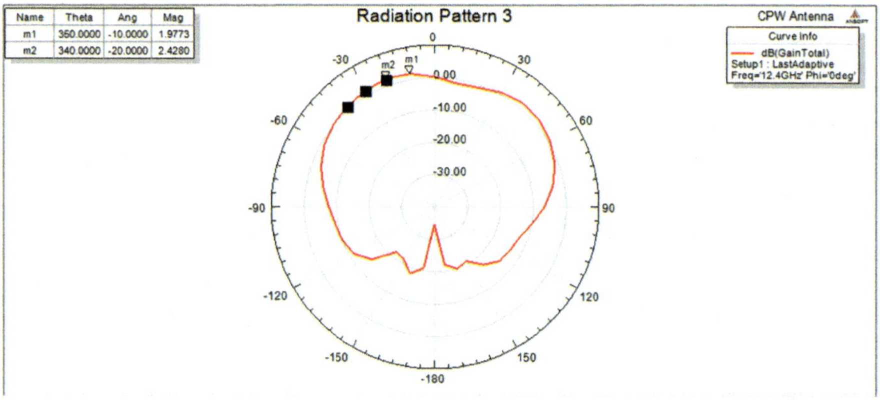

Fig. 4 Radiation pattern at 12.2 GHz

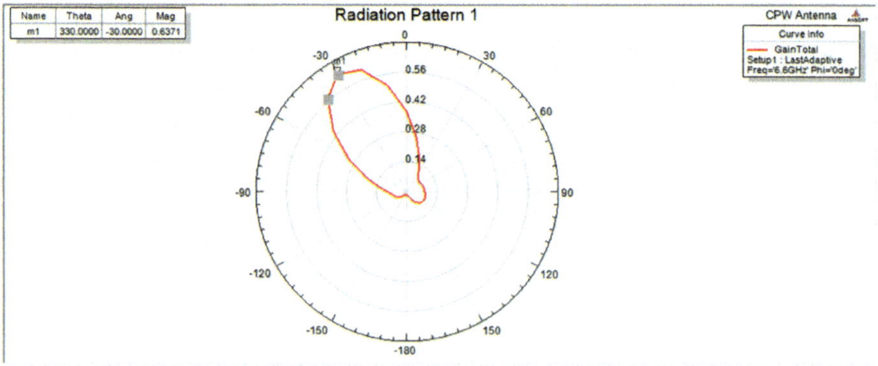

Fig. 5 Radiation pattern at 6.6 GHz

Fig. 6 Current distribution at 12.2 GHz

Fig. 7 Current distribution at 6.6 GHz

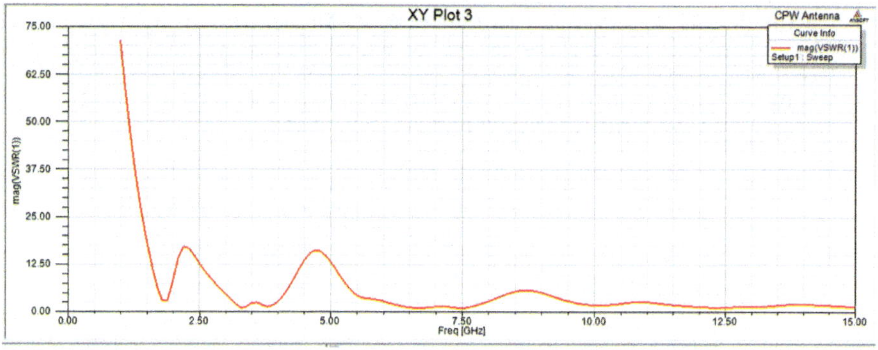

Fig. 8 VSWR plot of the proposed antenna

4 Conclusion

A CPW-fed U-slotted microstrip patch antenna has been proposed and analyzed. The obtained frequencies of operations are 3.2–4, 6.3–7.7 and 11.6–13.4 GHz which are suitable for handling WLAN, C and Ku band application-related applications and can be used extensively in the domain where there is the need for the integration of both these technologies. The proposed gain of the antenna is 2.42 and 0.6 dB, respectively, for the aforementioned frequency of operation. The overall antenna design is simple and easy to fabricate, and significant reduction is done for conduction loss as copper is etched out for making the U-shaped slots reducing the conduction loss; thus, it can be stated the proposed antenna is well suited for WLAN, C and Ku band applications for increasing the gain. Modification in the slot can be made to make the antenna resonate at Wi-Fi, Wi-Max and X band frequencies, and a margin has been provided as when fabricated additional solder leads to addition of inductance which makes the antenna resonate at a much lower frequency than the simulated structure.

References

1. Kumar G, Ray KP (2003) Broadband microstrip antenna (Chaps. 1, 2). Artech House Inc
2. Wadell BC.(1991) Transmission line design handbook, pp 1–5, 73–75. Artech House, Norwood
3. Simons Rainee N (2001) Coplanar waveguide circuits, components, and systems, pp 1–5, 15–21. Wiley, Inc., New York
4. Yoon J, Jeong GT, Kwak KS (2007) Fabrication and measurement of triangular-slot antenna for triple band (2.4/5.2/5.8 GHz) antenna with rectangular tuning stub. Microw Opt Technol Lett 49(8)
5. Pei J, Wang AG, Gao S, Leng W (2011) Miniaturized triple-band antenna with a defected ground plane for WLAN/WiMAX applications. Antennas Wirel Propag Lett IEEE 10:298–301
6. Azim R, Islam MT, Misran N (2011) Dual polarized microstrip patch antenna for Ku-band application. Informacije MIDEM 41(2):114–117 (Ljubljana)
7. Liu D, Gaucher B (2004) A branched inverted-F antenna for dual band WLAN applications. In: Proceedings of the IEEE antennas and propagation society international symposium, Monterey, CA, vol 3, pp 2623–2626
8. Raj RK, Joseph M, Aanandan CK, Vasudevan K, Mohanan P (2006) A new compact microstrip-fed dual-band coplanar antenna for WLAN applications. IEEE Trans Antennas Propag 54(12):3755–3762
9. Wu JW, Hsiao HM, Lu JH, Chang SH (2004) Dual broadband design of rectangular slot antenna for 2.4 and 5 GHz wireless. IEE Electron Lett 40(23)

Design and Parameter Extraction of Split Ring Resonator for Surface Crack Detection in Different Materials

Varun Seshadri and Ravi Prakash Dwivedi

Abstract In this paper, a simple approach has been proposed to detect crack using the changes in the resonant frequency of the resonator. Extensive simulations have been carried out using finite element analysis tool HFSS. SRR of various geometrical shapes and the corresponding result is also presented in this paper. A theoretical formulation to estimate the resonant frequency of the split ring resonator is also proposed. A step-by-step procedure is devised for analyzing the material parameters to facilitate and simplify the design procedure of metamaterial structures. The entire idea of this paper is based on the interpretation of S-parameters of SRR.

Keywords Split ring resonator · Metamaterial · Magnetic susceptibility
Resonant cavity method · Capacitive reactance · Inductive reactance
Resonance · LC resonator · Strip line · S-parameters

1 Introduction

The metamaterial is basically used to provide the existing patch antenna a wide frequency range. Meta in Greek means beyond. Since such materials exhibit some unnatural properties such as backward wave propagation and negative refractive index so they are termed as metamaterial. Such properties have been realized previously and it is not of the present interest of the paper [1]. In 1968, Veselago first proposed the theoretical concept of metamaterials. But these structures were

V. Seshadri · R. P. Dwivedi (✉)
School of Electronics Engineering, VIT University, Chennai, Tamil Nadu, India
e-mail: raviprakash.dwivedi@vit.ac.in

V. Seshadri
e-mail: varunseshadri.vatsala2015@vit.ac.in

G. Gnanagurunathan et al. (eds.), *Optical and Microwave Technologies*,
Lecture Notes in Electrical Engineering 468,
https://doi.org/10.1007/978-981-10-7293-2_17

experimentally realized and proved later in 2000 [2]. Some applications for metamaterial antennas are wireless communication, satellites, space vehicle navigation, space communication, aircrafts, global positioning systems (GPS).

As the primitive techniques such as acoustic sensing leads to the production of harmonics due the perturbations on the crack that interfere with the sensing capabilities of the acoustic sensors which do not yield proper results as a consequence. Even the image processing technique of sensing cracks using high resolution charged coupled device (CCD) cameras does not give the depth information of the crack being sensed on the surface of the material. So far for military applications and underwater installation of pipe lines, such methods do not provide better results. The equivalent circuit model analysis of the SRR shows that they behave like *LC* resonator that can be excited by an external magnetic flux [3].

The method on which the SRR working principle is based is known as the *Resonant Cavity Method*. In this method, a sample of small dimension compared to the dimension of the cavity is inserted into the cavity. This insertion of the sample changes the resonant frequency of the cavity. The resonant cavities resonate at certain frequencies, and they are structures of high-quality factors. This method supports high-impedance environment, and there is a possibility of making reasonable measurements with small samples. The measurement is realized at only one frequency or few frequencies. So, due to the above-mentioned advantages, the resonant cavity method is suitable for materials with low loss. The equivalent circuit model shall consist of the corresponding capacitive and inductive reactance. The negative refractive index would be the permeability and permittivity [4] (Fig. 1).

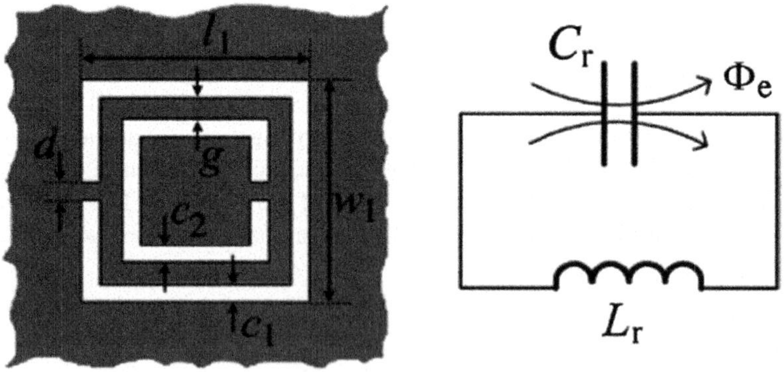

Fig. 1 Equivalent circuit model of the split ring resonator for single unit cell

2 Theoretical Analysis

The list of formulae for the design purpose is as follows:

$$Z_{0}=\frac{60}{\varepsilon_{\text{eff}}}\ln\left(\frac{8h}{w}+\frac{0.25w}{h}\right)$$

$$\varepsilon_{\text{eff}}=\left(\frac{\varepsilon_{\text{r}}+1}{2}\right)+\left[1+12\left(\frac{h}{w}\right)^{-0.5}+0.04\left\{1-\left(\frac{w}{h}\right)\right\}^{2}\right]$$

The resonant frequency of SRR is given by-

$$f_{0}=\frac{1}{2\pi\sqrt{L_{\text{T}}C_{\text{eq}}}}$$

$$C_{\text{eq}}=C_{\text{pul}}\left(2a_{\text{avg}}-\frac{g}{2}\right)+\frac{\varepsilon_{0}\text{ch}}{2g}$$

Here, C_{pul} is the per unit length capacitance between the rings which is given by the formula (Fig. 2):

$$C_{\text{pul}}=\sqrt{\varepsilon_{\text{e}}}/C_{0}Z_{0}$$

$$a_{\text{avg}}=a_{\text{ext}}-c-\frac{d}{2}$$

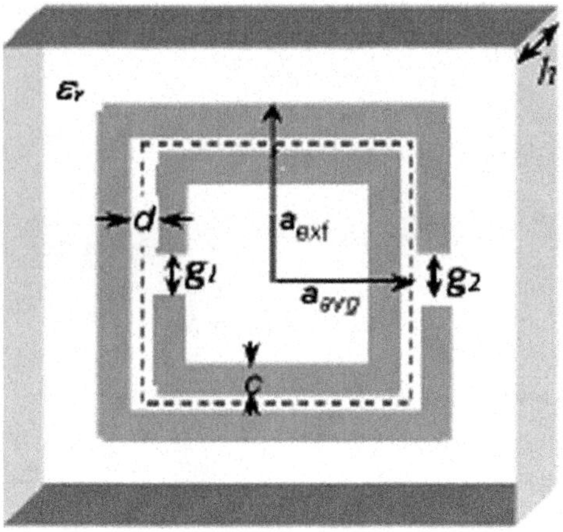

Fig. 2 Geometrically specified diagram of the square SRR structure used for the project

The above-mentioned formula is for the average distance.

$$L_T = 0.0002l\left[2.303\log_{10}\frac{4l}{c} - \gamma\right]\mu\mathrm{H}$$

Here, the constant γ has a value of 2.853 for designed wire loop of square geometry. All dimensions are in mm (c is thickness and l is the length given by $l = 8aex{-}g$) [5].

3 Unit Cell Design

The purpose of the split ring resonator (SRR) is to produce the desired magnetic susceptibility [7, 8]. The substrate used here is Fr-4 epoxy with a dielectric constant of 4.4. The sides are of 2.5 mm each. Copper has been used for the rings and the ground line (Fig. 3).

The dimensions and the specifications of the conventional unit cell are given in Table 1.

We have used lumped port as the radiation would be on both sides of prototype. Boundaries have been set for the outer and inner rings which is the perfect E boundary. Analysis setup has been given for the unit cell model. The radiation box would consist of air and that would have a length of 2.51 mm each in all the three axes of the rectangular coordinate system.

Depending on the basic square structure, as it is simple to fabricate the unit cell other variants by changing the geometry and keeping the above-mentioned specification for each shape the same have been also developed. These structures have been analyzed and simulated and results are also shown in Fig. 4.

All the above-given unit cells produce the same result which is given in Figs. 5 and 6.

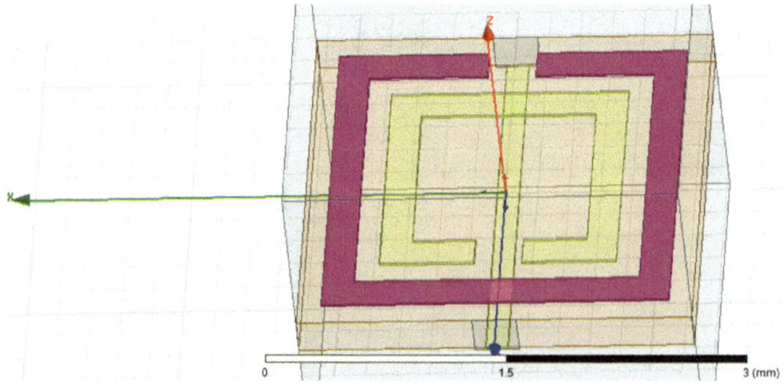

Fig. 3 Square SRR structure designed using ANSYS HFSS

Table 1 Specifications of the square SRR unit cell

Specifications	Magnitude	Unit
Outer ring perimeter	8.8	mm
Inner ring perimeter	6	mm
Ground line perimeter	5.28	mm
Gap perimeter ($g1 = g2$)	1	mm
Ring thickness	0.017	mm
Ground line thickness	0.017	mm
Gap thickness ($g1 = g2$)	0.017	mm
Port perimeter	1.094	mm
Substrate thickness (h)	0.25	mm
Spacing between the rings (d)	0.15	mm
Ring width (c)	0.2	mm

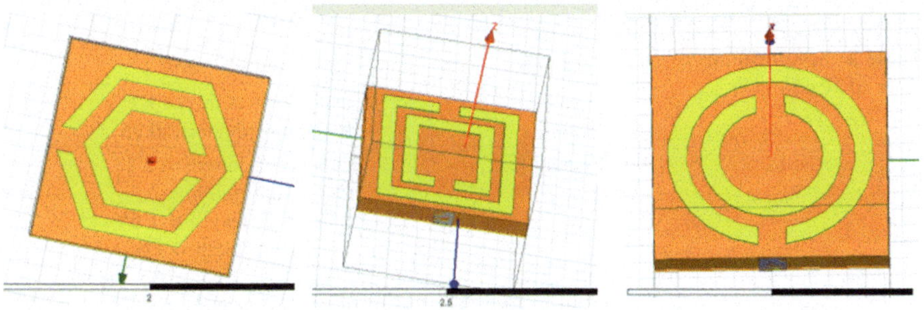

Fig. 4 Cylindrical, rectangular, and hexagonal unit cells of SRR

Fig. 5 Magnitude of S-parameters obtained from HFSS

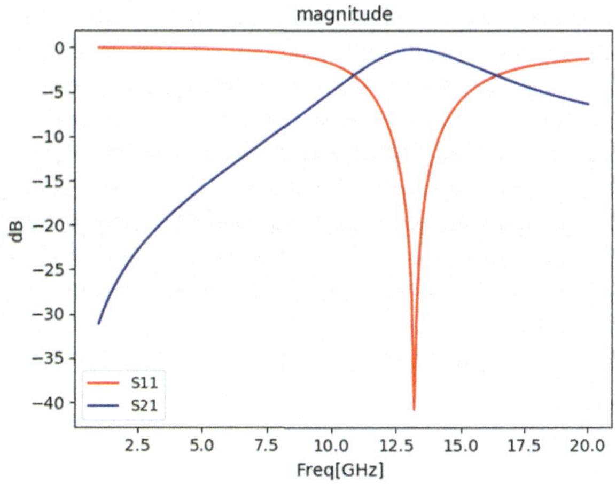

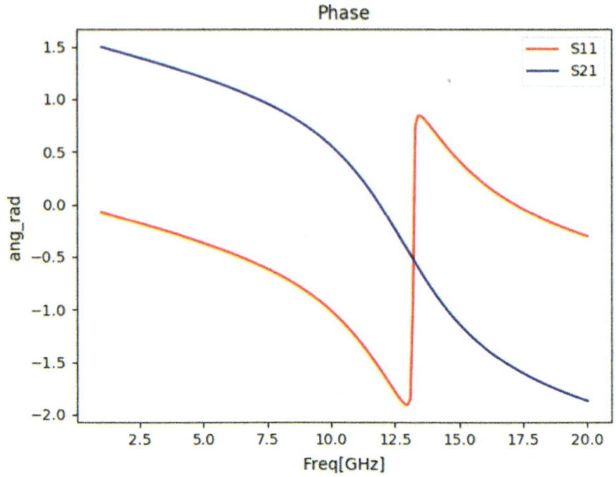

Fig. 6 Phase of S-parameters obtained from HFSS

Table 2 Specifications of other SRR unit cells

Structure	Specifications (mm)
Cylinder	Outer ring radius = 1.1 Inner ring radius = 0.75
Rectangle	Outer ring length = 2.2 Outer ring width = 2 Inner ring length = 1.5 Inner ring width = 1.3
Hexagon	Outer ring sides = 1.2 Inner ring sides = 0.8

Here, the substrate and gap dimensions, spacing between the rings, ring thickness, ring width, substrate dimensions, and the ground line dimensions are same as mentioned in Table 1 (Table 2).

Thus, we observe that the results obtained through HFSS by simulating the unit cell give the exact similar result. We have also proposed another model to make the response of the SRR for crack detection purpose more sensitive, wide range, and sharp [6]. The substrate dimensions have been changed keeping the material same which is Fr-4 epoxy. The length of the ground line is also increased accordingly. Rest all dimensions are kept same as mentioned in this paper. The substrate has sides of 3.5 mm and so the length of the ground line is 3.5 mm. The gap is of 0.2 mm (Fig. 7).

The simulated design and the corresponding results are given in Fig. 8.

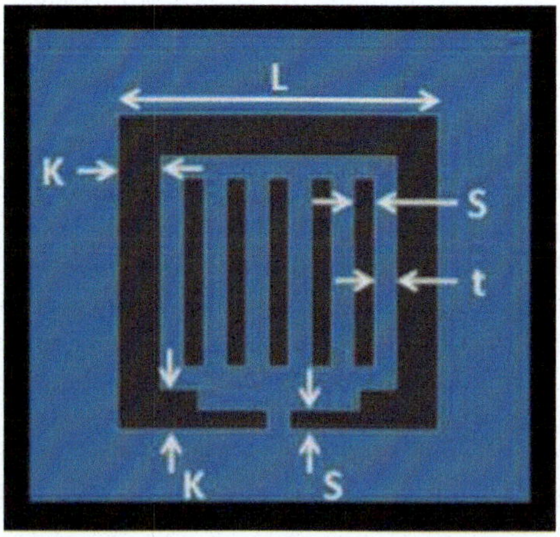

Fig. 7 Outline of the model showing dimensions, **L** = 3 mm, **S** = 0.2 mm, **t** = 0.2 mm, and **K** = 0.4 mm

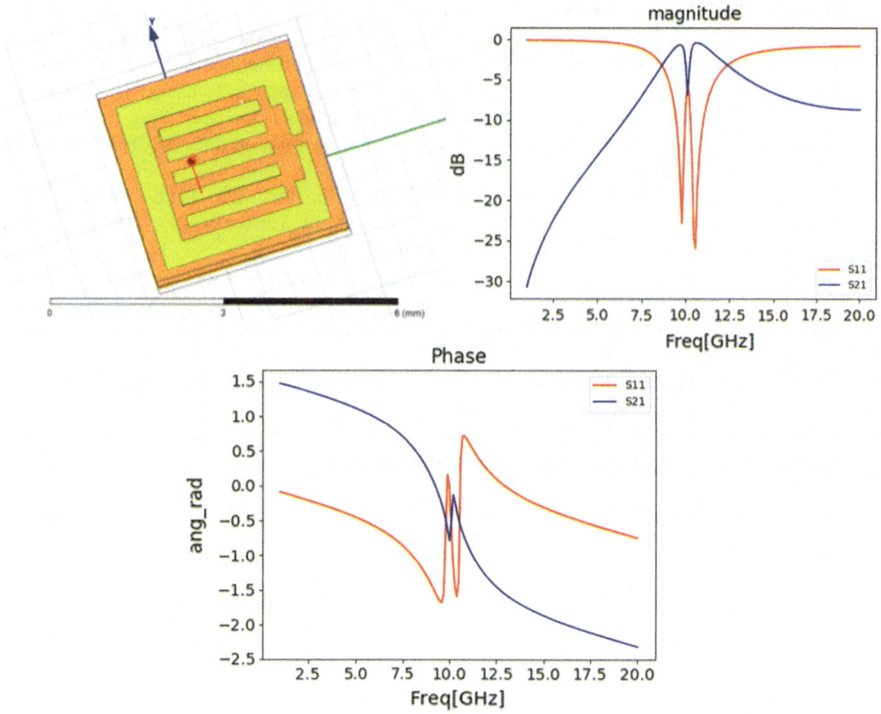

Fig. 8 Clockwise-SRR unit cell for high sensitivity simulated using HFSS. The magnitude of the S-parameters for the above-mentioned model obtained from HFSS. The phase of the S-parameters for the above-mentioned model obtained from HFSS

4 Conclusion

In this work, the analysis of various metamaterial structures was presented based on the S-parameters obtained from a full-wave simulator. The periodic boundary condition and excitation mode have been presented as well. The structural differences of different unit cells have been highlighted. The results obtained have also been compared with some published results. The basic idea is when different material blocks are placed under such cells the resonant frequency of the SRR shifts according to the dielectric constant of the material being sensed and used thus by revealing the presence of crack. This method would be compared with the results so obtained from an ideal block to determine the crack dimensions thereby generalizing the result further. The procedure will help researchers in the area of metamaterial property analysis with ease, using commercially available software.

References

1. Khushboo, UshaKiran K (2016) Design of metamaterial for wide stop band application. In: IEEE WiSPNET 2016 conference
2. Chandrasekhar K, Mishra D, Poddar DR, Mishra RK (2007) Design of left-handed metamaterials using hexagonal split ring resonator at S-band frequencies. In: 2007 IEEE
3. Baena JD, Bonahe J, Martin F, Sillero RM, Falcone F, Lopetegi T, Garcia-Garcia J, Gil I, Portillo MF, Sorolla M (2005) Equivalent-circuit models for split ring resonators & complimentary split ring resonators coupled to planar transmission lines. IEEE Microw Trans Microw Theor Tech 53(4)
4. Jabita AA (2013) Design of singly split single ring resonator for measurement of dielectric constant of materials using resonant method. Master Thesis, University of Gavle, Department of Engineering & Sustainable Development, June 2013
5. Saha C, Siddiqui JY, Antar YMM (2011) Square split ring resonator backed coplanar waveguide for filter applications. IEEE
6. Numan AB, Sharawi MS (2013) Extraction of material parameters for metamaterials using a full-wave simulator. IEEE Antennas Propag Mag 55(5)
7. Veselago V (1968) The electrodynamics of substances with simultaneously negative values of $\epsilon\&\mu$. Soviet Physics Uspekhi 10(4):504–514
8. Bandlow B, Schuhmann R (2008) Some constraints of the concept of effective material parameters in metamaterials, Universitat Paderbom

Optical Channel Analysis of Turbo Coded MIMO-OFDM System for Visible Light Communication

Sabitha Gauni, C. T. Manimegalai, K. Kalimuthu, V. C. S. Kaushik and T. Rama Rao

Abstract Optical wireless communication (OWC) system uses intensity modulation with direct detection (IMDD) to modulate the intensity of the signal in order to transmit information. There is an increase in the exploration of the modulation techniques used in visible light communications (VLC) in recent times. This paper deals with the comparison of power efficient modulation techniques such as dual-header pulse interval modulation (DH-PIM) and digital pulse interval modulation (DPIM) for VLC systems. Further, power allocation is studied under AWGN and Rayleigh channel. It is seen that the results prove to be spectrally efficient in the analysis of MIMO-OFDM VLC systems.

Keywords Optical wireless communication · Visible light communication DPIM · DH-PIM · MIMO-OFDM

1 Introduction

The research in the areas of optical wireless communication (OWC) and visible light communications (VLC) is unending as the bandwidth varying from 430 to 750 THz is very attractive [1–3]. VLC in essence has a dual role of illumination and

S. Gauni (✉) · C. T. Manimegalai · K. Kalimuthu · V. C. S. Kaushik · T. Rama Rao
SRM University, Kattankulathur, Chennai, India
e-mail: sabitha.g@ktr.srmuniv.ac.in; sabianup@gmail.com

C. T. Manimegalai
e-mail: manimegalai.c@ktr.srmuniv.ac.in

K. Kalimuthu
e-mail: kalimuthu.k@ktr.srmuniv.ac.in

V. C. S. Kaushik
e-mail: vcskaushik1@gmail.com

T. Rama Rao
e-mail: ramaraotr@gmail.com

G. Gnanagurunathan et al. (eds.), *Optical and Microwave Technologies*,
Lecture Notes in Electrical Engineering 468,
https://doi.org/10.1007/978-981-10-7293-2_18

communication. There is an added advantage of having cheap end devices such as LED and photo-detectors. The modulation schemes used for visible light communication are limited, for the reason that for intensity modulation with direct detection (IMDD), the modulated signal is to be unipolar and real. The analysis of OWC systems with modulation techniques like digital pulse interval modulation (DPIM) and dual-header pulse interval modulation (DH-PIM) has gained interest.

Though affected by strong distortions like terrestrial broadcasting, the coded orthogonal frequency division multiplexing (COFDM) technique provides reliable reception of signals. The limitation to such high bit-rate modulation schemes is multipath propagation. Therefore, using COFDM, the information which is to be transmitted is split over many carriers, to lower the signalling rate on each of them rather than lowering the assumed channel coherence bandwidth. The codes are developed by interleaving two or more component codes of the same information sequence. In a system with two component codes, turbo decoding is done by passing soft decisions between the input and output of the two decoders. This process has to be iterated several times for reliable decisions. Because of this advantage, turbo codes are now part of a large number of communication standards.

The paper gives a brief description of the system model for optical wireless communication (Sect. 2). In Sect. 3, the proposed system model is explained. Section 4 presents the performance results, followed by conclusion in Sect. 5.

2 System Model for Optical Wireless Communication

Optical wireless transceivers are economically high in cost for its use in mass-market applications [4]. Therefore, it is highly desirable to choose a better modulation technique which will be easy to implement, as achieving great bandwidth efficiency and/or power efficiency in a highly complex technique is of little use and is not feasible in terms of cost also. Thus, the two techniques of DPIM and DH-PIM are explored [5, 6].

2.1 Modulation Techniques for VLC

2.1.1 Differential Pulse Interval Modulation (DPIM)

Differential pulse interval modulation is a type of asynchronous modulation technique, where each input block of '*M*' data bits is mapped to one of the varying length of '*L*' possible symbols. Each symbol comprises a pulse which is of one time slot duration followed by a series of empty time slots. The number of empty time slots is determined by the decimal value of the *M*-bit data stream that is encoded. The length of the symbols in DPIM is unfixed. Also due to variation in symbol length detection at the receiver is done through clock synchronization at the receiver and not symbol synchronization. This simplifies the system and is an advantage over PPM [7].

2.1.2 Double Header Pulse Interval Modulation (DH-PIM)

The double header pulse interval modulation has more complexity because the included time slot in each symbol is also alternating between two types of starting pulse. The symbol s_k comprises of a head time slot, followed by empty time slots which are 'M' in number [8]. The head slot has (α + 1) time slots where α is an integer. Let the two types of heads be H1 and H2; the initial pulse width of H1 is α/2 time slots, accompanied by protected time slots which are (α/2) + 1 in number; H2 pulse is 'α' time slots wide, followed by a single time slot. The head time slot of symbol s_k is H1, when $k < 2M - 1$, else it is H2.

2.2 *Advantages of Differential Modulation Techniques*

2.2.1 Power Efficiency

The comparison of the modulation techniques is made with respect to average optical power required for a certain target bit error rate (BER) performance or signal-to-noise ratio (SNR) [9]. Power efficiency is given as the energy per pulse per average energy per bit.

DPIM's average power requirement and power efficiency is given by:

$$P_{\text{avgDPIM}} = \sqrt{\frac{2N_0R_b}{R^2L_{\text{DPIM}}\log_2 L}}Q^{-1}(P_{\text{ebitDPIM}}) \tag{1}$$

$$\mu_{\text{DPIM}} = \sqrt{\frac{8}{(L+1)\log_2 L}} \tag{2}$$

For DH-PIM, the average power requirement and the power efficiency is given by:

$$P_{\text{avgDH-PIM}} = \sqrt{\frac{2N_0R_b}{16R^2L_{\text{DH-PIM}}M}}Q^{-1}(P_{\text{seDH-PIM}}) \tag{3}$$

$$\mu_{\text{DH-PIM}} = \sqrt{\frac{9\alpha^2}{2M(2^{M-1}+2\alpha+1)}} \tag{4}$$

2.2.2 Bandwidth Efficiency

The optical bandwidth is practically limitless. The modulation schemes involving high bandwidth are more prone to inter symbol interference (ISI) and as result incurring more power penalty. Transmission bandwidth is defined as minimum bit slot $(T_{\min})$

$$B_{\text{req}} = \frac{1}{T_{\min}} \tag{5}$$

DPIM:

$$T_{\text{sDPIM}} = \frac{T_b M}{L_{\text{DIPM}}} \tag{6}$$

$$B_{\text{reqDPIM}} = \frac{L+1}{2M} R_b \tag{7}$$

DH-PIM:

$$T_{\text{sDH-PIM}} = \frac{2M}{(2^{M-1} + 2\alpha + 1)R_b} \tag{8}$$

$$B_{\text{reqDH-PIM}} = \frac{(2^{M-1} + 2\alpha + 1)}{\alpha M} R_b \tag{9}$$

2.2.3 Transmission Reliability

A study based on packet transmission rate is more convenient as the symbol length varies in most of the modulation techniques. The size of packet is assumed to be constant of N_{pkt} bits [10]. The average slot length is given by

$$L_{\text{pkt}} = \frac{\bar{L} N_{\text{pkt}}}{M} \tag{10}$$

$\bar{L}$ Therefore, the packet transmission rate R_{pkt} is given by:

$$R_{\text{pkt}} = \frac{R_s M}{\bar{L} N_{\text{pkt}}} \tag{11}$$

R_s

For DPIM,

$$R_{\text{pktDPIM}} = \frac{2MB_{\text{req}}}{N_{\text{pkt}}(2^M + 1)} \tag{12}$$

For PIM,

$$R_{\text{pktDH-PIL}} = \frac{\alpha MB_{\text{req}}}{N_{\text{pk}t}(2^{M-1} + 1 + 2\alpha)} \tag{13}$$

3 Proposed System Model

The proposed turbo coded OFDM system is presented in Fig. 1. The turbo encoder encodes the information bits that are generated by a random data generator. These bits are then interleaved and then demultiplexed into various sub-channels. Based on the given modulation scheme, these bits are then mapped to a single constellation point. This is then followed by dinary to decimal conversion in preparation for modulation. The modulator performs IFFT on the encoded decimal bits to convert from frequency domain to time domain signal. In accordance with OFDM technique, a cyclic prefix is appended to the head of each symbol. The previous two steps of IFFT conversion and the appendage of cyclic prefix form the core of OFDM modulation. After the mapping and modulation of OFDM symbols, the transmitter sends the bits over the channel to the receiver. The channel is assumed to be a frequency-selective additive white gaussian noise (AWGN) channel.

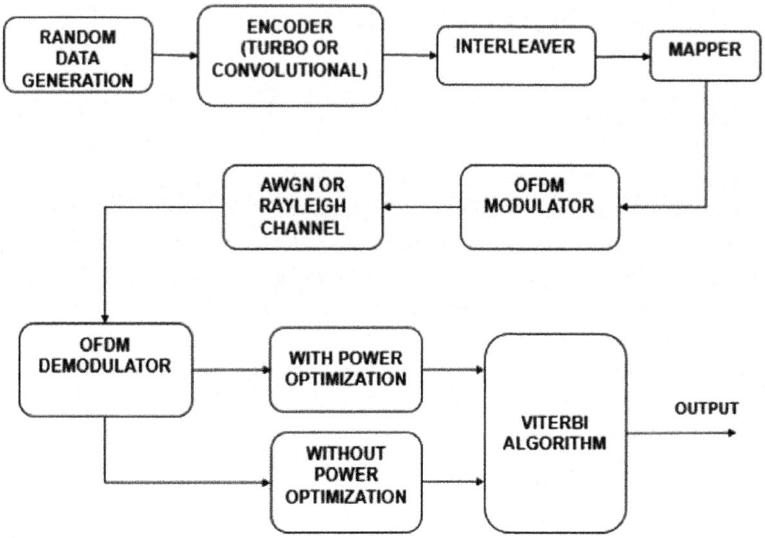

Fig. 1 System model

The received signal is demodulated at the receiver, converted back from decimal to binary, de-interleaved and decoded by a turbo decoder as follows: the time domain signals are converted into frequency domain signals using discrete Fourier transform (DFT). The soft metric of each coded bit in each sub-channel is computed by the demapper. The de-interleaver transmit the output of the de-mapper to the de-coder. For coherent detection, the gain and phase component estimates of the channel response are computed at the receiver. For the estimation, the pilot signals like pilot symbols and pilot tones are used. The feedback taken from the receiver to transmitter aids the transmitter in allocating different power levels to different sub-channels maintaining the total power transmitted constant. Hence, the coded OFDM system improves performance by making the power allocation efficient. The received data symbols at the decoder are used for computing the error rate in transmission and coding.

3.1 *Turbo Coded OFDM Systems*

Turbo coding is done by interleaving message with the convolutional codes. The turbo coded OFDM system shows better performance compared to convolutional codes as they have less power consumption due to reduced complexity. Unlike the block codes, the convolutional code does not break the message stream.

The turbo coded OFDM (TC-OFDM) system is a mixture of turbo code with OFDM transmission. This yields significant improvements in terms of transmission of data with less energy which is a remarkable move in personal communication devices [11, 12]. The performance of the system increases exponentially with the iterative property of turbo decoders, since both the soft bit estimates and the known pilot symbols are used. The performance of such systems improves the reception of signals in frequency-selective and time-selective channels.

3.1.1 Turbo Encoder

The block diagram of the encoder as shown in Fig. 2 for a turbo code is parallel concatenated convolutional code [13]. The input binary data sequence, $d_k = (d_1, \ldots d_N)$, is given as input to the convolutional encoder. The coded bit stream, Xk1, is generated as the output of the ENC1. The sequence of data Xk1 is then interleaved. Interleaving is the process by which the bits are converted into a matrix and input bit positions are spread out in a pseudo-random manner. The second convolutional encoder ENC2 takes this interleaved data sequence as input and generates Xk2 as the second coded bit stream. The systematic code bits Xks and parity bits from both the first and the second encoder Xk1, Xk2 are multiplexed together before being modulated and transmitted.

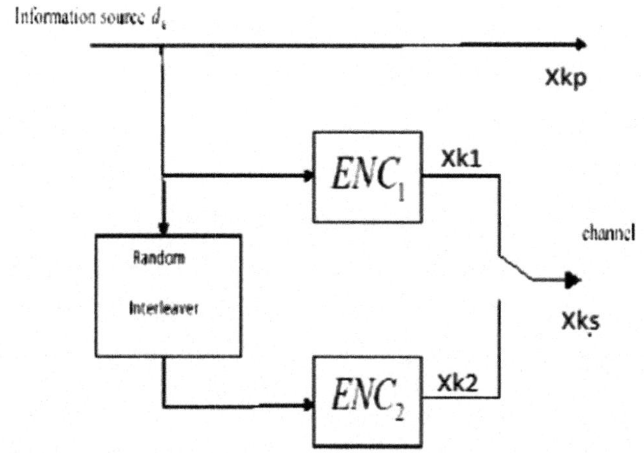

Fig. 2 Turbo encoder

3.1.2 Turbo Decoder

Figure 3 shows the block diagram of a turbo decoder. The received sequence of code values, Rk = {Ysk, Ypk}, from the demodulator is given as input to the turbo decoder. There are two maximum a posteriori (MAP) decoders in turbo decoder: DEC1 decodes the sequences coming from ENC1, and DEC2 decodes the sequences coming from ENC2. The input to each decoder is the received sequence of parity values Ypk1 of the first encoder ENC1 and the received sequence of systematic values Ysk. The output of DEC1 is the transmitted data d_k which is a sequence of soft estimates EXT1, is called extrinsic data. The extrinsic data does not have any data from any of the two decoders. The output is now interleaved identical to the interleaver in ENC1 and then given as input to the second decoder DEC2 [14]. Apart from the interleaved extrinsic information EXT1, the DEC2 also takes the interleaved output of the received sequence of parity values from ENC2 (the second encoder) and the received sequence of systematic values Ysk. Using an inverse form of interleaver, the output of DEC2 is interleaved. The output consists of d_k which is the transmitted sequence of data and EXT2 which is the soft estimates. This extrinsic data obtained as feedback from DEC1 does not have any parity bits from the first code. This method is repeated for a number of iterations.

The number of iterations exponentially increases the performance of the turbo codes, as the bit error rate decreases. The convergence of the estimates of the two decoders occurs after several iterations. The DEC2 produces the log-likelihood estimate of d_k as the output. The log-likelihood value considers the probability that a transmitted bit is '0' or '1' based on the parity and the systematic information obtained from both component codes [15]. The transmitted bit is decided to be '0' if

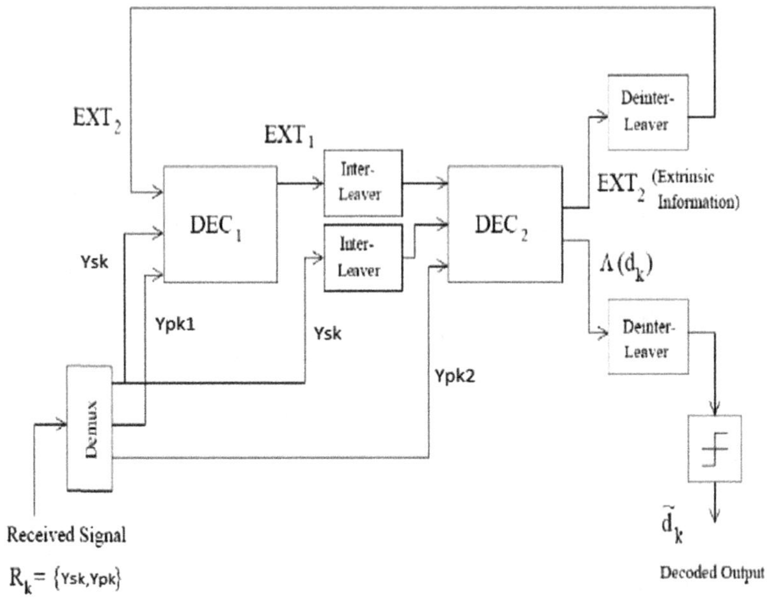

Fig. 3 Turbo decoder

the occurrence of the negative values (of d_k) is more; likewise, the transmitted bit is decided to be '1' if the occurrence of positive values is more. These bits are then de-interleaved so that the sequences of the systematic and parity streams coincide. Based on an operation using a simple threshold, the result is decided with hard decision estimates, d_k, for the transmitted bits.

The convergence of the decoding estimates EXT1 and EXT2 is misleading because of the set of corrupted code bits from the error sequence that cannot be corrected by either of the decoders. Therefore, the EXT1 and EXT2 outputs may converge or diverge to a false soft value.

3.2 *Soft Output Viterbi Algorithm (SOVA)*

The conventional Viterbi algorithm is used for finding the hidden states called the Viterbi path with dynamic programming algorithm. This algorithm requires a sequence of observed events. The encoder state is simulated with the received values.

The conventional Viterbi algorithm is modified to obtain the SOVA algorithm. The modified Viterbi algorithm provides a soft output for every decoded bit. In the proposed SOVA algorithm, to accept a priori information, the path metrics used for the selection of the maximum likelihood path through the trellis are modified.

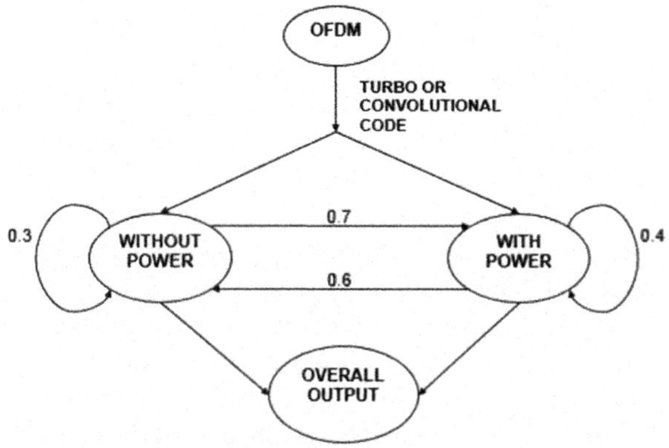

Fig. 4 The Viterbi algorithm

The algorithm followed by the Viterbi algorithm is shown in Fig. 4. The arrow marks indicate the probabilities from one state to another. There are two states considered in this system: with power optimization and without power optimization [6]. The transition probabilities from one state to another and itself are given in the figure and these values are preset by the programmer.

The arrows converging onto the state represent the start probabilities; the arrows between them indicate the transition probabilities, and finally, the arrows emerging from the two states are called the emerging probabilities. These values are then used by the Viterbi decoder to compute the error in the system.

3.3 Rayleigh Fading Channel

The transmitted signals are affected by complex Gaussian random variable of the wireless channel. There is a multiplicative effect on the transmitted signals.

The received signal amplitude is considered to be a Rayleigh random variable, so the mean of the channel coefficient is zero. The power density function of the signal amplitude received in a Rayleigh fading channel is given as:

$$f(x) = \frac{x}{\sigma^2 \alpha} \exp\left(-\frac{x^2}{2\sigma^2 \alpha^2}\right) \quad \forall x > 0. \tag{14}$$

In the above equation, x follows the Rayleigh distribution; α is the fading parameter; and σ^2 stands for the variance of the amplitude of the received signal.

4 Performance Results

The following are the simulation results of the proposed system.

Figure 5 shows the same plot but with the transmission capacities normalized to the transmission capacity of PPM. DPIM offers higher capacity at lower values of *M* than DH-PIM, but the difference between them is negligible at high orders. The transmission capacities of DPIM and DH-PIM are about four times as that of PPM for a particular value of *M* (Fig. 6).

Figures 7 and 8 show the performance of the SOVA algorithm in AWGN and Rayleigh channels, respectively. It is clear from the plot that Rayleigh channel encounters more BER than AWGN, due to the presence of multiple paths. Figures 9 and 10 depict the same algorithm performed in a MIMO-OFDM system

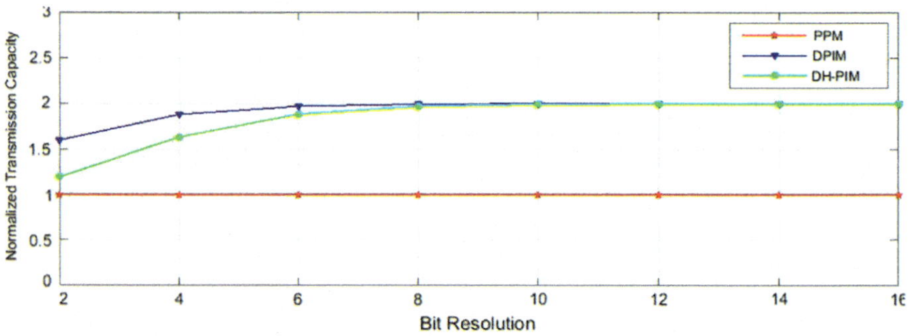

Fig. 5 Normalized transmission capacity versus '*M*'

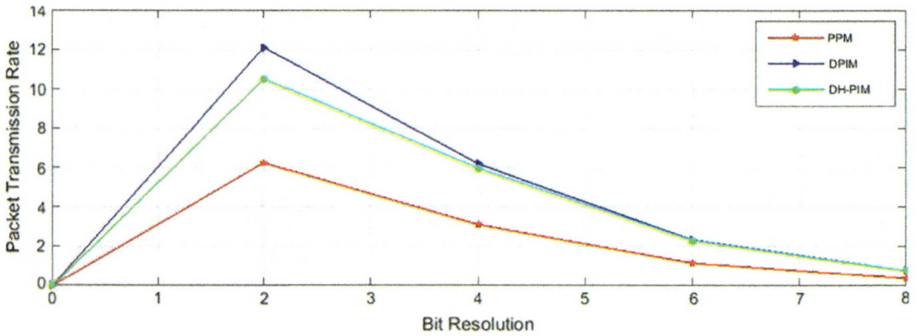

Fig. 6 Packet transmission rate versus '*M*'

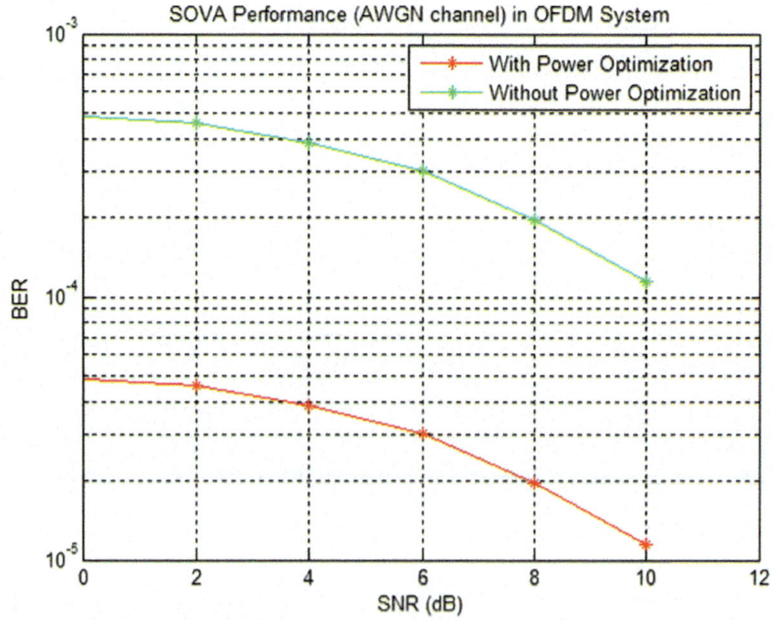

Fig. 7 Performance of SOVA (AWGN channel) in OFDM system

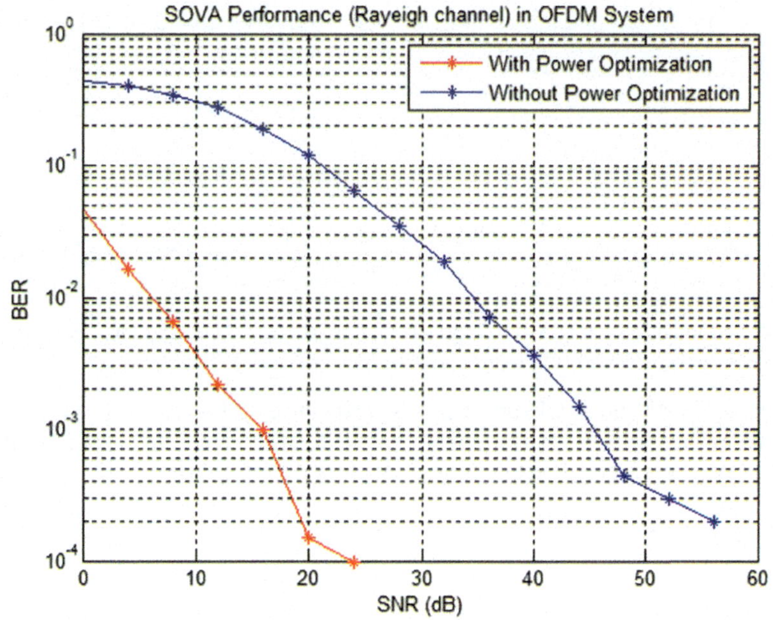

Fig. 8 Performance of SOVA (Rayleigh channel) in OFDM system

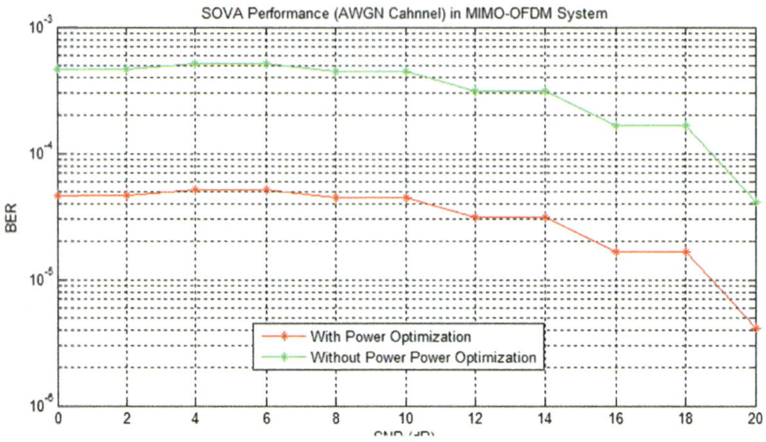

Fig. 9 Performance of SOVA (AWGN channel) in MIMO-OFDM system

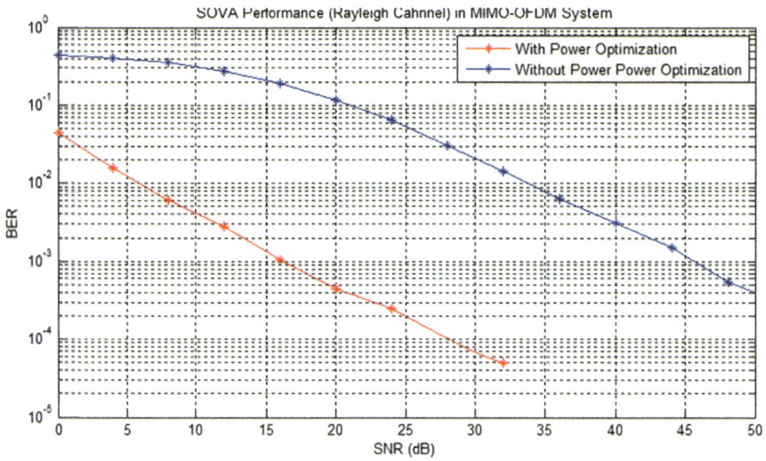

Fig. 10 Performance of SOVA (Rayleigh channel) in MIMO-OFDM system

for both AWGN and Rayleigh channels, respectively. Finally, Fig. 11 shows the comparison between the performances in Rayleigh channel and AWGN channel, when SOVA algorithm is applied to a MIMO-OFDM scenario.

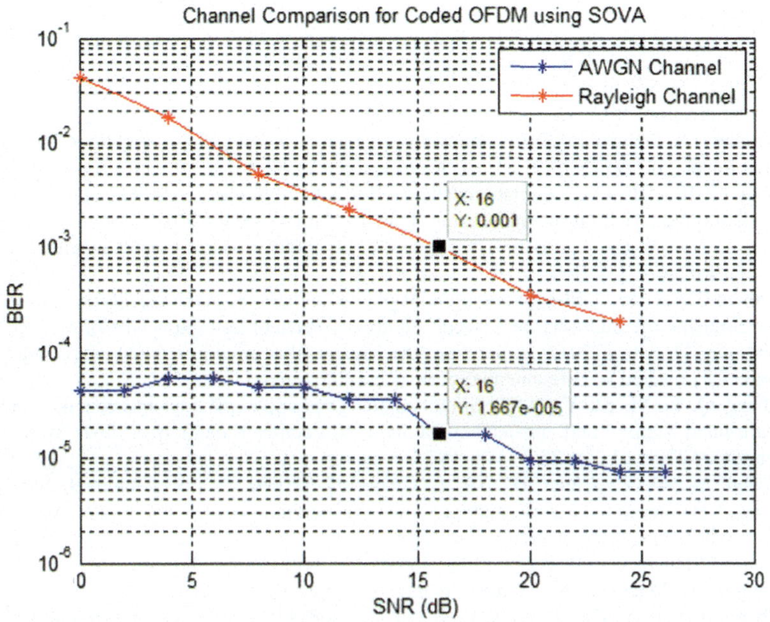

Fig. 11 Comparison of SOVA performance in Rayleigh and AWGN channels

5 Conclusions

The performance of the optical communication system was compared with respect to the efficiency of modulation techniques used. For lower values of '*L*' in DPIM and DH-PIM, they are not efficient in terms of bandwidth and power usage. Hence, MIMO-OFDM systems with turbo coding are incorporated to increase the performance of the system. Eventually, for higher value of M, the transmission rate for DH-PIM and DPIM are the same. By viewing the performance graphs of the algorithm in various scenarios, it is observed that when power is optimized, the results show reduced BER. When the existing system in OFDM is extended to MIMO, there is a significant increase in SNR (almost double the value when observed in an OFDM system). The SNR in an OFDM system is about 10 dB, whereas in a MIMO-OFDM system, it is 20 dB, observed in AWGN channel. In Rayleigh channel, the increase in SNR is about 2–3 dB. At around 15 dB, the error rate for AWGN is 0.00001, while the error rate for Rayleigh channel is 0.001, because of the multipath degradation in Rayleigh channel.

References

1. Burchardt H, Serafimovski N, Tsonev D, Videv S, Haas H (2014) VLC: beyond point to point communication. IEEE Commun Mag 98–105
2. Choubey A, Sharma D (2015) Analyzing of various modulation methods on basis of error performance for VLC. In: IEEE EESCO conference, pp 1–5
3. Yu Z (2010) Adaptive modulation schemes for optical wireless communication systems. Ph. D. dissertation, School of Engineering, University of Warwick 2010
4. Yan G, Min W (2010) Performance research of modulation for optical wireless communication. ISCSCT 357–360
5. Lethien C et al (2012) Energy-autonomous picosecond remote antenna unit for radio-over-fiber system using the multiservice concept. Photonics Technol Lett 24(8):649–651
6. Matsuura M, Sato J (2015) Bidirectional radio-over-fiber systems using double-clad fibers for optically powered remote antenna units. IEEE Photonics J 7(1)
7. Chang Q, Fu H, Su Y (2008) Simultaneous generation and transmission of downstream multi-band signals and upstream data in a bidirectional radio over fiber system. IEEE Photonics Technol Lett 20(3):181–183
8. Hsueh Y, Jia Z, Chien H, Yu J, Chang G (2009) A novel bidirectional 60-GHz radio-over-fiber scheme with multiband signal generation using a single intensity modulator. IEEE Photonics Technol Lett 21(18):1338–1340
9. Zhang L, Hu X, Cao P, Wang T, Su Y (2010) A multiband radio over fiber system using a single-drive Mach-Zehder Modulator. In: Proceeding of ACP 2010, FP5
10. Chang Q, Tian Y, Gao J, Ye T, Li Q, Su Y (2008) Generation and transmission of optical carrier suppressed-optical differential (Qudrature) phase-shift keying (OCS-OD(Q)PSK) signals in radio over fiber systems. J Lightwave Technol 26(15):2611–2618
11. Salim M, Yadav RP, Ravi Kanth S (2010) Performance analysis of log-map, SOVA and modified SOVA algorithm for turbo decoder. Int J Comput Appl 9(11):0975–8887
12. Yadav RP, Salim M, Ravikanth S (2011) Performance analysis of log-map, SOVA and modified SOVA algorithm for turbo decoder. Int J Comput Appl 9(11):0975–8887
13. Le Floch B, Alard M, Berrou C (2001) Coded orthogonal frequency division multiplex. Proc IEEE 83(6):982–996
14. Cluzeau M, Finiasz M, Tillich JP (2010) Methods for the reconstruction of parallel turbo codes. In: Proceedings of the IEEE international symposium on information theory (ISIT), Austin, pp. 2008–2012, 13–18 June 2010
15. Jia Z, Yu J, Hsueh Y-T, Chowdhury A, Chien H-C, Buck JA, Chang G-K (2008) Multiband signal generation and dispersion-tolerant transmission based on photonic frequency tripling technology for 60-GHz radio-over-fiber systems. IEEE Photonics Technol Lett 20(17):1 470–1472

Frequency Tuning Method for Small Profile Metamaterial Based on Tri-Ring Resonator

Maruti Tamrakar and K. Usha Kiran

Abstract The metamaterials are structures that show the different electromagnetic properties than usual transmission line/medium, such as negative values of permittivity, permeability and refraction. Due to unique behaviour of metamaterial, it gives passband or stopband response for a certain frequencies: a property called resonant. Typical response of metamaterial is narrow band and resonant at particular frequency based on its dimension. A method is proposed in this paper to control or shift the response frequency of tri-ring resonator without altering its dimension.

Keywords Metamaterial · Frequency tuning · Band rejection · Resonator

1 Introduction

Current research is focused on metamaterials with unique properties of negative values of permeability and/or permittivity for wideband or multiband frequency. Metamaterial structures are used with antennas for gain and beam-width improvement and with passive RF circuits for better passband insertion loss and stopband rejection [1, 2]. Among multiple kinds of metamaterial structures, split ring resonator (SRR) is most popular, which shows negative permeability for finite frequency band [3–5]. SRR is found to be very useful for narrowband applications [1, 6]. Nowadays, research approach is shifted towards multiband, wideband frequency response and/or frequency tuning [1, 2, 6, 7]. A frequency tuning method is proposed in this paper for narrow band small profile metamaterial.

It is a well-known fact that metamaterials are structures which resonate at a particular frequency, and by varying the dimension of these structures, resonant

M. Tamrakar · K. Usha Kiran (✉)
School of Electronics Engineering, VIT University, Chennai 600127, Tamil Nadu, India
e-mail: usha.kiran@vit.ac.in

M. Tamrakar
e-mail: maruti.tamrakar2013@vit.ac.in

G. Gnanagurunathan et al. (eds.), *Optical and Microwave Technologies*,
Lecture Notes in Electrical Engineering 468,
https://doi.org/10.1007/978-981-10-7293-2_19

frequency could be shifted [8]. Under this paper, unique technique is proposed to shift the resonant frequency without changing the physical dimensions, which is substantiated with different metamaterial structures.

In this work, a complementary tri-ring resonator is studied for resonance frequency tuning. The tri-ring is designed in FR4 substrate by optimizing design parameters like ring radius, width of ring, slot in ring and gap between rings.

2 Design and Simulation

Presented metamaterial structures are designed and simulated in CST microwave studio for validation of proposed method. The effective medium parameters permittivity (ε), permeability (μ) and refractive index (n) are estimated using MATLAB by given input of complex s-parameters S11 and S21 (refection and transmission characteristics) obtained from simulation [9–12].

2.1 Single Ring Resonator

Ring resonator has discontinuity which referred as 'slot' in this paper. Conventional ring resonator structures are much smaller than the resonance wave length. In the conventional ring resonator, external magnetic field penetrates through the ring and currents are induced in the structure, but slot prevents currents from flowing around the ring, which considerably increases the resonance frequency of the structure [1]. Single-complementary ring resonator with 50 Ω transmission line shown in Fig. 1 is simulated in FR4 substrate of dimension 50 mm × 30 mm, in which dielectric constant is 4.4, substrate thickness is 0.762 mm, and copper thickness is 0.035 mm. Figures 2 and 3 are showing the s-parameter magnitude and phase results of single-complementary ring resonator of ring outer radius of 3.6 mm, inner radius of 3.4 mm and slot width of 0.2 mm.

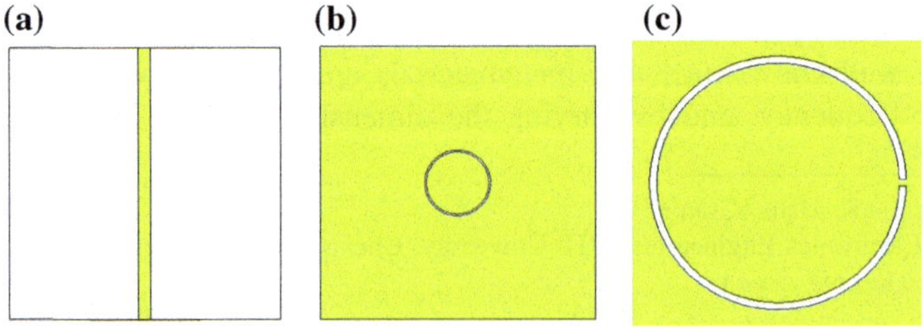

Fig. 1 Single ring resonator structure, **a** transmission line, **b** complementary ring in ground plane, **c** close view of complementary ring in ground plane

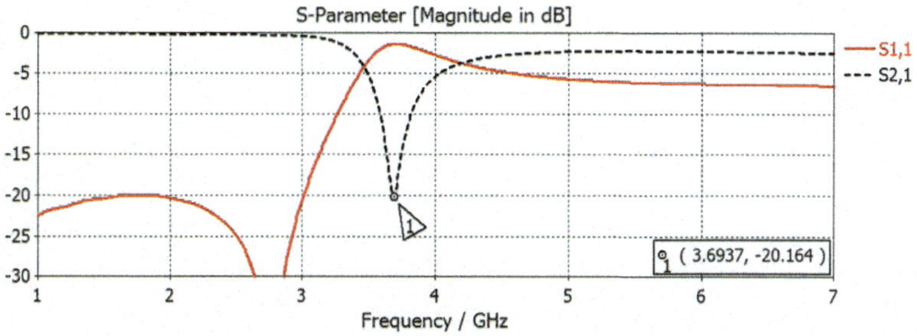

Fig. 2 S-parameter magnitude (dB) plot of complementary single ring resonator

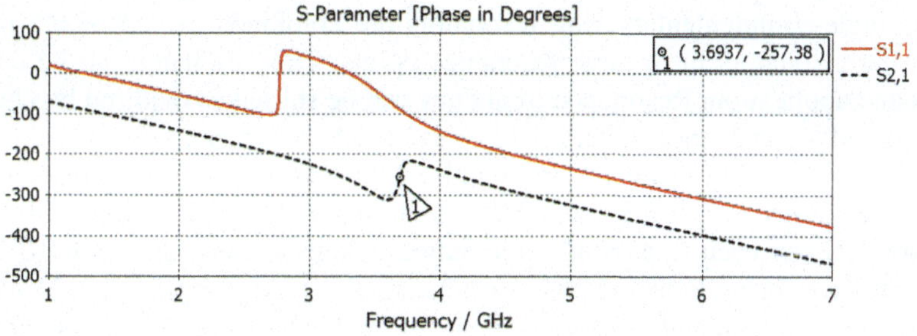

Fig. 3 S-parameter phase (degree) plot of complementary single ring resonator

In Figs. 2 and 3, it can be seen that metamaterial structure of Fig. 1 gives passband for lower frequency 1.0–3.0 GHz and stopband at 3.7 GHz for finite bandwidth. Further simulation is carried out for three different ring radii, and each ring gives different resonant frequencies for its respective radius. Ring of outer radius of 3.6, 3.2 and 2.8 mm have resonance at 3.7, 4.17 and 4.66 GHz, respectively; same is presented in Fig. 4.

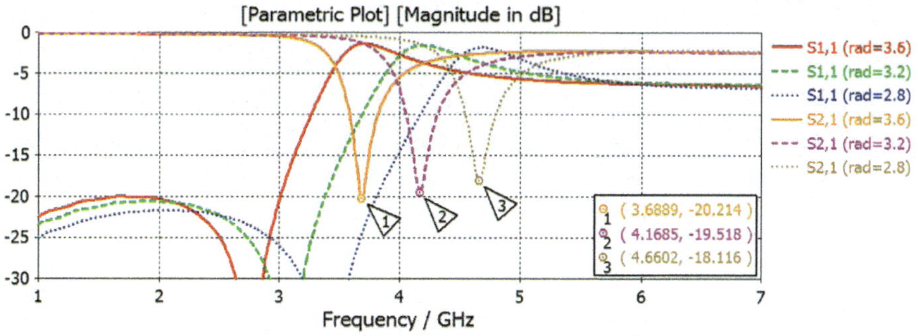

Fig. 4 S-parameter magnitude (dB) plot of complementary single ring resonator for three different ring radii

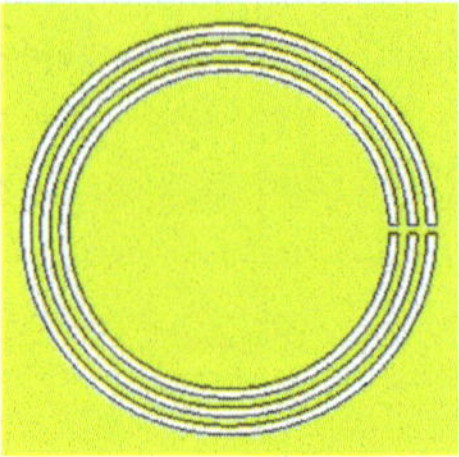

Fig. 5 Close view of complementary tri-ring resonator structure with slot aligned in same axis

2.2 *Tri-Ring Resonator*

A new 3-ring resonator unit cell design, called here as tri-ring, is introduced here. Three single-complementary ring resonators are combined at gap of 0.2 mm, as these ring resonators have three resonance points. This resonator can be used for multiband application. Resonance frequency can be shifted as required by changing the ring radius, and rejection level is controlled by changing substrate thickness, gap between two ring resonator and width of slots. In Fig 5, tri-ring resonator has outer ring radius of 3.6 mm, and innermost ring radius is 2.6 mm with gap of 0.2 mm between each ring. Slots' orientation is kept in same direction.

S-parameter magnitude and phase results are shown in Figs. 6 and 7. Minor shift in resonance points and rejection level for higher frequency can be observed between Figs. 4 and 6 for s-parameter magnitude. First/dominating resonance has fractional frequency bandwidth 4.25% for −15 dB rejection level, and 8.4% for −10 dB rejection level.

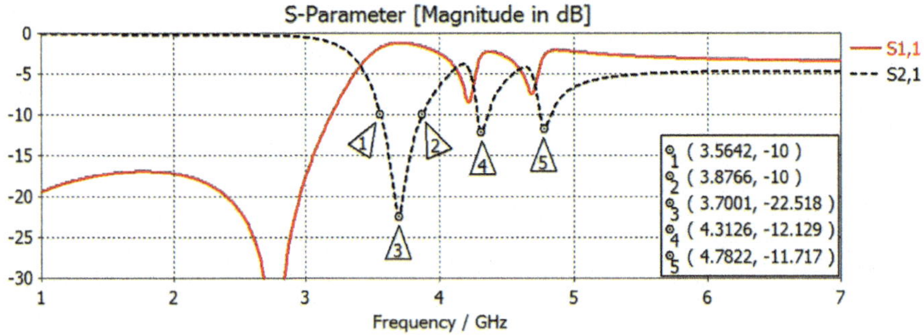

Fig. 6 S-parameter magnitude (dB) plot of complementary tri-ring resonator with slot aligned in same axis

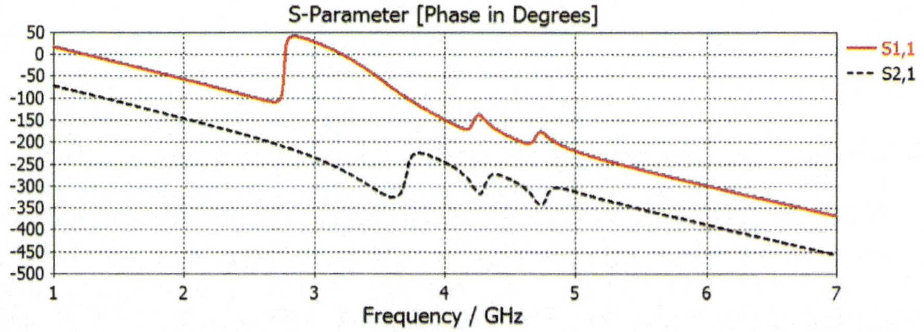

Fig. 7 S-parameter phase (degree) plot of complementary tri-ring resonator with slot aligned in same axis

2.3 *Metamaterial Properties of Tri-Ring Resonator*

The electromagnetic response of a material is defined by its electromagnetic properties: permittivity (ε) and permeability (μ). The effective medium parameters —permittivity (ε_{eff}), permeability (μ_{eff}) and refractive index (n_{eff})—are estimated using MATLAB by given input of complex s-parameters S11 and S21 (transmission and refection characteristics) obtained from simulation.

As per the result presented in Fig. 8, the tri-ring resonator has three distinct cases in terms of values of ε_{eff} and μ_{eff} over the frequency range from 1 to 4 GHz.

Case 1: Double positive, in which both parameters ε_{eff} and μ_{eff} are positive. The effective refractive index (Eq. 1) is also positive, and the metamaterial gives passband response with forward wave propagation. In Fig. 8, the response can be seen for frequency 1.0–2.45 GHz.

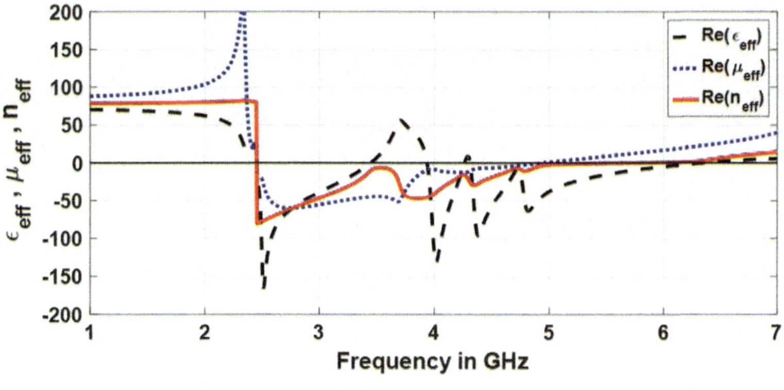

Fig. 8 Effective permeability (μ_{eff}), effective permittivity (ε_{eff}) and effective refractive index (n_{eff}) plot of complementary tri-ring resonator with slot aligned in same axis

$$n_{\text{eff}} = \pm\sqrt{(\mu_{\text{eff}} \cdot \varepsilon_{\text{eff}})} \tag{1}$$

Case 2: Double Negative, in which both parameters ε_{eff} and μ_{eff} are negative. The n_{eff} is negative, and the metamaterial gives passband response with reverse wave propagation. The response can be seen in Fig. 8 for frequency 2.45–3.45 GHz.

Case 3: Single negative, in which both parameters ε_{eff} and μ_{eff} are having opposite polarity at particular frequency; if one is positive polarity, then other one is negative polarity. The n_{eff} is complex value, and the metamaterial gives stopband response. The stopband response is due to positive value of ε_{eff} and negative value of μ_{eff}, and it can be seen in Fig. 8 for frequency 3.45–3.93 GHz. The ε_{eff} is 56, and μ_{eff} is −50 at 3.7 GHz resonance point. This indicates that tri-ring resonator has magnetic response at 3.7 GHz for narrow bandwidth.

3 Frequency Tuning Methods

To shift the frequency towards lower side, following two methods are investigated: first one is by increasing the dimensions of resonator, and the other one is by changing the orientation of slots of three resonators.

3.1 By Varying Ring Diameter

The tri-ring resonator is scaled the ring outer radius to 3.6, 4 and 4.4 mm, while gap between rings and slot width remain constant. Slot of the rings is aligned in same axis as shown in Fig. 5 for this simulation. It can be observed that first/dominate resonance is shifted towards lower frequency from 3.7 to 3.378 and 3.148 GHz with increasing ring radius (Fig. 9).

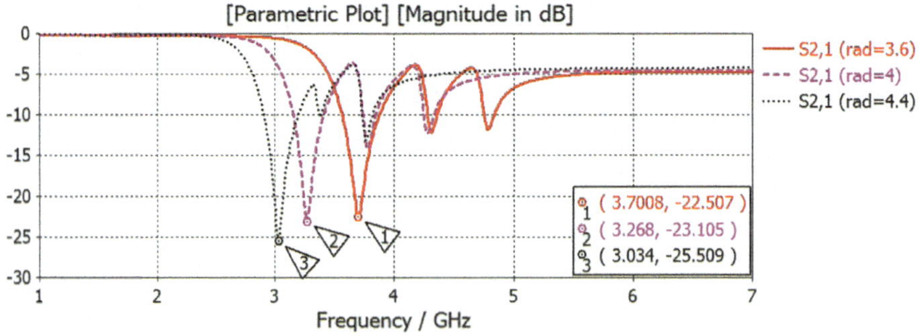

Fig. 9 S-parameter magnitude (dB) plot of complementary tri-ring resonator with slot aligned in same axis for three different ring radii

3.2 By Rotating the Inner Ring

A new and unique method is proposed here to shift/tune frequency without altering resonator dimension. To achieve frequency shift, the inner rings of tri-ring resonator rotated, which generates a large capacitance, lower the resonant frequency and concentrate the electric field. Combination of three rings also increases the magnetic field at the centre of resonator.

The middle ring is rotated by 30°, 60° and 90°, and inner ring is rotated by double of middle ring rotation. Figure 10 represents the inner rings rotation with respect to outer ring. No rotation or 0° rotation is presented in Fig. 5.

From Fig. 11, it can be observed that first/dominant resonance frequency is shifted form 3.7 to 3.4, 3.06 and 2.818 GHz for 30°, 60° and 90° rotation of middle ring of tri-ring resonator, while rejection level at resonance is approximately same.

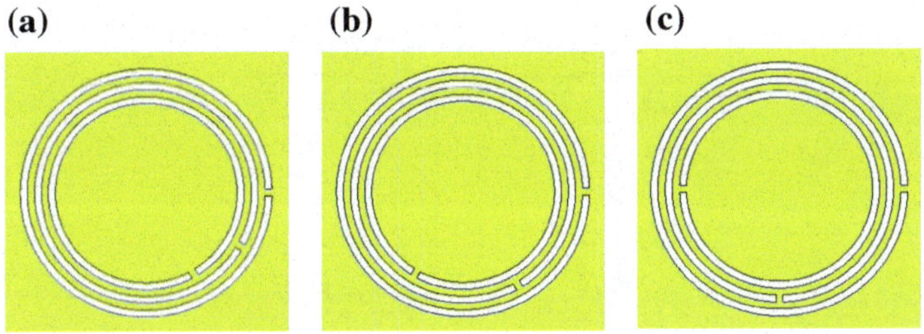

Fig. 10 Close view of complementary tri-ring resonator structure with inner slot rotation by **a** angle 30°, **b** angle 60°, **c** angle 90°

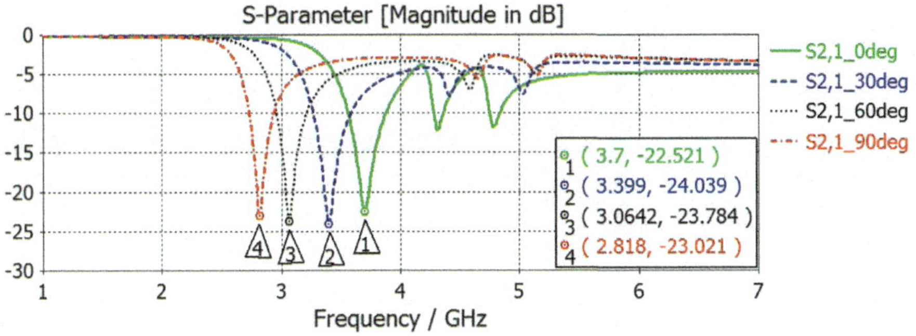

Fig. 11 S-parameter, S21 magnitude (dB) plot of complementary tri-ring resonator with inner slot rotation by angle 0°, 30°, 60° and 90°

Table 1 Simulation results summary for tri-ring resonator

Outer ring diameter (mm)	Middle ring diameter (mm)	Inner ring diameter (mm)	Outer ring rotation (deg)	Middle ring rotation (deg)	Inner ring rotation (deg)	Resonance frequency (GHz)	Rejection level (dB)	Frequency shift (GHz)
3.6	3.2	2.8	0	0	0	3.7	−22.521	Reference
4	3.6	3.2	0	0	0	3.376	−16.948	0.324
4.4	4	3.6	0	0	0	3.148	−19.194	0.552
3.6	3.2	2.8	0	30	60	3.399	−24.039	0.301
3.6	3.2	2.8	0	60	120	3.064	−23.784	0.635
3.6	3.2	2.8	0	90	180	2.818	−23.021	0.882

4 Results and Discussion

The proposed frequency tuning method is also simulated and validated with other structures like rectangular and polygonal resonator. The MATLAB simulation results in Fig. 8 show that the passband response for frequency form 1.0 to 3.45 GHz is due to combined effect of double positive and double negative of ε_{eff} and μ_{eff}. The ε_{eff} and μ_{eff} have opposite polarity for frequency 3.45–3.93 GHz, and negative value of μ_{eff} indicates the strong magnetic resonance, causing stopband response.

The summary of simulated results is given in Table 1. Frequency shift is calculated with reference to first design criteria, presented in Table 1. It is observed in Fig. 9 that resonance frequency is shifted from 3.7 to 3.148 GHz by increasing the ring radius from 3.6 to 4.4 mm. The shift in frequency is 0.55 GHz while ring radius increased by 22%.

The tri-ring complementary resonator of 3.6 mm outer radius with slots aligned is giving multi-resonance frequencies with first resonance point at 3.7 GHz and by rotating the inner rings by 90°; the frequency resonant point is shifted to 2.8 GHz. The frequency shift of 0.882 GHz is noted without altering the resonator dimensions. The shift in resonant frequency decreases by increasing the rotation angle/misalignment. The results are similar for clockwise and anticlockwise rotation of inner rings. The effect of inner ring rotation on frequency shift is more compared to increase/decrease of ring radius.

5 Conclusion

The metamaterial simulated and analysed in this paper has three complementary ring resonators in a single structure, and each resonator resonates to its respective frequency. Depending on the orientation of slots in ring resonator, the resonant frequency of the structure is shifted while maintaining the same dimension. Rotation of inner ring gives better control of frequency for tuning compared to changing the ring radius. The structure can also be used for multiband rejection if slot of rings is aligned in an axis. The study shows that by rotating/misaligning the inner ring of proposed metamaterial structure is effective method to tune and shift the resonance frequency while rejection level remains the same for tuned frequency.

References

1. Ekmekci E, Turhan-Sayan G (2009) Comparative investigation of resonance characteristics and electrical size of the double-sided SRR, BC-SRR and conventional SRR type metamaterials for varying substrate parameters. Prog Electromagnet Res B 12:35–62
2. Ekmekci E et al (2011) Frequency-tunable metamaterials using broadside-coupled split ring resonators. Phys Rev B 83:193103

3. Rigi-Tamandani A et al (2013) Wideband planar split ring resonator based metamaterials. Prog Electromagnet Res M 28:115–128
4. Duran-Sindreu M et al (2012) Electrically small resonators for planar metamaterial, microwave circuit and antenna design: a comparative analysis. Appl Sci 2:375–395
5. Islam SS et al (2014) Design and analysis of a new double negative metamaterial. J Microelectron, Electron Compon Mater 44:218–223
6. Turkmen O, Ekmekci E, Turhan-Sayan G (2011) A new multi-ring SRR type metamaterial design with multiple magnetic resonances. Prog Electromagnet Res Symp Proc 20–23
7. Ekmeki E, Strikwerda AC, Fan K (2011) Frequency tunable metamaterial designs using near field coupled SRR structures in the terahertz region. In: International conference on infrared, millimeter, and terahertz waves, pp 1–2
8. Sassi I, Talbi L, Hettak K (2015) Compact multi-band filter based on multi-ring complementary split ring resonators. Prog Electromagnet Res 57:127–135
9. Kadam NT, Janwalkar KS, Odhekar AA (2015) Parameter extraction for negative index metamaterials. Int J Comput Appl 0975–8887
10. Si Y, Zhang Z (2015) Extraction of metamaterial parameters. In: IEEE 6th international symposium on microwave, antenna, propagation, and EMC technologies, pp 608–610
11. Szabo Z, Park GH, Hedge R, Li EP (2010) A unique extraction of metamaterial parameters based on Kramers–Kronig relationship. IEEE Trans Microw Theor Tech 2646–2653
12. Chen X, Grzegorczyk TM, Wu BI, Pacheco J Jr, Kong JA (2004) Robust method to retrieve the constitutive parameters of metamaterials. Phys Rev E70 016608

Redesigning Mach-Zehnder Modulator with Ring Resonators

R. G. Jesuwanth Sugesh and A. Sivasubramanian

Abstract Recent technological advances have allowed network users to transfer large files. This has increased the network traffic in Telecom and Data centres. Optical fibre communication has large bandwidth and channel capacity to meet the demands. Silicon photonics aims in improving the data transfer with efficiency on chip. Modulators play potent role in the process. The work deals in the analysis on improving the favoured Mach-Zehnder modulator (MZM). The results show the number of ring resonators that can be used with respect to the wavelength to improve the extinction ratio. Extinction ratio is very important when it comes to processing valuable data and that is improved with the help of the ring resonators in MZM.

1 Introduction

The network traffic is growing at an alarming rate. The advancement in network technology such as cloud computing and big data has put a stress on the Telecom and Data centres. When compared to electrical transmission, photonic transmission is known for its high speed and is the efficient way in terms of, energy to transfer high data, heat when compared to electrical transmission and distance whether it is for short distance or long. The processing capacity of the processors is submersed by the electrical wiring in the chip. In order to obtain the full potential of the processing chip, silicon photonics has been introduced. Silicon photonics is using the same route as followed by semiconductor industry aiming at smaller, cheaper, energy efficient and high data rate circuits. Silicon photonics uses the same CMOS fabrication process which adds to its benefit. Producing cheap and high efficient photonic devices with silicon is one of the challenges that need to be addressed in Silicon Photonics. Silicon has an indirect band gap that is not suitable for lasers. The

R. G. Jesuwanth Sugesh (✉) · A. Sivasubramanian
School of Electronics Engineering, VIT University, Chennai, Tamil Nadu, India
e-mail: rgjsugesh@gmail.com

A. Sivasubramanian
e-mail: sivasubramanian.a@vit.ac.in

G. Gnanagurunathan et al. (eds.), *Optical and Microwave Technologies*,
Lecture Notes in Electrical Engineering 468,
https://doi.org/10.1007/978-981-10-7293-2_20

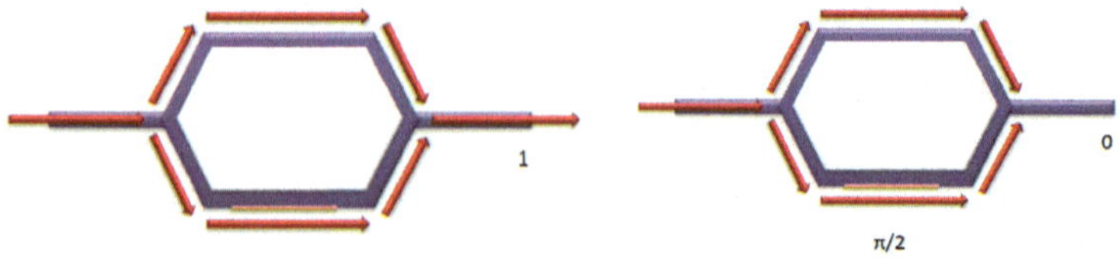

a. Constructive interference b. Destructive interference

Fig. 1 Working of MZM

important building blocks of silicon photonics are the generating, modulating and detecting the light. Modulators are the key components as it converts the light source into a beam of information and determines the data transmission capacity [1]. External modulators are more preferred as they can provide larger extinction ratio, higher data speed and low modulation distortion. Among them, Mach-Zehnder modulator (MZM) is the standard modulator being used in the recent days. MZM does not require any special tuning to work at large optical operating bandwidth [2]. Insensitive to fabrication tolerance and temperature variations makes it not necessary to require any thermal controllers which is used for wavelength stabilization thus reducing the power consumption [3–5]. This makes it an appropriate nominee for commercial systems [6]. MZM works on the basis of the constructive and destructive interference due to the phase shift in one or both the arms which consequently leads to intensity modulation [7]. The working of MZM is explained in Fig. 1.

MZMs exhibiting high modulation bandwidth have been demonstrated with the use of travelling wave structures [8]. But the drawback of these travelling wave MZMs is that they require high voltage swings to obtain higher extinction ratio (ER). Extinction ratio is very important in processing the received bits. This high-power need limits the process scalability and integration with other components [9].

The main objective of this work is to improve the ER of MZM. Section 2 introduces a method of improving the performance of MZM. The results of the experiment are discussed in Sect. 3. Conclusion of the paper is provided in Sect. 4.

2 Ring Resonators in MZM

Ring modulators are being preferred in Photonic Integrated Circuits for their low power consumption [10] and compact size [11], but they have bandwidth limitation and are less temperature tolerant [12]. Ring resonator (RR) being a passive element consists of two waveguides among which one is bent to form a circle, and another one is a straight waveguide. Between these two waveguides, coupling occurs. The working of RR is explained with the help of Fig. 2.

The bidirectional RR allows signal transmission clockwise or anticlockwise independently in the ring waveguide and backward or forward in the straight

Fig. 2 Ring resonators (RR)

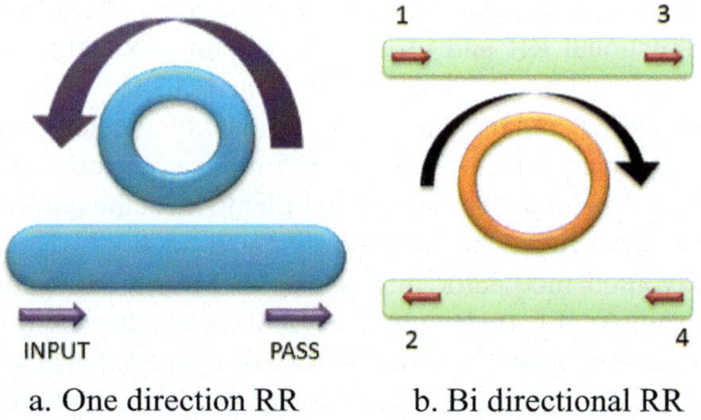

waveguide. The resonant frequencies are calculated based on the length of the waveguide which is used to form the ring structure [13].

As RR improves the modulation efficiency, introducing them to MZM's structure will improve the modulator's performance. The improvement in the transmission is also based on the wavelength. The length of the RR leads to extra phase shift, and in order to have proper phase shift balance, care should be taken on adding the RR on both the arms of the modulator. The paper describes the number of RRs needed to have an increase in the extinction ratio when compared with the traditional MZM.

3 Results

The circuit examined in the experiments was done using the simulation tool Rsoft OptSim is given in Fig. 3.

The circuit comprises of a continuous laser source, the internal structure of MZM and the ring resonators that are added to each arm for estimating the number of resonators necessary to improve the performance and the photodetector. The input bits for modulation are generated from pseudorandom bit sequence generator (PRBS). The laser source produces a continuous laser beam output which is split into equal between two arms. The phase shift occurs based on the input from the random bits generated from the PRBS. In both the arms, equal number of RR should be placed so as to keep the length in each arm equal otherwise it will lead to extra length

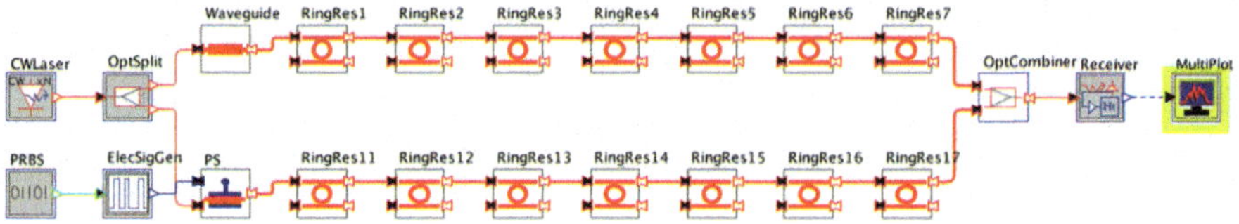

Fig. 3 Circuit diagram with 7 RRs in MZM

in one arm which will alter the phase shift. The bidirectional RR is converted to one directional RR and used in the circuit. The ring radius of the resonators is set as 11.14×10^{-5} m. The output from both the arms is combined using the optical combiner, and the output is detected with the help of photodetector. The wavelength was changed from 1150 to 1565 nm, and experiments were carried out for 10 Gbps with 7 as bit pattern length. In order to meet the purpose of real-world application, the main objective of this experiment is to improve the extinction ratio of the most favoured modulator that is the MZM with considering the wavelength used. An example of wavelength (λ = 1550 nm) is discussed in Fig. 4.

The graphs point out that there is an increase in the ER after the inclusion of RRs. The cascaded nature of the RR leads to the variation in the extinction ratio thus an optimum number of RR is needed. The RR inclusion is done on both the arms in order to have phase shift balance in both the arms. Figure 5 provides the eye diagram obtained for wavelength 1550 nm when adding seven RR to the arms of MZM. An improvement in the eye was obtained when adding seven RR. It can be seen that there is a clear opening in the eye which illustrates that the receiver will have lower likelihood of differentiating between a logical 1 and logical 0.

The steep slope ensures less timing errors. In real time, there is a possibility of variation in zero and one level time duration or continuous 1's and 0's so eye crossing is also to be taken into account. The eye crossing percentage is used for the diagnosis of amplitude distortion and pulse symmetry problems. The ideal crossing of a perfect eye is 50% and when it deviates, the eye closes and degrades the extinction ratio. Eye-crossing percentage is calculated using (1),

$$\text{Eye crossing} = 100 \times \frac{(\text{crossing level} - \text{zero level})}{(\text{one level} - \text{zero level})} \tag{1}$$

The eye crossing for Fig. 5 is calculated to be approximately 50%. The slope in the figure suggests that there will be less chances of eye closing leading to mistake in bits differentiation. Free spectral range is obtained as 10 nm.

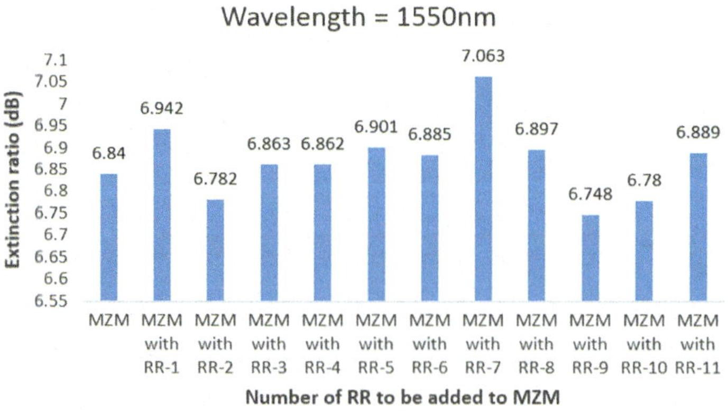

Fig. 4 Extinction ratio for λ = 1550 nm

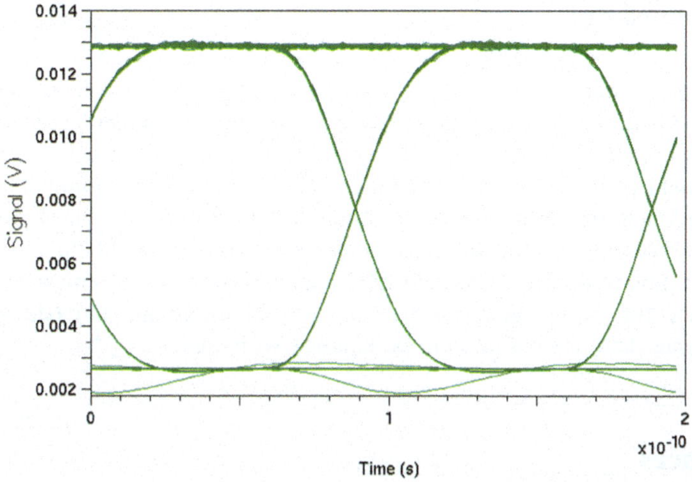

Fig. 5 Eye diagram for λ = 1550 nm and with RR-7

$$\mathrm{DCD} = 100 \times \frac{\text{time difference between rising and falling edges at 50 percent level}}{\text{bit period}} \tag{2}$$

The duty-cycle distortion (DCD) also reveals the pulse symmetry problem. The DCD has to be lower for better extinction ratio. The DCD is calculated using (2). The DCD is calculated to be almost 0% for the eye diagram in Fig. 5. The eye is wider so there is less intersymbol interference for this circuit.

Table 1 lists out the number of RRs to be added in the arms of MZM with respect to wavelength. The RRs do not consume power to work and are compact in size. This gives the leverage of including them to MZM.

Table 1 Number of RR needed in MZM to improve ER per wavelength

Wavelength (nm)	Conventional MZM (ER-dB)	MZM with RR (ER-dB)	Number of RR for peak
1150	6.9	7.289	5
1250	6.885	7.06	1
1350	6.8581	6.9823	7
1400	6.87	7.07	9
1450	6.642	7.16	5
1500	6.90	7.10	1
1550	6.84	7.063	7
1565	6.886	7.006	1

4 Conclusion

Each photonic transmission within the chip has to be done on separate wavelength, and the extinction ratio is very important when it comes to processing valuable data. The work concludes the number of ring resonators used for a wavelength for the improvement of extinction ratio in MZM. The dependence of number of RR and wavelength is to be studied further. Extinction ratio ensures that the receiver is able to differentiate the bits transmitted, and this proposed work will help to improve the extinction ratio of the conventional MZM with inclusion of RR. This work will help to ease the time consumption for researchers in improving the performance of MZM where data processing is to be handled with care.

References

1. Liu A, Liao L, Chetrit Y, Basak J, Nguyen H, Rubin D, Paniccia M (2008) 200 Gbps photonic integrated chip on silicon platform. In: 2008 5th international conference on group IV photonics, GFP, pp 368–370
2. Cignoli M, Minoia G, Repossi M, Baldi D, Ghilioni A, Temporiti E, Svelto F (2015) A 1310 nm 3D-integrated silicon photonics Mach-Zehnder-based transmitter with 275 mW multistage CMOS driver achieving 6 dB extinction ratio at 25 Gb/s. In: Digital technical paper—IEEE international solid-state circuits conference, vol 58. pp 416–417
3. Lentine AL, Derose CT, Davids PS, Nicolas JD, Zortman WA, Cox JA, Jones A, Douglas C, Pomerene AT, Starbuck AL, Savignon DJ, Wiwi M, Chu PB (2015) Silicon photonics platform for national security applications. In: IEEE Aerospace Conference, 2015, pp 1–9
4. Reed GT, Jason Png CE (2005) Silicon optical modulators. Mater Today 8(1):40–50
5. Li G, Zheng X, Yao J, Thacker H, Shubin I, Luo Y, Raj K, Cunningham JE, Krishnamoorthy AV (2011) High-efficiency 25 Gb/s CMOS ring modulator with integrated thermal tuning. In: IEEE international conference group IV photonics GFP, Sept 2016, pp 8–10
6. Sepehrian H, Yekani A, Rusch LA, Shi W (2016) CMOS-photonics codesign of an integrated DAC-less PAM-4 silicon photonic transmitter. IEEE Trans Circ Sys 1–11
7. Chen HW, Kuo YH, Bowers JE (2009) High speed silicon modulators. In: 2009 14th optoelectronics communication conference OECC, vol 805. pp 9–10
8. Zhou Y, Zhou L, Zhu H, Wang J, Feiran S, Zhiren W, Liu L, Wang T, Xinwan L, Chen J (2015) Optimized silicon MZI modulators for 50 Gbit/s OOK and 40 Gbit/s BPSK modulation. In: 2015 Opto-electronics communication conference OECC, pp 5–7
9. Qi N, Li X, Li H, Xiao X, Wang L, Li Z, Gao Z, Yu Y, Moyal M, Chiang PY (2015) A 25 Gb/s, 520 mW, 6.4 Vpp silicon-photonic mach-zehnder modulator with distributed driver in CMOS. In: Optical fiber communication conference, Nov 2015, p W1B.3
10. Bhowmik BB, Gupta S (2013) Silicon micro-ring resonator based 16-QAM modulator for WDM transmission systems. In: international conference on microwave and photonics (ICMAP), pp 1–5

11. Li J, Li G, Zheng X, Raj K, Krishnamoorthy AV, Buckwalter JF (2013) A 25-Gb/s monolithic optical transmitter with micro-ring modulator in 130-nm SoI CMOS. IEEE Photonics Technol Lett 25(19):1901–1903
12. Buckwalter JF, Zheng X, Li G, Raj K, Krishnamoorthy AV (2012) A monolithic 25-Gb/s transceiver with photonic ring modulators and ge detectors in a 130-nm CMOS SOI process. IEEE J Solid-State Circ 47(6):1309–1322
13. Arellano C, Mingaleev S (2011) Efficient design of photonic integrated circuits. In: International conference transparent optical networks (ICTON), pp 4–8

Study on Gain Enhancement of the Antenna Using Planar Small Metasurface Lens

Prakhar D. Vyas and K. Usha Kiran

Abstract This paper proposes the analysis of unit cell of different shapes which would be in future used to form a metasurface lens for enhancing the gain. In this paper, various shapes such as rectangular, hexagonal, etc. of the unit cell are designed to operate at about 5 GHz frequency and are analyzed and simulated using HFSS 16.0. The relative permittivity and the relative permeability is calculated and analyzed for the designed unit cell.

Keywords Metamaterial · Gain · Negative refractive index · Directivity

1 Introduction

In today's technology, the very basic element is used in optics and microwave application. The most fundamental function of any lens would be imaging and focusing. One type of lens that is capable to focus the source field with flat geometry is metamaterial negative refractive lens. Metamaterial can be defined as a material possessing properties that generally cannot be found in other natural materials, one such property being negative refractive index. The efficiency of any lens antenna can be generally improved by using the MNRI property by controlling the refractive index, polarization, and wave impedance of the antenna. In this study, the property of negative refractive index is used to study the properties of the proposed antenna. The spherical wavefront could be transformed efficiently into planar wavefront using the proposed MNRI lens to obtain efficient directivity. Such high-directivity antennas would include applications such as wireless communication, radar communication, and satellite communication systems.

P. D. Vyas (✉) · K. Usha Kiran (✉)
School of Electronics Engineering, VIT University Chennai, Chennai 600127, Tamil Nadu, India
e-mail: prakhard.vyas2015@vit.ac.in

K. Usha Kiran
e-mail: usha.kiran@vit.ac

G. Gnanagurunathan et al. (eds.), *Optical and Microwave Technologies*, Lecture Notes in Electrical Engineering 468,
https://doi.org/10.1007/978-981-10-7293-2_21

In this work, different shapes like rectangle, triangle, and hexagonal shape are considered for unit cell design. It is seen that these shapes possess the double negative properties within the operating frequency range.

2 Proposed Work

The two-dimensional structure of metamaterials is the metasurface. The metamaterials can be designed to obtain any permittivity and permeability according to the principle. In this work, first a single cell, an unit cell, is designed for different structures. The unit cell for all shapes is designed on the top of the substrate [1–3] using HFSS as shown in Fig. 1a–f. The dimensions of each unit cell are changed as per the requirement so as to obtain the required operating frequency range. The designed unit cell was then analyzed, and the relative permittivity and permeability was calculated using the theoretical formula. Comparison of the calculated permittivity and permeability values was tabulated, and study was done on the gain enhancement. Section 3 presents the design of the unit cell, Sect. 4 presents the analysis of the unit cell, and concluding remarks is presented in the Sect. 5.

3 Design of the Unit Cell

The design consists of various shapes of unit cell as shown in Fig. 1. The unit cell is designed having the substrate at the top and the ground plane at the bottom [1]. The dielectric substrate used here is FR4 epoxy with a thickness of 1.16 mm, relative permittivity of 4.4, and a loss tangent of 0.02. The unit cell designed for each shape has equal areas. The dimensions of the unit cell of each shape are tabulated in Table 1.

The dimensions of each of the unit cell are listed in Table 1. The gap between the inner and outer cell is maintained to be 1 mm (Fig. 2).

The substrate used is FR4 epoxy with a thickness of 1.6 mm. Here to obtain the required operating frequency range, the dimensions of the unit cell were changed.

4 Analysis of the Unit Cell

The designed unit cell is symmetrical about x and y axis. In the analysis, appropriate boundary conditions are given to the unit cell [4]. The analysis of the unit cell is done to show the double negative nature of the material with negative permittivity and permeability. In the simulation of the unit cell, the boundary walls are a pair perfect electric conductor (PEC) and perfect magnetic conductor (PMC) having a two-port waveguide. Output port is given at the side of unit cell.

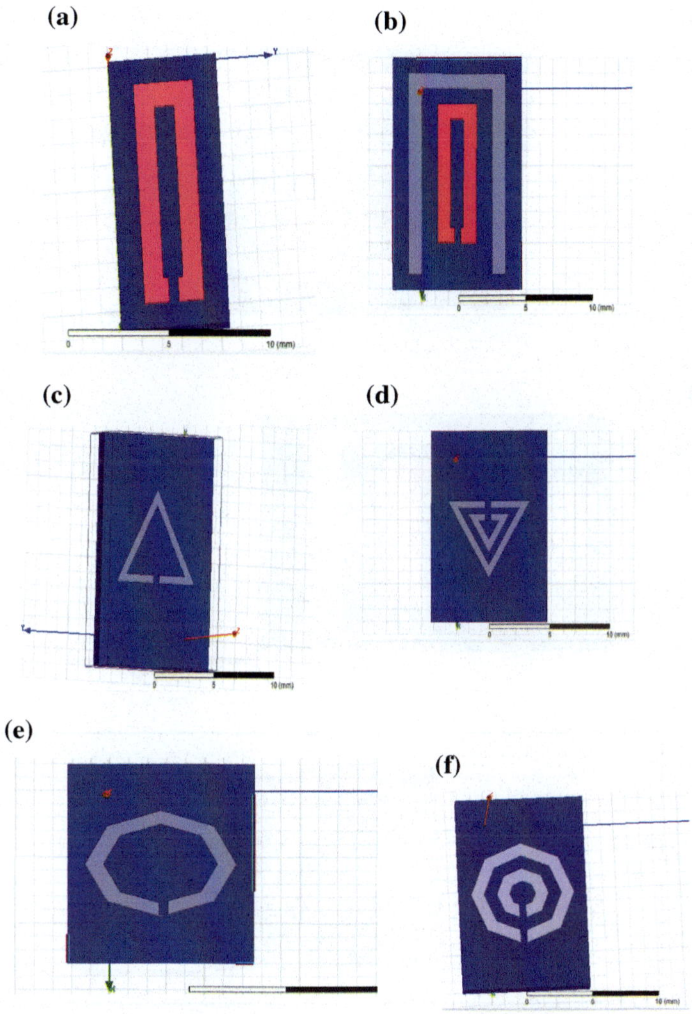

Fig. 1 **a** Rectangular unit cell, **b** inverted U unit cell, **c** triangular unit cell, **d** inverted triangle unit cell, **e** hexagonal unit cell, **f** double hexagonal

Table 1 Dimensions of unit cell

Shape of unit cell	Dimension in mm
Rectangular	$a = 13$, $b = 7.4$
Inverted U	$s = 6.92$
Triangular	$s = 6.92$
Inverted triangle	$s = 3$
Hexagonal	$s = 3$
Double hexagonal	$a = 13$, $b = 7.4$

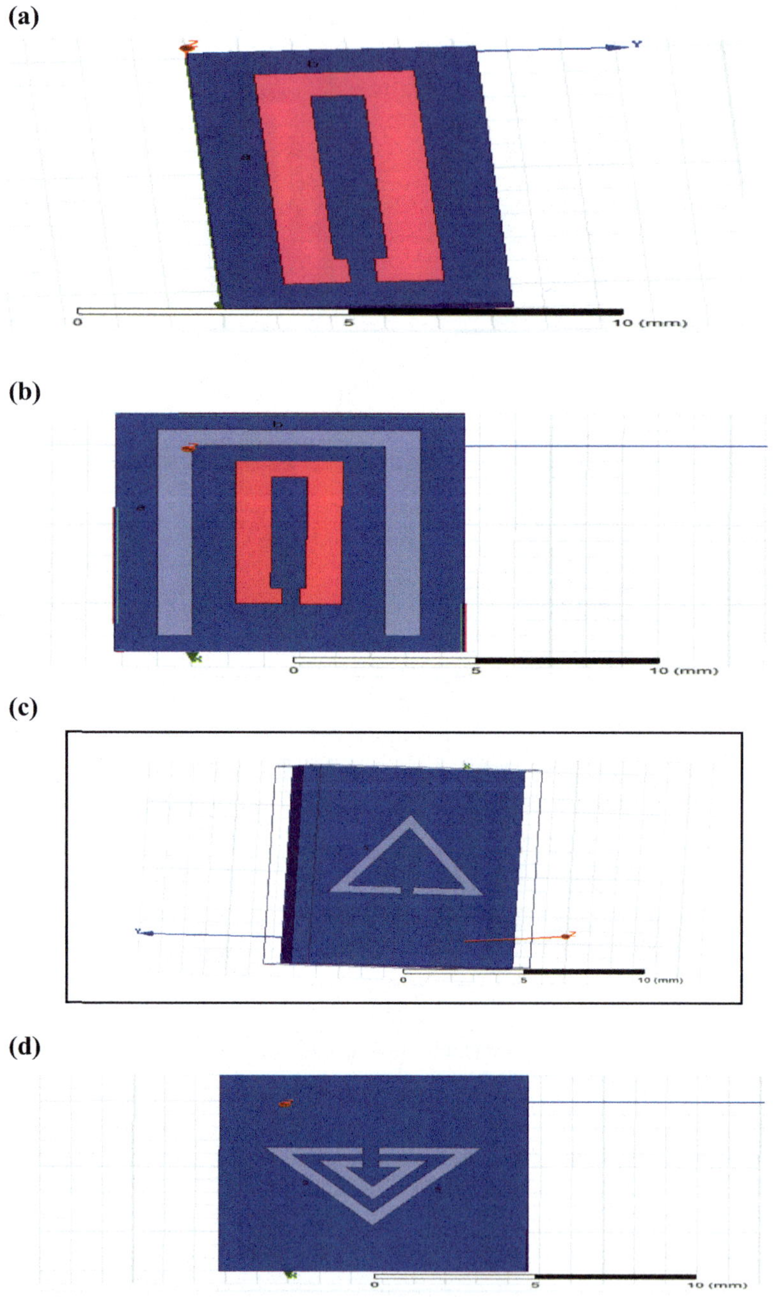

Fig. 2 **a** Dimension of rectangular unit cell, **b** dimension of inverted U unit cell, **c** Dimension of triangular unit cell, **d** dimension of inverted triangular unit cell, **e** dimension of hexagonal unit cell, **f** dimension of double hexagonal unit cell

(e)

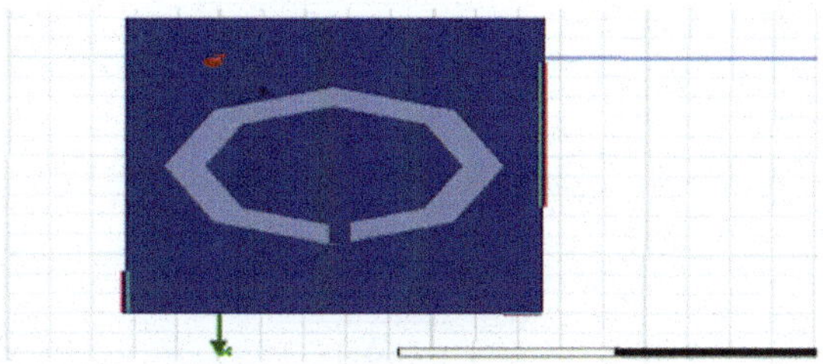

(f)

Fig. 2 (continued)

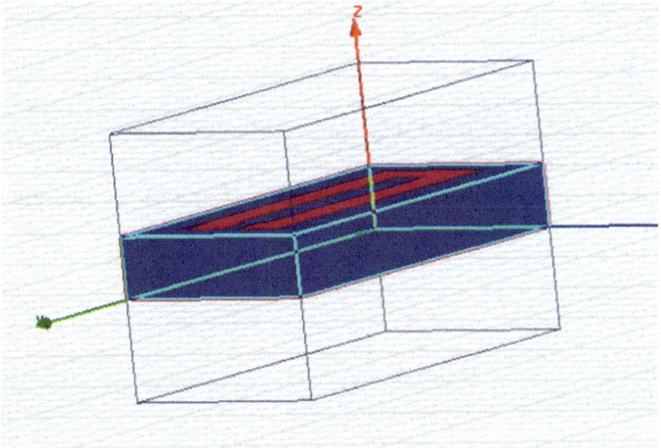

Fig. 3 Simulation of unit cell

In this paper, we use the *S*-parameter method to analyze the unit cell [5–8]. There are also other methods which can be used to analyze the unit cell. The S_{11} and S_{21} parameter for the unit cell are obtained from the arrangement shown in Fig. 3. Then the obtained values of S_{11} and S_{21} are to be exported to Microsoft Excel, and those values are used to obtain the equivalent permittivity and permeability values using the given Eqs. (1)–(3)

$$Z = \pm\sqrt{\frac{(1+S_{11})^2 - S_{21}^2}{(1-S_{11})^2 - S_{21}^2}} \tag{1}$$

$$e^{jnk_0 d} = X \pm j\sqrt{1-X^2} \tag{2}$$

$$X = \frac{(S_{11}^2 + S_{21}^2)}{2S_{21}} \tag{3}$$

Here, Z indicates the equivalent impedance, and n indicates the refractive index. The calculated values of Z and n then can be substituted in the Eqs. (4) and (5) below to find the permittivity and permeability.

$$\mu_r = nZ \tag{4}$$

$$\varepsilon_r = {}^{n}/_{Z} \tag{5}$$

The values obtained from (4) and (5) showed that the unit cell has negative refractive index and acts as a metamaterial in the required frequency band.

5 Result

The permittivity and the permeability values obtained in the resonating frequency range are listed in Table 2.

The simulated results of S_{11} and S_{21} for each shape of the designed unit cell are shown in Figs. 4, 5, 6, 7, 8, and 9.

Table 2 Calculated permittivity and permeability values

Shape of unit cell	Frequency in GHz	ε_r	μ_r
Rectangular	4.9	1.4	1.75
Inverted U	4.9	4.6	1.91
Triangle	5.9	1.41	3.33
Inverted triangle	5.9	1.44	3.5
Hexagonal	4.9	6.7	1.4
Double hexagonal	4.9	4.7	6.7

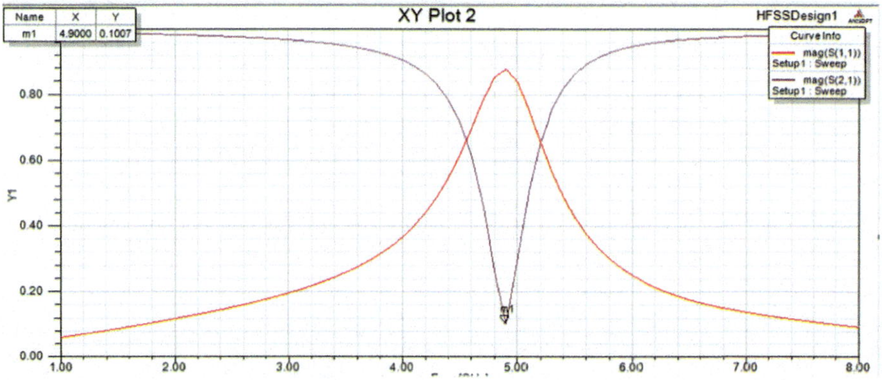

Fig. 4 S_{11}, S_{21} of rectangular unit cell in dB

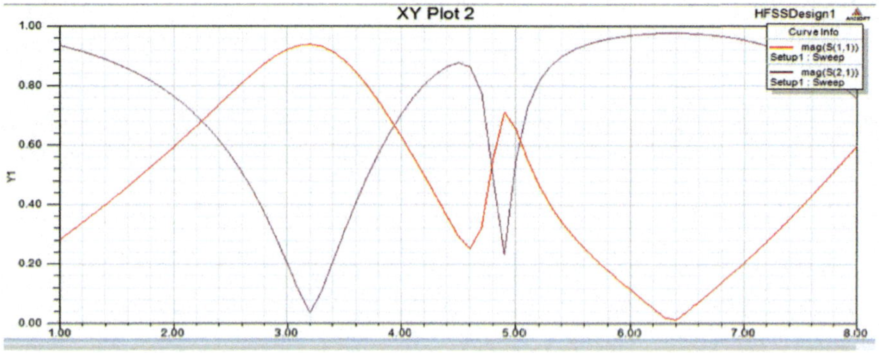

Fig. 5 S_{11}, S_{21} of inverted U unit cell in dB

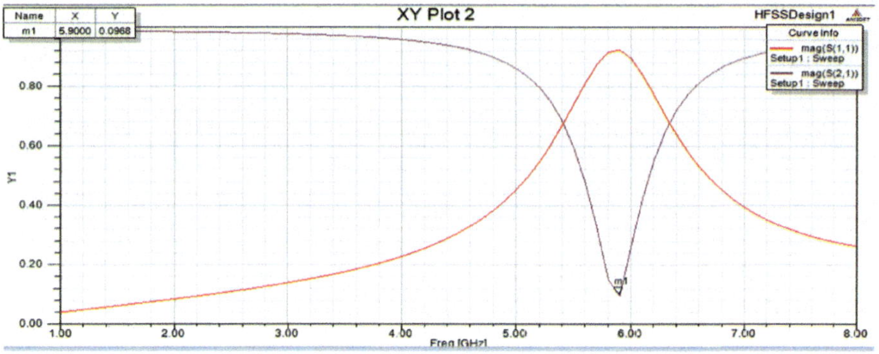

Fig. 6 S_{11}, S_{21} of triangular unit cell in dB

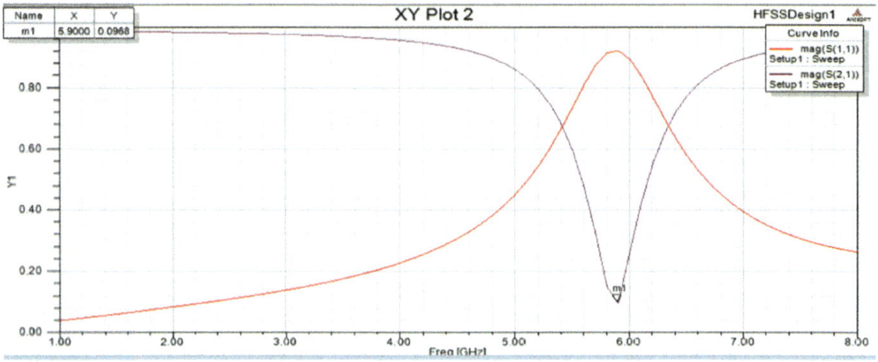

Fig. 7 S_{11}, S_{21} of inverted triangle unit cell in dB

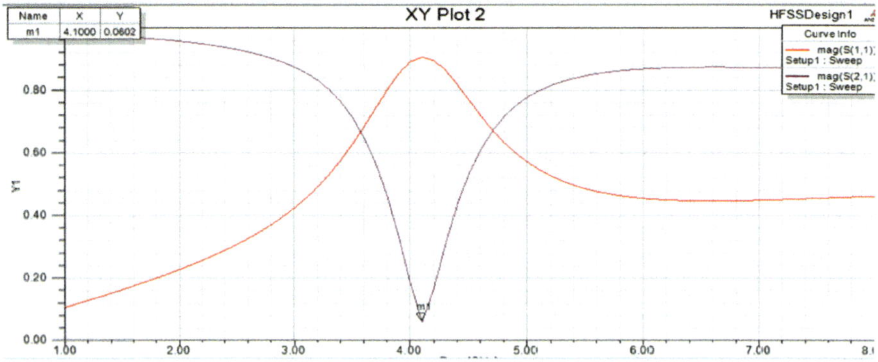

Fig. 8 S_{11}, S_{21} of hexagonal unit cell in dB

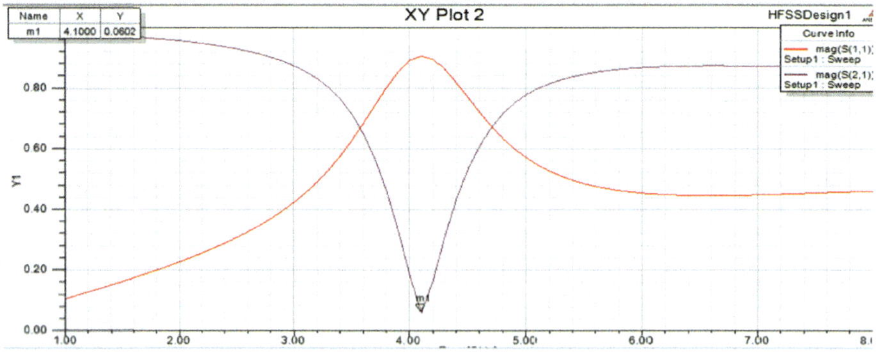

Fig. 9 S_{11}, S_{21} of double hexagonal unit cell in dB

6 Conclusion and Future Work

On the basis of the analysis of the above unit cell design, it is seen that all of the simulated unit cell is seen to posses negative permittivity and negative permeability and they are studied to be double negative material and hence metamaterials.

The future work would include optimizing the above designed unit cell to simulate at about 5.2 GHz. Later, the optimized values would be compared and studied to choose the better unit cell profile and using it as a metasurface to increase the gain of the antenna.

References

1. Pavan MN, Chattoraj N (2015) Design and analysis of a frequency reconfigurable antenna using metasurface for wireless applications. In: IEEE sponsored 2nd international conference on innovations in information, embedded and communication systems (ICIIECS)
2. Zhu HL, Liu XH, Cheung SW, Senior Member, IEEE, Yuk TL (2014) Frequency-reconfigurable antenna using metasurface. IEEE Trans Antennas Propag 62(1)
3. Panda AK, Sahu A (2011) International conference on computational intelligence and communication systems
4. Ziolkowski RW, Fellow, IEEE (2003) Design, fabrication, and testing of double negative metamaterials. IEEE Trans Antennas Propog 51(7)
5. Numan AB, Sharawi MS (2013) Extraction of material parameters for metamaterials using a full-wave simulator. IEEE Antennas Propag Mag 55(5)
6. Smith DR, Vier DC, Koschny T, Soukoulis CM (2005) Electromagnetic parameter retrieval from inhomogeneous metamaterials. Phys Rev E 71
7. Koschny T, Markos P, Smith DR, Soukoulis CM (2003) Resonant and antiresonant frequency dependence of the effective parameters of metameaterials. Phys Rev E68
8. Chen X, Grzegorczyk TM, Wu BI, Pacheco J, Kong JA (2004) Robust method to retrieve the constitutive effective parameters of metamaterials. Phys Rev E 70(1):016608.1–016608.7

Design of an Internal Multi-resonant PIFA Antenna for Mobile Telecommunication Networks

Subathra Thavakumar and M. Susila

Abstract A PIFA antenna which supports hexa bands of various wireless standards in this current era is designed. The proposed antenna is configured with the concept of loop and slot structure in order to acquire wider bandwidth, multiple resonance, and good efficiency and to diminish the size of conventional PIFA antenna. This multiband internal PIFA antenna is having the ability of operating in GSM (900 MHz), Standard Positioning Service (1.176 GHz) of Indian Regional Navigational Satellite System (IRNSS) L5 band UMTS-DCS (1710–1880 MHz) and PCS (1850–1990 MHz), LTE 2300 (2305–2400 MHz) and ISM bands of Wi-Fi/Bluetooth (2.4 GHz). The proposed multiband antenna is designed using the FR4 epoxy substrate with the thickness of 1.6 mm. The simulation results are analyzed and optimized using 3D electromagnetic simulation software which is based on the finite element methodology. Return loss, voltage standing wave ratio (VSWR), radiation pattern, and gain are the attributes obtained and validated for the designed PIFA antenna.

Keywords PIFA · Multiband · GSM · UMTS · LTE · ISM · IRNSS

1 Introduction

Nowadays, the growth of telecommunication standards is increasing in an enormous manner. So, the wireless devices are manufactured with the demand of supporting all wireless proliferations. Planar inverted-F antenna (PIFA) is the type of internal antenna which was derived from the linear inverted-F antenna by replacing the wire element with the planar element. It is widely used as a mobile antenna in this current scenario and particularly in the slim smartphones by its characteristics such as small in size, omnidirectional radiation pattern, cost-effective, supports multiband operation in low and high frequencies, minimum

S. Thavakumar (✉) · M. Susila
Department of Telecommunication Engineering, SRM University, Chennai, India
e-mail: subathra.t.94@gmail.com

G. Gnanagurunathan et al. (eds.), *Optical and Microwave Technologies*,
Lecture Notes in Electrical Engineering 468,
https://doi.org/10.1007/978-981-10-7293-2_22

specific absorption ratio (SAR) and easy to manufacture. More than that PIFA is having low backward radiation toward the user's head or any part of a body which in turn increases the performance of the antenna. Indian Regional Navigational Satellite System (IRNSS) is the recently launched regional-based navigational system by the Indian Space Research Organization (ISRO) under the operation of Navigation with Indian Constellation (NAVIC). This autonomous system is implemented to dispense the standard positioning service instead of GPS in order to prevent the unguaranteed positioning service during a hostile situation. IRNSS is designed to support two services, namely standard positioning service which is open to a civilian purpose and precision service for a military purpose which is highly encrypted one.

An internal mobile antenna was designed based on the meandered concept with a coaxial feed. It was capable of covering WiMAX, WLAN, GPS, and UMTS frequency bands [1]. FR4 dielectric substrate was used to fabricate the multiband patch antenna with the permittivity of 4.4. This substrate is cheaper and provides flexibility while designing reconfigurable and tunable patch antenna which has multiple resonance [2]. An autonomous and regional-based navigational system of IRNSS was described with the operational services stipulated for Indian subcontinent [3]. Indian GPS Satellite System of IRNSS was described based on the constellation of IRNSS satellites, positional service, and segmental program. It was named as our "own GPS system" [4]. Bandwidth coverage of multiband PIFA was increased by inserting the extra branch line along meandered structure and folded loop structure. That particular PIFA supported high-frequency bands of LTE and UMTS and the low-frequency bands of GSM [5]. The optimization concept of antenna characteristics was made easier by performing the modification in size, position, and shape of a particular antenna with the help of inserted slots [6]. Two different structured slots were introduced in a single patch to acquire the multiple resonance. That slotted technique provided the way to overcome the PIFA antenna limitation of narrow bandwidth [7]. The combination of meandered strip and two monopole strip was implemented to enhance the impedance matching at high frequency as well as low frequency in an internal PIFA antenna [8]. The IRNSS operation was defined in space segment, ground segment, and user segment with the orbital operation of IRNSS satellites. An architecture of the IRNSS satellites and its deployment in Geostationary and Synchronous orbit were described along timing group delay [9]. A comparative analysis among the antenna size reduction techniques such as stub loading, capacitive loading, introducing slots, and addition of parasitic element was made on the basis of antenna application [10, 11]. The combination of all bandwidth enhancing techniques was implemented in a single PIFA structure with the support of two different antenna designing tool was analyzed and optimized [12]. The proposed antenna enhances the bandwidth coverage and dispenses multiple resonance between 0.859 GHz (low-frequency operational band) and 2.809 GHz (high-frequency operational band) by its design methodology.

2 Configuration of Proposed Antenna

The proposed multiband PIFA antenna consists of two structures, namely loop and slotted structure. FR4 substrate which has dielectric constant (ε_r) 4.4, loss tangent (δ) 0.02, and height (h) 1.6 mm is used as a base material to design the proposed antenna. This PIFA structure consists of a ground plane, radiating patch, inserted feeds, and shorting pins. It is designed on the ground plane which has the total dimension of 100 × 35 × 1.6 mm. The dimension of loop antenna structure is 30 × 35 × 1.6 mm and slotted antenna structure is 18.4 × 34 × 1.6 mm that of a ground plane. The structural and geometric description of the multiband PIFA antenna is given in Figs. 1 and 2.

The loop structure is placed at the left side of the PIFA design to cover the frequency bands of GSM, UMTS, LTE, and Wi-Fi/Bluetooth standards, and the Slotted structure is placed at the right side to cover the standard positioning service of IRNSS at the L5 band. In order to design the miniaturized PIFA antenna, the concept of a loop and slotted methodology is implemented in the design of multi-resonant PIFA structure using the 3D electromagnetic tool. The detailed dimensions based on the length and width of proposed PIFA are specified in Table 1.

This PIFA has two inserted feeds with 50 Ω impedance matching and two shorting pins. The shorting effect is provided to the antenna by connecting a

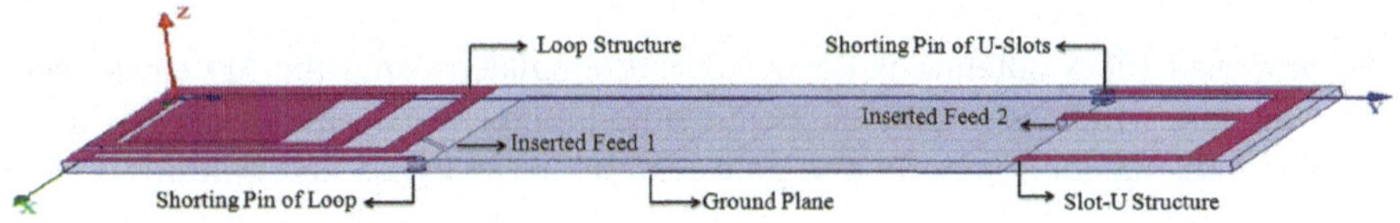

Fig. 1 Structure of proposed multiband PIFA

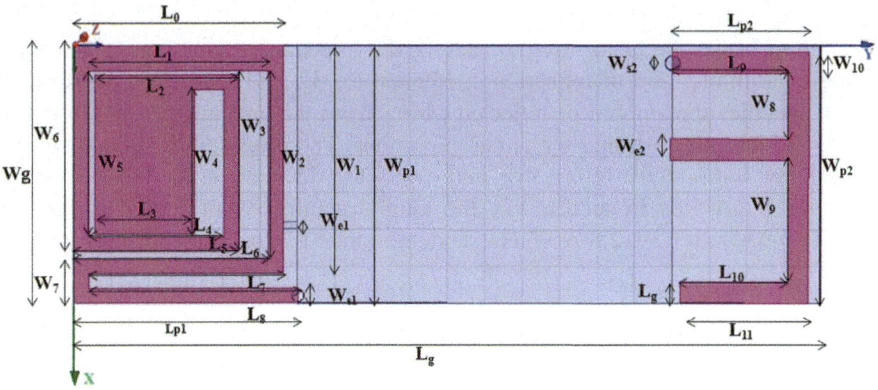

Fig. 2 Geometry of proposed antenna

Table 1 Dimensions of proposed PIFA

Parameter	Value (mm)	Parameter	Value (mm)	Parameter	Value (mm)	Parameter	Value (mm)
L_g	100	W_{s1}	1.6	L_6	26	W_3	24
W_g	35	W_{s2}	2	L_7	26	W_4	19.6
L_{p1}	30	L_0	28	L_8	28	W_5	22.5
L_{p2}	18.4	L_1	24	L_9	15.5	W_6	28
W_{p1}	35	L_2	19.3	L_{10}	14.5	W_7	6
W_{p2}	34	L_3	13.3	L_{11}	17.4	W_8	8.84
W_{e1}	1	L_4	18	W_1	31	W_9	16.46
W_{e2}	2.9	L_5	22	W_2	25.5	W_{10}	2.9

radiating patch and ground plane via shorting pins. The shorting pin width (W_{s1}) of loop structure is 1.6 and 2 mm for slotted-U structure (W_{s2}). The width of shorting pins is stipulated after the optimization process in order to attain good impedance matching for desired frequencies. This internal multi-resonant PIFA resonates at 900 MHz, 1.37 MHz, 1.65 GHz, and 2.0625 GHz. At 2.69 GHz, proposed PIFA achieves the maximum radiation efficiency of 92% with the peak gain of 2.19 (dB).

3 Validation of Simulated Results

The proposed PIFA antenna is designed and simulated using the 3D electromagnetic software which uses finite element methodology. The attributes such as return loss, VSWR, radiation pattern, and gain pattern are analyzed and optimized.

3.1 *Return Loss*

The return loss is the factor used to analyze the impedance matching. In general, it is the logarithmic ratio of reflection coefficient (r), i.e., S11 which is measured in dB. It specifies the amount of reflected power from the antenna element to that of power fed into the antenna for excitation purpose. The proposed multiband antenna is designed with—6 dB return loss and it met the desired results. Figure 3 represents the return loss in such a way the loop structure covers 2 bandwidths, i.e., 0.8590–0.93 and 1.39–2.8090 GHz, and the slotted-U structure covers 1.0670–1.4867 GHz.

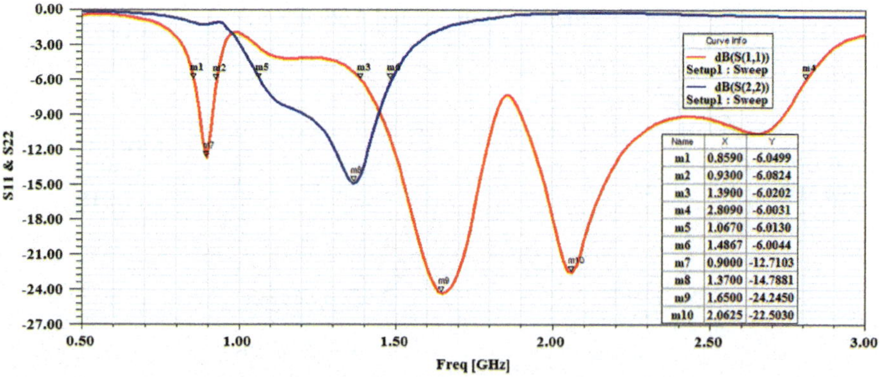

Fig. 3 Frequency versus reflection coefficient plot for return loss

3.2 *Voltage Standing Wave Ratio*

Figure 4 shows the obtained VSWR at the ratio of 3:1 of the proposed PIFA. This provides good impedance matching with respect to return loss.

3.3 *Radiation Pattern*

The 2D radiation pattern of proposed multiband PIFA antenna is determined in the far-field region with respect to *x–y* plane. The following radiation patterns are obtained at $\phi = 0°$ and 90° of total gain in decibel for all possible operating frequency bands of a proposed multiband PIFA antenna. Figure 5 clearly states that the proposed PIFA has attained the omnidirectional pattern which is essential for internal and multi-resonant mobile antenna.

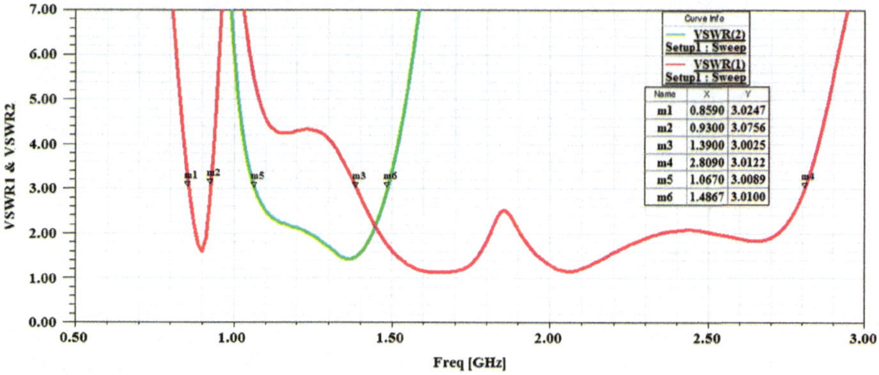

Fig. 4 Frequency versus VSWR plot

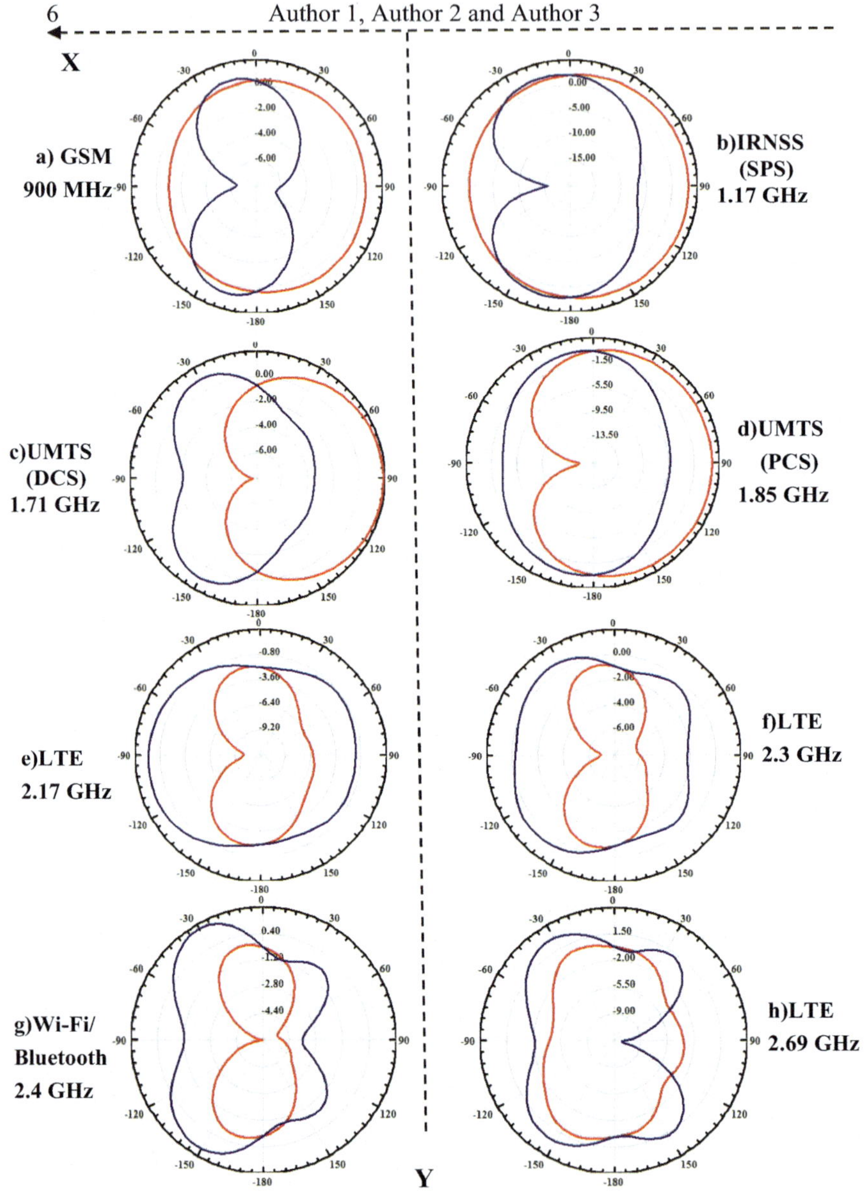

Fig. 5 2D radiation pattern of all operational frequency bands

Table 2 provides the gist of proposed antenna attributes such as radiation efficiency, peak gain, and peak directivity for the corresponding frequency bands. Here, the radiation efficiency is the ratio of peak gain (dB) and peak directivity (dB).

Table 2 Frequency versus proposed PIFA attributes

Attributes	Operating frequencies (GHz)							
	0.9	1.17	1.71	1.85	2.17	2.3	2.4	2.69
Radiation efficiency (%)	68	98	80	71	56	64	70	92
Peak gain (dB)	1.28	2.17	2.23	1.84	1.14	1.25	1.54	2.19
Peak directivity (dB)	1.87	2.21	2.77	2.58	2.03	1.96	2.18	2.37

Acknowledgements The authors would like to thank Dr. R. Rama Rao, Professor and Head Department of Electronics and Communication Engineering/Telecommunication Engineering, SRM University for the valuable guidance during this work.

References

1. Verma A, Chauhan A (2016) Compact slotted meandered PIFA versus conventional PIFA antenna for DCS, GPS, bluetooth/WLAN, 4 G LTE, WiMAX, UMTS, GLONASS applications. In: 3rd international conference on computing for sustainable global development. IEEE Publications, pp 951–954. ISSN: 0973-7529
2. Ramteke SM, Prabha S (2015) Multiband patch antenna for wireless communication system. Int J Electr Electron Eng 7(1):204–210. ISSN: 2321-2055
3. Khatri RR, Chauhan S (2016) Indian regional navigation satellite system. Int J Innovative Res Technol 2(11):380–384. ISSN: 2349-6002
4. Saini M, Gupta U (2014) Indian GPS satellite navigation system: an overview. Int J Enhanced Res Manag Comput Appl 3(6):32–37
5. Hong Y, Tak J, Baek J, Myeong B, Choi J (2014) Design of a multiband antenna for LTE/ GSM/UMTS band operation, Article ID 548160, vol 2014, 9 p. Hindawi Publishing Corporation, IJAP
6. Gokul R., Geetha Ramani J (2016) A novel tri band planar inverted F antenna for wireless communication. Middle-East J Sci Res (IDOSI Publications): 695–700. ISSN 1990-9233
7. Saini S, Kaur R, Singh S, Kumar N (2015) A compact T-slot multiband planar inverted-F antenna for handheld devices, pp 4209–4214. IEEE Publications. ISSN: 4673-8253
8. Zhang T, Li RL, Jin G, Wei G, Tentzeris MM (2015) A novel multiband planar antenna for GSM/ UMTS/ LTE/ Zigbee/RFID mobile devices. IEEE Trans Antenn Propag 59(11):4209–4214
9. Saikiran B, Vikram V (2013) IRNSS architecture and applications. KIET Int J Commun Electron 1(3):21–27. ISSN: 2320-8996
10. Loharia N, Rana SB (2016) Review of PIFA. Int J Electron Commun Technol 7(2): 66-68. ISSN: 2230-7109|ISSN: 2230-9543
11. Hosseini SE, Attari AR, Pourzadi A (2013) Multiband PIFA with a slot on the ground plane for wireless applications. Int J Inf Electron Eng 3(4):349–352
12. Gangwar RPS, Rani P (2015) Multiband planar inverted-F antenna (PIFA) for mobile phones. Int J Electr Electron Eng Res 5(6):71–78. ISSN(P): 2250-155X

Design of Substrate Integrated Waveguide Back to Back π-Shaped Slot Antenna for 60 GHz Applications

M. Nanda Kumar and T. Shanmuganantham

Abstract In this paper, proposed back to back π-shaped slot antenna with SIW cavity, which is connected to 50 Ω microstrip line feed and material Rogers RT/Duroid 5880 with $\in_r = 2.2$, is used to develop proposed structure. CST software is used to simulate the proposed antenna and observed the results like reflection coefficient, VSWR, gain, and radiation patterns. This structure will provide bandwidth of 3.3 GHz with respect to −10 dB reference line and resonant frequency is 60 GHz and preserves 47.14% in between the frequency range 57–64 GHz with respect to impedance bandwidth and VSWR (2:1). The reflection coefficient, gain, directivity, transmission efficiency, and radiation efficiency of proposed antenna at 60 GHz are −35.644 dB, 5.85 dBi, 5.68 dB, 78%, and 96%.

Keywords Substrate integrated waveguide (SIW) · Millimeter waves
Wireless LAN (WLAN) · Waveguides

1 Introduction

The frequency of millimeter wave technology is 30–300 GHz and plays a very important role in communications, attracted due to increasing the growth of academia and industry applications. The fixed unlicenced frequencies of millimeter wave technologies are 60GHz (wireless communication network) [1, 2], 79 GHz (Automotive radar systems), and 94 GHz (millimeter wave imaging). For these applications, investigation on broadband and high-gain antenna is very important and growing attention in last decades. 57−64 GHz band (7 GHz) is a unlicensed band which is assigned by federal communication commission (FCC) to access unlicensed devices and resonant frequency is 60 GHz. This frequency band is used

M. Nanda Kumar (✉) · T. Shanmuganantham
Department of Electronics Engineering, Pondicherry University, Pondicherry, India
e-mail: nanda.mkumar12@gmail.com

T. Shanmuganantham
e-mail: Shanmuga.dee@pondiuni.edu.in

G. Gnanagurunathan et al. (eds.), *Optical and Microwave Technologies*,
Lecture Notes in Electrical Engineering 468,
https://doi.org/10.1007/978-981-10-7293-2_23

for high data rates [3], short-range applications such as wireless local area network (WLAN), automotive applications, and GIFI.

The necessity of radio communications applications requires high data rates, prompted attention in use of the extra high frequency (EHF) bands. The 60 GHz band is more attracted for short-range communications, becomes available for use and offers the more frequency reuse and therefore makes it striking for short-range communication applications.

Substrate integrated waveguide [4] is one technology which is used to implement high-frequency applications and designed with help of rectangular waveguide. It is one type of substrate integrated circuits (SICs). Other names of SIW are laminated waveguide and post-wall waveguide. The structure of SIW is a waveguide, integrating with help of two rows of metallic holes or slots or via joining bottom and top planes of a dielectric substrate [5, 6] and displayed in Fig. 1. The main advantage of SIW is used as a planar technology which includes active, passive components, and antennas. SIW-based slot antennas are attracted for wireless applications due to small profile, easy integration with planar circuits, and good isolation from feed network.

This paper proposes back to back π-shaped slot antenna based on substrate integrated waveguide fed by microstrip line for 60 GHz applications. Here, we firstly designed SIW followed by microstrip used as feed to the SIW back to back π-shaped slot antenna.

This paper is ordered as follows. Section 2 describes microstrip design; Sect 3 describes design of SIW; Sect. 4 outlines SIW back to back π-shaped slot antenna and their results and finally, conclusion can be described in Sect. 5.

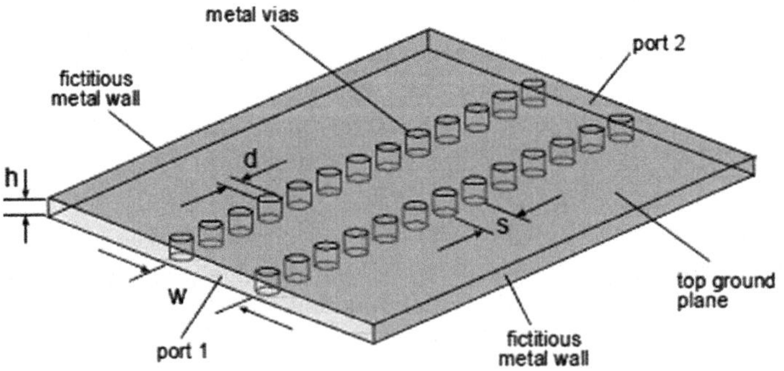

Fig. 1 Substrate integrated waveguide

2 Microstrip Design

This section first describes about the design of microstrip and followed by tapering of microstrip to rectangular waveguide.

2.1 *Microstrip*

The general representation of the microstrip is shown in Fig. 2a. The microstrip line width W depends on the dielectric substrate height *h* and the *W*/*h* can be determined with help of Eqs. 1–3.

$$\frac{W}{h} = \begin{cases} \frac{2}{\pi}\left\{ \begin{array}{c} A - 1 - \ln(2A - 1) + \ldots \\ \frac{\varepsilon_r - 1}{2\varepsilon_r}\left[\ln(A - 1 + 0.39) - \frac{0.1}{\varepsilon_r}\right] \end{array} \right\}, \frac{W}{h} > 2 \\ \frac{8e^A}{e^A - 2}, W/h < 2 \end{cases} \quad (1)$$

where

$$A = \frac{377\pi}{2Z_o\sqrt{\varepsilon_r}} \quad (2)$$

$$B = \frac{Z_o}{60\sqrt{\varepsilon_r}}\sqrt{\frac{\sqrt{\varepsilon_r} + 1}{2}} + \frac{\varepsilon_r - 1}{\varepsilon_r + 1}\left(0.23 + \frac{0.11}{\varepsilon_r}\right) \quad (3)$$

2.2 *Microstrip Tapering to Rectangular Waveguide*

Generally, to overcome the mismatch problem, tapering is used. The length and width of tapering can be determined with help of Eqs. 4–5.

$$W_{ef}/W_e = 4.38e^{0.627\frac{\varepsilon_r}{\varepsilon_e}} \quad (4)$$

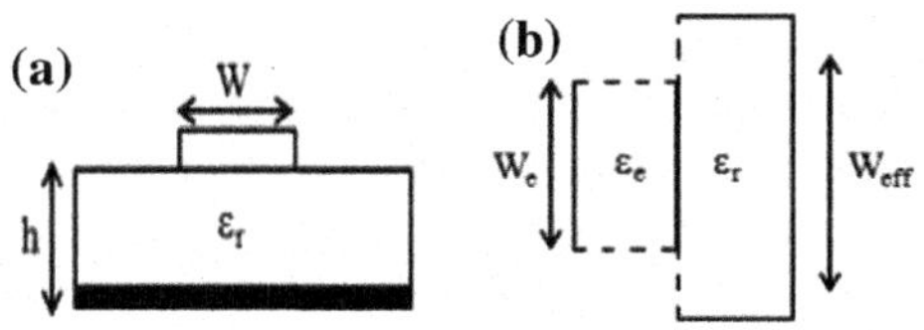

Fig. 2 **a** Microstrip line **b** Microstrip tapering to rectangular waveguide

The length of tapering can be determined by below equation

$$L_m = \frac{n\lambda_g}{4}, n = 1, 2, 3, 4. \ldots .. \quad (5)$$

where

$$\lambda_g = \frac{c}{f\sqrt{\varepsilon_r}} \quad (6)$$

3 Substrate Integrated Waveguide

It is one type of substrate integrated circuits (SICs), mainly intended for centimeter and millimeter applications [7] and Fig. 1 shows simple diagram of SIW. Assume "**s**" signifies distance between two holes; "**h**" represents the height of the SIW, and "**d**" represents diameter of the holes. It is constrained from rectangular waveguide and dominant mode is TE_{10} [8, 9], the width of dielectric-filled rectangular waveguide is

$$a_d = \frac{C}{2 * f_c * \sqrt{\in_r}} \quad (7)$$

where $\mathbf{a_d}$ is the width of dielectric-filled rectangular waveguide (DRW). The standard equation of SIW as mentioned in eq. 8. For better performance, choose $d \leq \frac{\lambda_g}{5}$ and $p \leq 2d$ [8, 9].

$$a_s = a_d + \frac{d^2}{0.95s} \quad (8)$$

4 Results and Discussion

The proposed structure was shown in Fig. 3, introduced for 60 GHz applications and designed using Rogers RT/Duroid with $\in_r = 2.2$ height is 0.381 mm and metallization thickness of 25 μm. The parameters described in Table 1 are used for designing proposed antenna. The length of pi shape (w3) value is greater than half of guided wavelength, and width of pi slot (L3) is less than a length of pi slot (w3). The length of microstrip and tapered microstrip is multiple odd numbers of quarter guided wavelength. The size of proposed structure is 4 * 10.5 * 0.381 mm.

The frequency versus reflection coefficient (S_{11}) of proposed structure observed the result between 57 and 64 GHz applications in Fig. 4. The impedance bandwidth is 3.31 GHz for a proposed antenna with respect to −10 dB reference line,

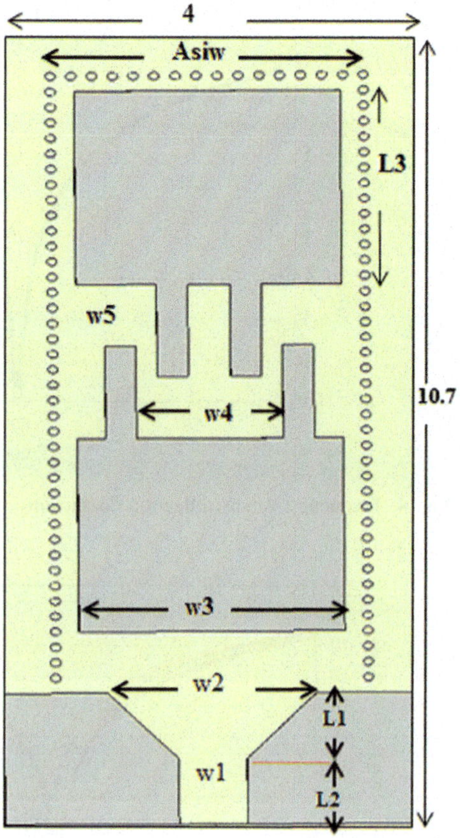

Fig. 3 Proposed structure (All dimensions are in mm)

Table 1 Parameters used for design

Parameter	Value (mm)	Parameter	Value (mm)
s	0.2	w1	0.65
d	0.1	w2	2
L1 = L2	0.842	w3	2.55
Asiw	3	w4	2
L3	2.5	w5	0.3

preserves 47.28% of a band in between 57 and 64 GHz frequency and also observed that the −43.9 dB accuracy at 60 GHz.

As per the scenario of slot, width will affect the resonant frequency and reflection coefficient. Figure 5 represents the frequency versus reflection coefficient for different values of slot widths (w3). Slot width changes from low to high (2.45mm to 2.65mm), the resonant frequency (60.2–59.8 GHz) is decreased and also observed that reflection coefficient is also decreases from -48dB to −42 dB.

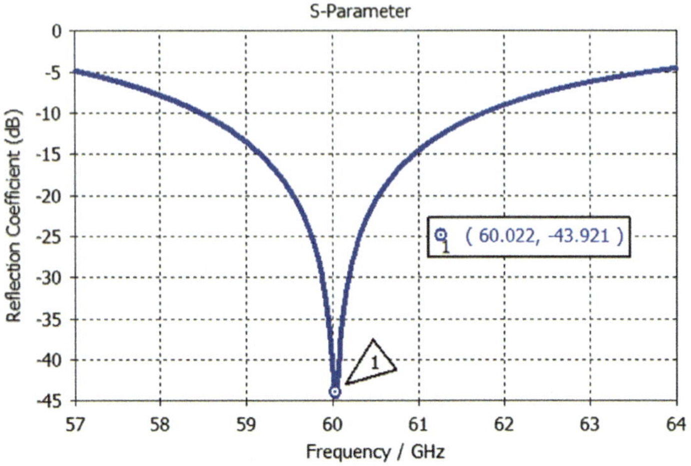

Fig. 4 Frequency versus reflection coefficient

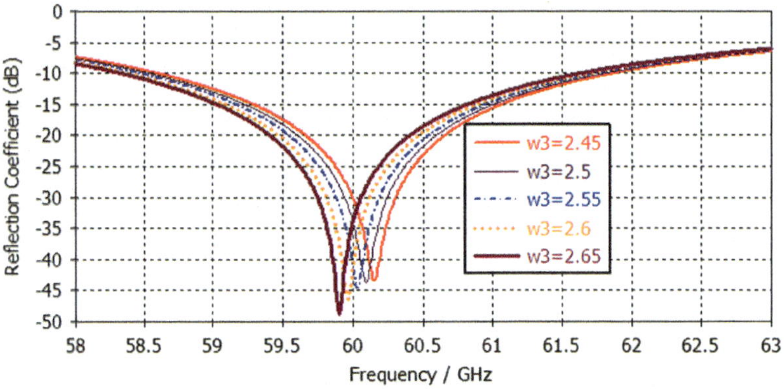

Fig. 5 Frequency versus reflection coefficient (dB) for different values of slot widths (w3)

The length of slot will affect the resonant frequency as shown in Fig. 6. The length of slot is proportional to reflection coefficient but resonant frequency is constant, that is, length of the slot decreases from 2.6 to 2.4 mm and the corresponding reflection coefficient also decreses from −45.429 to −43 dB. Fig. 7 represents the efficiencies of proposed antenna of 60–62.5 GHz frequency and noticed that radiation efficiency is 92.5 % at 60 GHz and 95.5 % at 62 GHz and transmission efficiency is 84.2 % at 60 GHz and 86.2 % at 62 GHz.

The VSWR for proposed antenna is shown in Fig. 8 and also achieves around 3.3 GHz bandwidth with respect to VSWR (2:1) and also observed that 1.0182 at

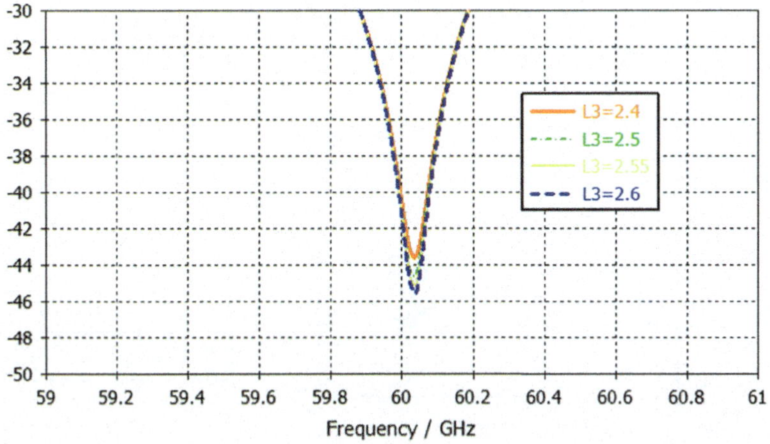

Fig. 6 Frequency versus reflection coefficient (dB) for different values of slot length (L3)

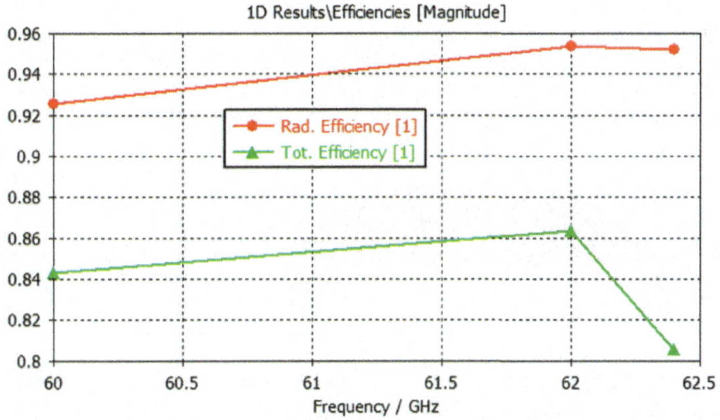

Fig. 7 Efficiency of proposed antenna

60 GHz. Figure 9 represents the three-dimensional radiation pattern of proposed antenna at 60 GHz and observed that the main lobe direction is along broadside (*z*-axis).

Table 2 shows the comparison of proposed antenna with existing antennas with different parameters like bandwidth, reflection coefficient, gain, and size of an antenna. The accuracy and bandwidth improvement of proposed structure were more, a size of antenna is moderate but the gain was average compared to existing works.

The two-dimensional radiation pattern (E- and H-fields) of antenna at 60 GHz is shown in Fig. 10. In that red color indicates H-field and green color indicates E-field and also observed that the antenna will propagate in bidirectional.

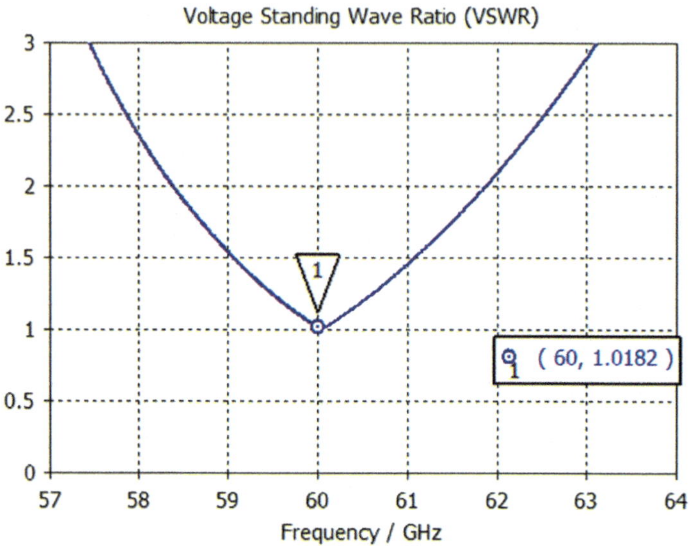

Fig. 8 Voltage standing wave ratio (VSWR)

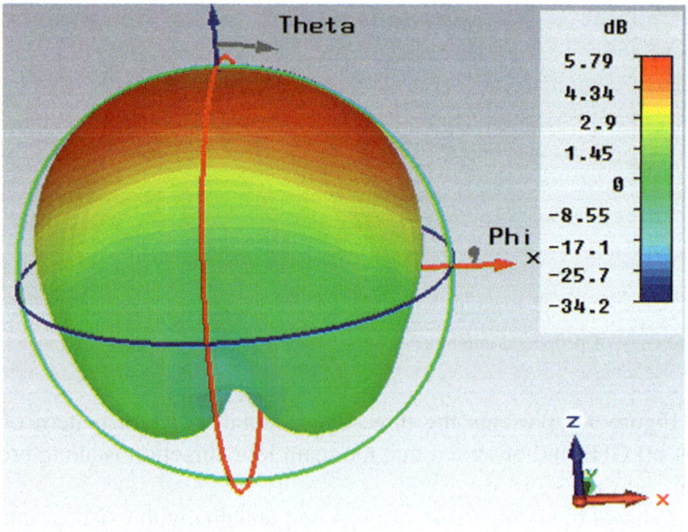

Fig. 9 Three-dimensional gain pattern

Table 2 Different result comparison of proposed structure with existing structure

S. No.	Resonant frequency (GHz)	Bandwidth (GHz)	Reflection coefficient (dB)	Gain (dBi)	Size (mm)
Proposed paper	60	**3.3**	−43.921	5.79	4 * 10.7 * 0.381
[1]	61	2.144	−41.1	8.97	4 * 20.26 * 0.381
[2]	60	1.85	−35	4	4 * 4 * 0.381
[4]	60	1.5	−40	16.5	9.93 * 43.29 * 0.381
[5]	60	0.8	−12.07	7.2	4 * 10.7 * 0.381

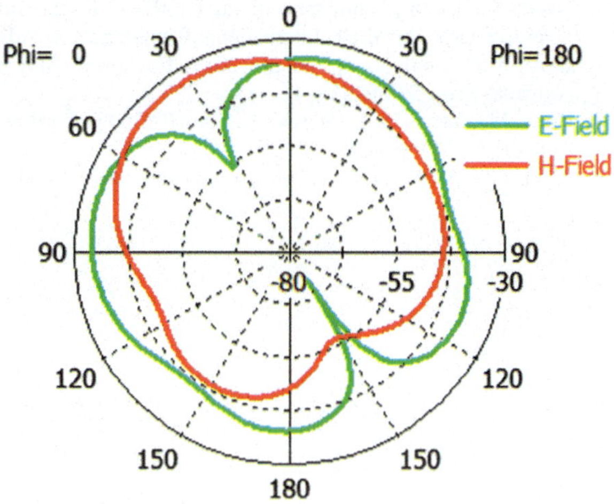

Fig. 10 E-plane and H-plane pattern at 60 GHz

5 Conclusion

This paper, invented substrate integrated waveguide back to back π-shaped slot antenna connected with 50 Ω microstrip feed for 57–64 GHz band applications, choosed Rogers RT/Duroid 5880 with dielectric substrate thickness 0.381 mm for implementing proposed structure and obtained 3.3 GHz bandwidth with respect to reflection coefficient (below −10 dB is the reference line) and VSWR and also observes the gain of proposed antenna is 5.79 dBi at 60 GHz. This antenna is suitable for broadband and high data rate applications like GIFI and automotive applications and covers the maximum range of 10 m.

References

1. Nanda Kumar M, Shanmuganantham T (2016) Substrate integrated waveguide cavity backed with U and V shaped slot antenna for 60 GHz applications. In: International conference on smart engineering materials
2. Nanda kumar M, Shanmuganantham T (2016) Substrate integrated waveguide cavity backed bowtie slot antenna for 60 GHz applications. In: IEEE international conference on emerging technology trends
3. Lockie D, Peck D (2009) High data rate Mmw radios. IEEE Microw Mage 10(5):75–83
4. Shrivastava P, Rama Rao T (2015) Performance investigations with ATLSA on 60 GHz radio link in a narrow hallway environment. Progr Electromagn Res 58:69–77
5. Ramesh S, Rama Rao T (2015) Planar high gain dielectric loaded exponentially TSA for millimeter wave wireless communications. Wirel Press Commun 84:3179–3192
6. Nanda Kumar M Shanmuganantham T (2016) Current and future challenges in SIW antennas —an overview. In: IEEE International Conference on Advanced Computing (Feb-2016)
7. Bozzi M, Georgiadis A, Wu K (2011) Review of SIW circuits and antennas. IET Microw Antennas Prop 5:909–920
8. Bozzi M, Perregrini L, Wu K, Arcioni P (2009) Current and future research trends in SIW technology. Radioengg 18(2):201–207
9. Zhang XC, Xu J, Yu ZY (2006) Novel band pass SIW filter based on CSRRS. PIER 72:39–46

Performance Improvement of Gain in Distributed Raman Amplifier Using Forward and Backward Pumps

S. Shameem and A. Sivasubramanian

Abstract Optical amplifiers have played a revolutionary role in fiber communication systems from early 1990s itself. An optical signal can be amplified within the optical domain without converting to electrical domain using optical amplifiers in an inexpensive way without electronic regenerators. The fiber itself acts as the gain medium for distributed Raman amplifier. The working principle of Raman fiber amplifier is called stimulated Raman scattering (SRS), which is a nonlinear effect and it requires an optical power higher than a threshold of at least 500 mW [1]. An input signal can be amplified while traveling through the fiber in the same direction or in opposite direction with a pump wavelength of appropriate power. By carefully adjusting the forward and backward pump power, the transmission performance can be improved. In this paper, the gain and SNR of the amplifier are studied for different pumping schemes and found that the transmission performance is improved by operating on low pump power ratio which does not require any changes in the existing system.

Keywords Distributed Raman amplifier · Raman pump power
EDFA · OptSim

1 Introduction

Research on Raman amplification for optical fiber communication systems started in early 1970s [1] itself, and its benefits were found by mid-1980. But the requirement of high-energy pump laser sources to get sufficient amplifier gain prevented the use of Raman Fiber Amplifiers from commercially accepting and use of lumped amplifiers like EDFA increased widely. No dopants are required and the transmission fiber itself turns to gain medium and give a distributed amplification.

S. Shameem (✉) · A. Sivasubramanian
School of Electronics Engineering, VIT University, Chennai, Tamil Nadu, India
e-mail: shameem.s2015@vit.ac.in

G. Gnanagurunathan et al. (eds.), *Optical and Microwave Technologies*,
Lecture Notes in Electrical Engineering 468,
https://doi.org/10.1007/978-981-10-7293-2_24

High-power pumps were invented in 1990s, and this fueled more advances in the use of Raman amplifiers commercially. Researchers found that distributed Raman fiber amplifier can provide better signal quality than EDFAs, [2] which enabled the use of higher order modulation formats. The system designer can select the desired frequency or bandwidth by selecting the appropriate pump power. After 1990s, lots of researches and studies were carried out in this topic, and by early 2000s, the commercial use of Raman amplifiers started in ultra-long and very long-haul optical communication systems. When compared with EDFA which are considered as lumped amplifiers, viewed from the transmission line point, distributed Raman amplifiers can retain the gain level of the signal over a long distance throughout the transmission line [3].

One of the main factors which decides the amplifier performance is noise figure. Raman amplifiers can be used to improve the noise figure and so it is widely used in very long-haul optical communication systems. Noise figure and its spectrum depend on noise sources such as Rayleigh Backscattering, Relative Intensity Noise, and ASE. Other parameters include length of the fiber, spectral density, gain, dispersion and attenuation characteristics of fiber as well as of signal and pump. Noise figure and noise factor are used to measure the degradation of the signal to noise ratio which is caused by the different components connected to the system as well as the channel through which the signal is transmitted. For amplifying the signal, a specific gain is required and to achieve high gain, in most cases, more pump power is required which results in high ASE and other noises which will again result in lower signal power of the amplifier.

In this paper, Raman amplifier using bidirectional pumping scheme is proposed. A flat gain spectrum along with improved noise figure is attained using this setup. The performance is analyzed with multiple backward pumps and single forward pump.

2 Raman Fiber Amplifier

Raman effect is the result of third-order nonlinearity. Raman scattering occurs when a pump photon with ω_p frequency scatters a photon and leads to transfer of energy and momentum by which the incoming photon frequency gets converted to a new frequency ω_s, where $\omega_s = \omega_p + \Omega$, where $\Omega \ll \omega_p$. If $\Omega < 0$, it is called Stokes wave and anti-stoke if $\Omega > 0$. Raman gain is the result of the Stokes shift corresponding to the difference in the pump and signal wave which is to be amplified. An input signal can be amplified while traveling through the fiber in the same direction or in opposite direction with a pump wavelength of appropriate power [4]. With high optical power, this nonlinear effect can transfer most of the pump power to the signal.

3 Forward and Backward Pumping for Flat Gain and Better Noise

For amplifying a signal using fiber amplifiers, there are different options for pumping. In forward pumping, the pump and the signal wave travel in same direction while in backward pumping, the signal and pump travel in opposite direction. If both forward and backward pumps are used simultaneously in the fiber amplifier, it is called bidirectional pumping.

In 2001, Emori.et al. experimentally demonstrated the bidirectional scheme that realized both flat gain and noise figure [5]. The direction of the pump will not affect the power conversion efficiency of the amplifier, and the overall gain of the amplifier depends extremely on the average excitation density along the length of the fiber [6].

The pump power required to achieve a particular Raman gain is dependent on components like Raman gain coefficient, polarization, length of the fiber, fiber losses due to pump depletion, and wavelength [7]. Care must be taken to use forward pump of low relative intensity noise (RIN) to avoid degradation of transmission performance of data due to the high power pumps [8]. Advantages of using higher order pumps have been studied by [9, 10]. Raman amplifier with forward pump is found to have better noise performance when compared to the backward pumped amplifier because the pump wave is applied prior to the transmission loss. Also, the backward pumping helps to attain higher gain.

4 Results

In this paper, distributed multi pump Raman amplifier is analyzed. The setup consists of 40 Gbs DWDM link with an inter-channel spacing of 50 GHz. Six CW laser pumps with wavelengths 1425, 1420, 1435, 1450, 1465, and 1495 nm are used as pump wavelengths. An ideal optical mux is used to give the pump wave to the amplifier. The bandwidth considered is for a 70 nm, 70 km distributed Raman fiber amplifier. The forward and backward gain and forward and backward SNR have been investigated for the above mentioned different pump wavelengths.

The SNR for low pump values is higher when compared to high pump values. Thus, we got a pump wavelength depended gain. In other words, the Raman amplifier gain is dependent on the input wavelength of the pump.

In Fig. 1, the basic schematic diagram is shown in which the distributed fiber is provided with multiple pumping, one forward and five backward pumps which are considered to be the best pumping method [6]. The system setup done in OptSim simulation is shown in Fig. 2. Figure 3 shows the signal power plot in which we can see that the backward noise is zero. Spectral density plot for the different pump wavelength is shown in Fig. 4.

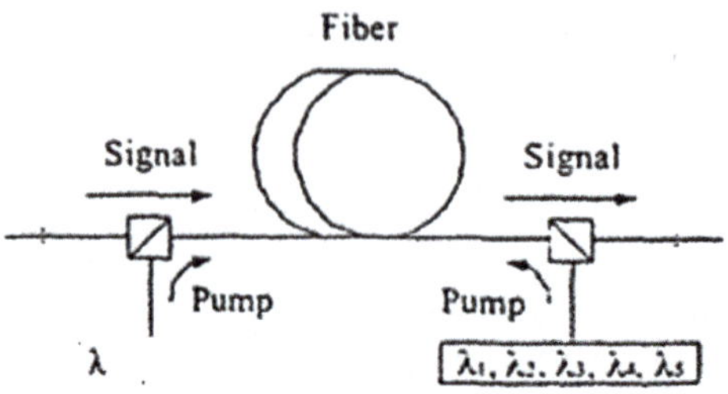

Fig. 1 Schematic block diagram of the proposed model [4]

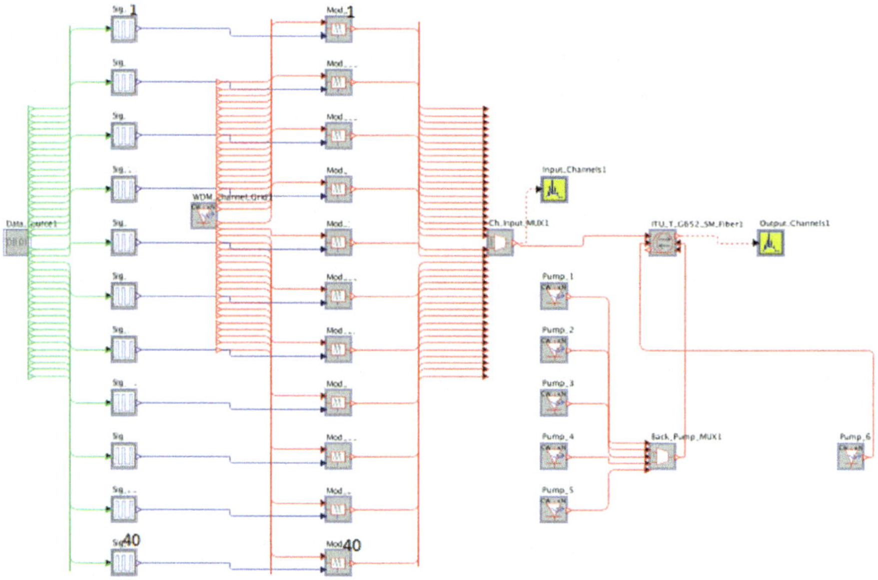

Fig. 2 Block diagram for multi pump distributed Raman amplifier

Forward gain, backward gain, and SNR for the different pump values are shown in the tabular column. Compared to the other pumping methods, the overall SNR is high for low pump wavelengths than higher wavelengths considered for the setup. The setup presented here shows that it is possible to improve the performance by adopting the bidirectional pumping than to use the conventional backward only pumping to increase the explained with reference to different pump wavelengths (Table 1).

Figure 4 Illustrates about the spectrum of the various pump wavelengths given to the amplifier session.

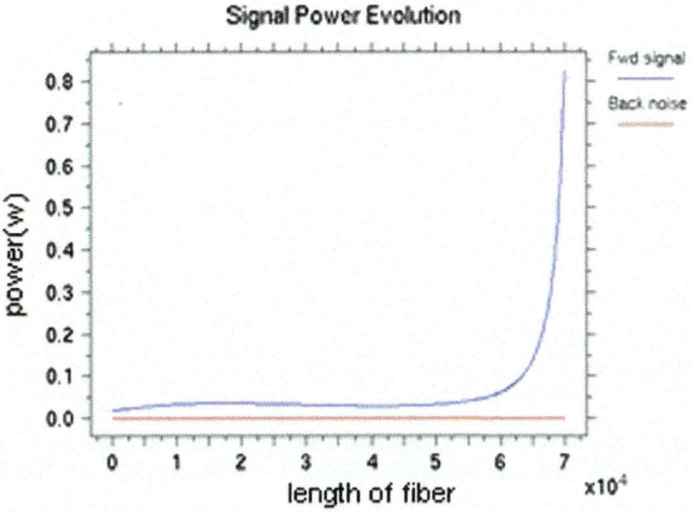

Fig. 3 Signal power for forward and backward pumps

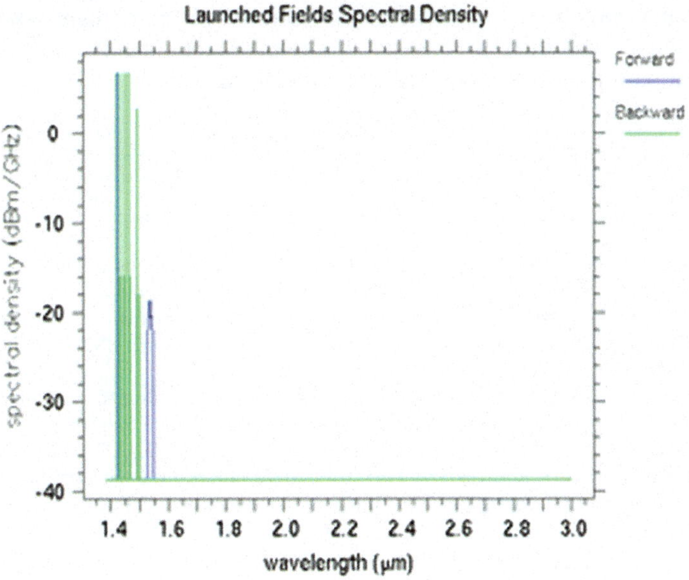

Fig. 4 Spectral density

Table 1 Forward and backward gain and SNR for different pump wavelengths

Pump wavelength (nm)	Backward gain (dB)	Forward gain (dB)	Forward SNR (dB)	Backward SNR (dB)
1420	−53	−50	32	90
1425	−44	–	–	40
1435	−45	–	–	45
1450	−32	–	–	50
1465	−22	–	–	52
1495	−35	20	44	60

5 Conclusion

In this paper, we have investigated and explained various aspects of improving SNR and gain for different pumping values for a Raman fiber amplifier. It was found that bidirectional pumping offers better noise performance when compared to forward only and backward only pumping schemes. Our study needs more extensive review of the methods to improve the noise figure which includes higher order pumping and even clad pumped Raman amplifiers. In this paper, we found that Raman gain is produced where there is an optical fiber in the light wave communication system, and thus, we can say that Raman amplification is flexible. Thus, the system performance is increased at a lower cost.

References

1. Bromage J (2004) Raman amplification for fiber communications systems. J Light wave Technol 22(1):79–93
2. Kaur R, Kaur K Analysis and investigation of noise figure of fiber raman amplifier. Int J Electron Comput Sci Eng. Available Online at www.ijecse.org ISSN-2277-1956, ISSN 2277-1956/V1N3-1743-1747
3. Namiki S, Emori Y (2001) Ultra broad-band Raman amplifiers pumped and gain-equalized by wavelength-division-multiplexed high-power laser diodes. IEEE J Sel Top Quantum Electron 7(1):3–16
4. Mustafa FM, Khalaf AA, Elgeldawy FA (2013) Multi-pumped Raman amplifier for long-haul UW—WDM optical communication systems: gain flatness and bandwidth enhancements, ICACT 2013
5. Kado S, Emori Y, Namiki S, Tsukiji K, Yoshida J, Kimura T (2001) Broadband flat-noise Raman amplifier using low-noise bi-directionally pumping sources. In ECOC'01. 27th European conference on optical communication, vol 6. IEEE
6. Iqbal MA, Martella GR, Gallazzi F, Tan M, Harper P, Ania-Castañón JD (2016) Performance improvement of broadband distributed raman amplifier using bidirectional pumping with first and dual order forward pumps. In ICTON 2016
7. Isoe GM, Rotich EK, Boiyo DK, Gamatham RRG, Leitch AWR, Gibbon TB, Muguro KM, Waswa DW, (2015) Noise figure and pump reflection power in SMF-reach optical fibre for Raman amplification, In: IEEE AFRICON 2015

8. Tan M, Rosa P, Le S, Phillips I, Harper P (2015) Evaluation of 100G DP-QPSK long-haul transmission performance using second order co-pumped Raman laser based amplification. Opt Express 23:22181–22189
9. Ania-Castañón JD, Pustovskikh AA, Kobtsev SM, Turitsyn SK (2007) Simple design method for gain-flattened three-pump Raman amplifiers. Opt Quantum Electron 39(3):213
10. Tan M, Rosa P, Phillips ID, Harper P (2014) Long-haul transmission performance evaluation of ultra-long Raman fiber laser based amplification influenced by second order co-pumping. In Proceedings of the Asia communications and photonics conference, 2014, paper ATh1E

A Comparative Study on Asymmetric, Triangular, and Rectangular Core Large-Mode-Area PCF Designs

Netra Dalvi, Radhika Ramesh and Pravin Joshi

Abstract The main limiting factor of a high-power and high-bit-rate device performance is the onset of nonlinear effects which includes stimulated Raman scattering (SRS), stimulated Brillouin scattering (SBS), four-wave mixing (FWM), and self-phase modulation (SPM). The solution to these limiting factors is large mode area (LMA) fiber. In this paper, asymmetric, triangular, and rectangular core LMA designs are compared and analyzed in terms of confinement loss, effective mode area, dispersion, and mode field diameter (MFD). The structural parameter and mode details were analyzed using the commercially available RSoft full vectorial finite element method (FemSIM™) from Synopsys.

Keywords Nonlinear effects · LMA · Fiber modes · Effective mode area
Confinement loss · Mode field diameter

1 Introduction

Recently, photonic crystal fiber (PCF) has gained significant attention due to their wide range of optical properties like high nonlinearity, endlessly large-mode-area single-mode (SM) operation [1], tailorable chromatic dispersion, and high birefringence [2]. PCF is proven technology which is competing with conventional fiber in many applications and has ability to vary the size and position of the air holes. Recently, high-bit-rate and high-power devices are gaining importance. However, in high-power devices, the main limitation in the performance of PCF is because of nonlinear effect such as stimulated Raman scattering (SRS), stimulated

N. Dalvi (✉) · R. Ramesh · P. Joshi
Fiber Optic Services (FOS), Mumbai, India
e-mail: netra.dalvi@foservice.com

R. Ramesh
e-mail: radhika.ramesh@foservice.com

P. Joshi
e-mail: phj@foservice.com

G. Gnanagurunathan et al. (eds.), *Optical and Microwave Technologies*,
Lecture Notes in Electrical Engineering 468,
https://doi.org/10.1007/978-981-10-7293-2_25

Brillouin scattering (SBS), four-wave mixing (FWM), and self-phase modulation (SPM) [3]. In high-bit-rate transmission systems, there is a change in the dispersive behavior due to SPM; SRS and SBS increase the noise to signal ratio, also FWM and SRS increase the cross talk between different WDM systems. These nonlinear effects can be overcome by using large-mode-area (LMA) fiber. Large-mode-area fiber support high-power transmission from lasers and amplifiers. LMA fiber has potential applications such as high-power delivery, multi-wavelength guidance, short-pulse delivery, and laser pig tailing.

In LMA fibers, due to increase in numerical aperture, modal dispersion is introduced. Hence, there is a trade-off between LMA and SM operation. Single-mode operation of LMA fiber can be obtained by (i) scaling down the numerical aperture of conventional step-index fiber [4] and (ii) by changing the geometry of cladding or index profiles to differentiate higher-order modes [5]. A graded index cladding design with a radial increase of refractive index profile, coaxial dual cores, and leaky cladding has been proposed by Dussardier et al. [6] for LMA SM operation. However, the extended SM operation is impossible in conventional LMA fiber; a few higher-order modes are always supported.

Single-mode operation in LMA PCFs is achieved by considering air holes in cladding region. Also, the confinement loss can be reduced by doping the center core of silica PCF. A novel technique has been applied to improve the properties in LMA PCF by employing a different level of doping concentration around the core region [7]. Novel LMA PCF structures have been designed based on selective material-filled technology. More recently, photonic bandgap fiber with SM operation and multiple cladding regions has been designed with a very large-mode-area of $\sim 2650\ \mu m^2$ [8].

In this paper, asymmetric, triangular, and rectangular core LMA designs are compared [9–11] and analyzed in terms of confinement loss, effective mode area, dispersion, and MFD. The effective indices and confinement loss of the first three modes—FM, FHOM, and SHOM—are analyzed at a wavelength of 1.064 μm. The structural parameter and mode details were analyzed using the commercially available RSoft full vectorial finite element method (FemSIM™) from Synopsys, a widely used tool for designing and analyzing complex fiber structure.

2 Structural Parameter, Simulation Method, Results, and Discussion

2.1 Structural Parameter

The cross-sectional view of LMA PCF designs is shown in Figs. 1, 2, and 3 [9–11]. In the structures, fluorine-doped rods and air holes are placed in the silica substrate. The refractive index of silica substrate is taken as 1.45 at wavelength 1.064 μm. This design is used to deliver high laser power from Nd:YAG or fiber laser.

A perfectly matched layer is employed as a boundary condition and is added to the edge of the simulation domain.

In Fig. 1, PCF design is characterized by the rectangular array of air holes drawn in the silica along the length of the fiber. One air hole has been removed from the center to make a core. There are four rectangular rings of air holes around the central core. In the second ring of the structure, the air holes are replaced with fluorine rods. The rods in the second ring have diameter D1 (2 × R1, where R1 is the radius of the larger rods in the second ring). Air holes diameter in the first, third, and fourth rings are taken as D2 (2 × R2, where R2 is the radius of the air holes in the third and fourth rings). The distance between the centers of two adjoining air holes, i.e., period, has been taken as a constant value Λ. In Fig. 2, the design is characterized by the triangular core in the center, created by removing air holes in the center region. Core is surrounded by triangular ring of fluorine rod with a diameter D0 (7.5 μm). There are four rings around the central core. The first triangular ring has fluorine rods, and other three rings have air holes of diameter D1 (8 μm). Λ is the hole pitch. In Fig. 3, the design is characterized by the asymmetric array of air holes drawn in the silica. The first hexagonal ring contains fluorine rod,

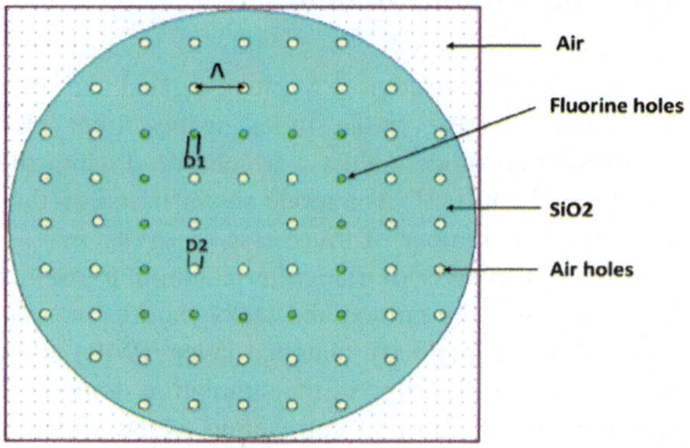

Fig. 1 Rectangular core LMA PCF design

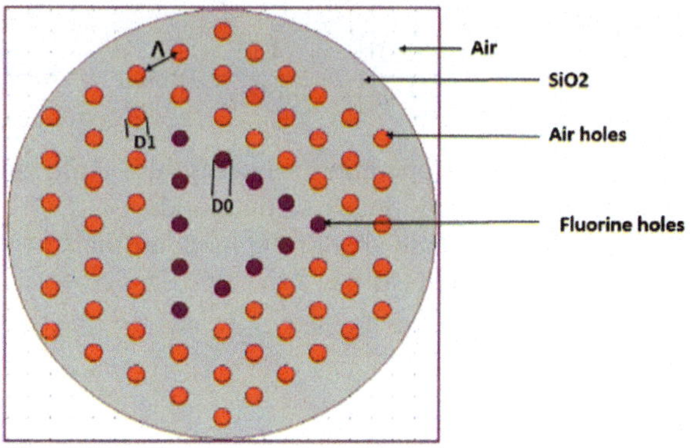

Fig. 2 Triangle core LMA PCF design

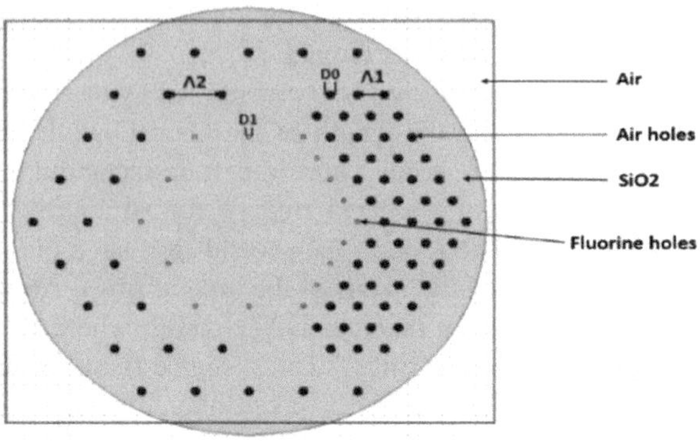

Fig. 3 Asymmetric core LMA PCF design

diameter D2 and other four rings contain air holes, diameter D1. In this structure, one side of air holes is closely placed to each other having period 12 μm and other side has a period of 24 μm.

2.2 *Simulation Method*

In the proposed structure, we investigated the effective mode area, confinement loss, and effective mode index using finite element method. The structural parameter and mode details were analyzed using the RSoft FemSIM™ Tool from Synopsys. FemSIM™ is a mode solver based on the finite element method that can calculate any number of transverse or cavity modes of an arbitrary structure [12]. A nonuniform hybrid triangular/rectangular mesh is used to solve for all field components of the guided and leaky modes. In order to evaluate confinement loss of the mode, a perfectly matched layer (PML) is employed. PML has a thickness and is added to the edge of the simulation domain. The real and imaginary parts of the propagation constant correspond to the effective index of the mode and confinement loss of the propagating mode, respectively. The confinement loss can be calculated using the following relation [13].

$$L\left[\frac{\mathrm{d}B}{m}\right] = \frac{40\pi}{\ln(10)\lambda}\mathrm{Im}(n_{\mathrm{eff}}) = 8.686k_o\mathrm{Im}(n_{\mathrm{eff}})$$

where λ represents operating wavelength of light, and $\mathrm{Im}(n_{\mathrm{eff}})$ is the imaginary part of the effective mode index of that mode

The effective mode area of the mode of the LMA PCF was calculated using the following equation [13]

$$A_{\mathrm{eff}} = \frac{\left(\int\int E^2 \mathrm{d}_x \mathrm{d}_y\right)^2}{\int E^4 \mathrm{d}_x \mathrm{d}_y}$$

where E represents amplitude of the transverse electric field. The group-velocity dispersion parameter D is [14]:

$$D = -\frac{2\pi c}{\lambda^2}\frac{\mathrm{d}^2\beta}{\mathrm{d}w^2} = \frac{2\pi c}{v_{\mathrm{g}}^2\lambda^2}\frac{\mathrm{d}v_{\mathrm{g}}}{\mathrm{d}w}$$

where $\lambda = \frac{2\pi c}{w}$ in vacuum and $v_{\mathrm{g}} = \frac{\mathrm{d}w}{\mathrm{d}\beta}$ the group velocity.

The effective mode field diameter (eff. MFD) is [12]:

$$\mathrm{d}_{\mathrm{eff}} = \frac{2\sqrt{2}\int E_i^2 r\mathrm{d}r}{\left[\int E_i^4 r\mathrm{d}r\right]^{\frac{1}{2}}}$$

$$\mathrm{d}_{\mathrm{eff}} = \frac{2}{\sqrt{\pi}}\sqrt{A_{\mathrm{eff}}}$$

where $E(r)$ represents mode field distribution.

2.3 *Result and Discussion*

The effective mode index, confinement loss, and effective mode area were investigated using full vectorial finite element method. Figure 4 shows the mode field distribution of three different structures under consideration at a wavelength of 1.064 μm.

The refractive index of fused silica has been taken as 1.45. The confinement losses of the modes increase on increasing mode order. To confirm this, we have considered the first three modes, i.e., fundamental mode (FM), the first higher-order mode (FHOM), and the second higher-order mode (SHOM).

Table 1 shows the variation of confinement loss and effective index of FM, FHOM, and SHOM modes of the proposed PCF structures.

The confinement loss of the higher-order modes in all three structures [9–11] was observed to be greater than FM. This has led to increase in confinement loss of higher modes enabling SM operation. The three structures were analyzed in terms of dispersion, MFD, and effective mode area of the fundamental mode at wavelength of 1.064 μm as shown in Table 2.

Effective mode area is inversely proportional to nonlinearity. Asymmetric structure has a larger effective mode area ~5307.54 μm^2 compared to other two structures. The rectangular structure showed a negative dispersion of −26 ps/km-nm hence can find application as a dispersion compensator. At 1.064 μm wavelength, MFD value of 82.1891 μm is recorded for the asymmetric structure.

Fig. 4 Mode field distribution, **a** rectangle, **b** triangle, and **c** asymmetric

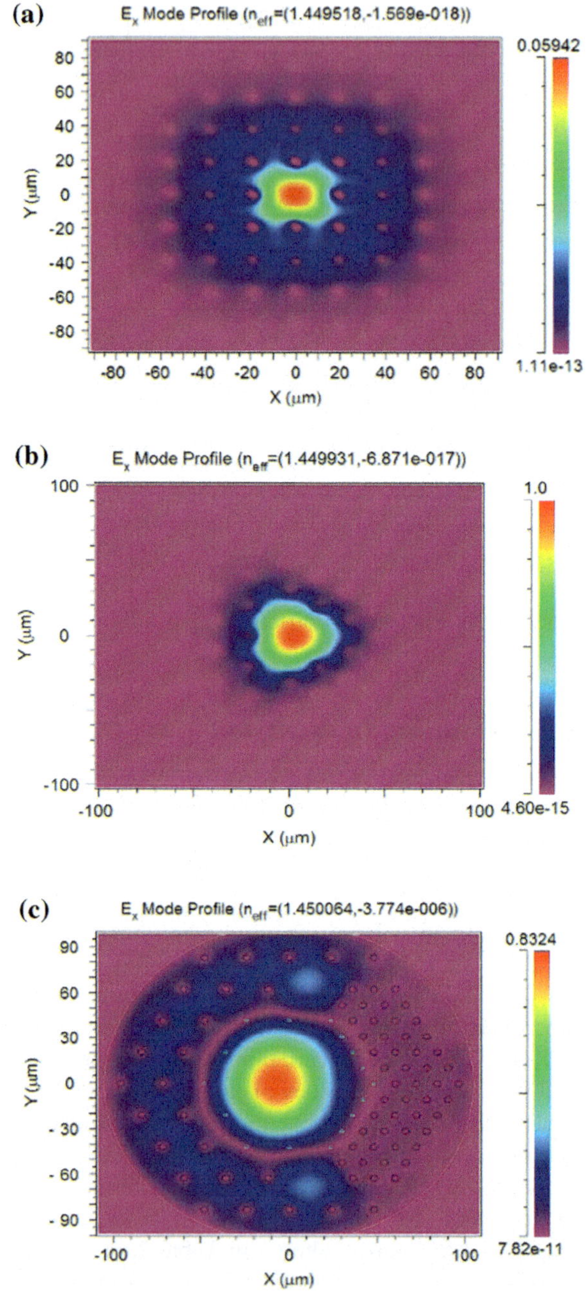

Table 1 Effective index and confinement loss of three modes, i.e., FM, FHOM, and SHOM @ 1.064 μm

Structures	Effective index		
	FM	FHOM	SHOM
Rectangle	1.4495420460	1.4495420460	1.4495214220
Triangle	1.4499309060	1.4499309060	1.4498546120
Asymmetric	1.4499825240	1.4499825240	1.4499613050
Structures	Confinement loss (dB/cm)		
	FM	FHOM	SHOM
Rectangle	$8.2 * 10^{-14}$	$3.41 * 10^{-13}$	$1.26 * 10^{-12}$
Triangle	$7.9 * 10^{-14}$	$1.72 * 10^{-13}$	$1.28 * 10^{-13}$
Asymmetric	$4.2 * 10^{-01}$	$3.3 * 10^{-02}$	$3.3 * 10^{-02}$

Table 2 Dispersion, EMA, and MFD of FM @ 1.064 μm

Structures	Dispersion (ps/km-nm)	Effective mode area (μm^2)	Mode field diameter (μm)
Rectangle	−26	1878.82	48.90
Triangle	−0.1	1393.13	42.10789
Asymmetric	0	5307.54	82.1891

3 Conclusion

We have compared three fluorine-doped large-mode-area PCF structures—rectangular [9], triangular [10], and asymmetric [11] core. The large effective mode area will help to mitigate the nonlinear effect at high-power and high-bit-rate applications. In all three structures [9–11], creation of a center defect and replacing the air holes by fluorine rod on a silica substrate has resulted in leaky modes. This has increased the confinement loss of higher-order modes enabling a single-mode operation. However, strangely the asymmetric LMA PCF design shows a higher loss in fundamental mode, which needs to be investigated later on. The asymmetric PCF design offers an effective mode area larger compared to the rectangular and triangular core geometries.

References

1. Kishore K, Sinha RK, Varshney AD (2012) Experimental verification of improved effective index method for endlessly single mode photonic crystal fiber. Opt Lasers Eng 50:182–186
2. Jiang G, Fu Y, Huang Y (2015) High birefringence rectangular-hole photonic crystal fiber. Opt Fiber Technol 26:163–171

3. Ferreira MFS (2008) Nonlinear effects in optical fibers: limitations and benefits. Proc SPIE 6793:679302
4. Broderick NGR, Offerhause HL, Richardson DJ, Sammut RA, Caplen JE, Dong L (1999) Large mode area fibers for high power applications. Opt Fiber Technol 5:185–196
5. Rastogi V, Chiang KS (2003) Leaky optical fiber for large mode area single mode operation. Electron Lett 39:1110–1112
6. Dussardier B, Rastogi V, Kumar A, Monnom G (2011) Large mode—area leaky optical fiber fabricated by MCVD. Appl Opt 50:3118–3122
7. Ademgil H, Haxha S (2011) Bending insensitive large mode area photonic crystal fiber. Optik 122:1950–1956
8. Gu G, Kong F, Hawkins TW, Jones M, Dong L (2015) Extending mode areas of single-mode all-solid photonic bandgap fibers. Opt Express 23:9147–9156
9. Reena, Saini TS, Kumar A, Kalra Y, Sinha RK (2016) Rectangular-core large-mode-area photonic crystal fiber for high power applications: design and analysis. Appl Opt
10. Wang X, Lou S, Lu W (2013) Bending orientation insensitive large mode area photonic crystal fiber with triangular core. IEEE Photonics J 5(4)
11. Napierała M, Nasilowski T, Bereś-Pawlik E, Mergo P, Berghmans F, Thienpont H (2011) Large-mode-area photonic crystal fiber with double lattice constant structure and low bending loss. Opt Express 19(23):22628–22636
12. Synopsys RSoft Component Device tool manual
13. Haxha S, Ademgil H (2008) Novel design of photonic crystal fibres with low confinement losses, nearly zero ultra-flatted chromatic dispersion, negative chromatic dispersion and improved effective mode area. Opt Commun 281:278–286
14. https://en.wikipedia.org/wiki/Dispersion_(optics)#Group_and_phase_velocity

Miniaturized High-Gain UWB Monopole Antenna with Dual Band Rejection Using CSRR

Ravi Prakash Dwivedi and K. Usha Kiran

Abstract Ultra-wideband technology (UWB) is used for transmitting large amount of data over wide frequency range. There is always a trade-off between gain and bandwidth of micro-strip patch antenna for UWB. It is difficult to obtain wideband and high gain in an antenna simultaneously. In the proposed design, a compact, high-gain ultra-wideband tree-shaped radiating patch antenna with band notch characteristics is designed. The radiating patch is printed on FR4 substrate with permittivity 4.4. Tree-shaped radiating patch and defected ground structure (DGS) provides a wideband over 2.8–11.6 GHz frequency range. For providing stop-band notches in the existing wireless technologies, two complementary split-ring resonators (CSRRs) are used. By etching square complementary split-ring resonator (CSRR) in the feed, a stop-band notch at WLAN (5.1–5.9 GHz) frequency range is obtained, and a single CSRR in the patch provides a notch at WiMAX (3.3–3.6 GHz) frequency range. The antenna has varied positive gain over all the frequencies other than notched bands. Maximum gain of 5 dB at 9 GHz and minimum of 0.4 dB at 4 GHz is obtained. Antenna is simulated using high frequency structure simulator (HFSS) software. The designed antenna is validated by comparing simulated results with measured results.

Keywords CSRR · DGS · Dual band · Stop-band notches · UWB

1 Introduction

Communication system demands low cost, less weight, compactable antenna which can be operated over wide range of frequencies. This objective can be achieved by using ultra-wideband (UWB) technologies. UWB technology has attractive char-

---

R. P. Dwivedi (✉) · K. Usha Kiran
SENSE, VIT University, Chennai Campus, Chennai 600127, Tamil Nadu, India
e-mail: raviprakash.dwivedi@vit.ac.in

K. Usha Kiran
e-mail: usha.kiran@vit.ac.in

G. Gnanagurunathan et al. (eds.), *Optical and Microwave Technologies*,
Lecture Notes in Electrical Engineering 468,
https://doi.org/10.1007/978-981-10-7293-2_26

acteristics such as high data rate of more than 100 Mbps, low power and multiple access communication. The frequency band allocated by Federal Communications Commission (FCC) for UWB is from 3.1 to 10.6 GHz. Since it has wider bandwidth, it may cause interferences with coexisting users such as WiMAX (3.4–3.7 GHz) and WLAN (5.1–5.9 GHz). This problem can be solved by providing band notches on UWB in these frequency ranges. The UWB can be obtained by using several techniques such as providing steps in the radiating patch with DGS [1, 2], parallel dual patches along with modified ground structure [3]. By connecting parallel identical monopoles on the substrate using feed line [4], bevel-shaped rectangular radiating patch, and trident-shaped feeding structure [5].

The notches can be provided in UWB in many ways. Electronic band gap (EGB) slots in the patch, slits in the ground plane, SRR, CSRR are some methods which can be adopted. EBG can provide only one stop-band notch. In another technique, suitable shapes are etched on the radiating patch to provide both single and dual band-notches [1]. Inserting slit in either radiating patch or ground plane can provide band notch. The length of the embedded slit must be quarter wavelength of notch frequency [3]. A strip bar which is placed in between two monopoles and connected to the feed line can cause notch in WLAN frequency [4]. A parasitic strip can be inserted to the antenna for achieving frequency band-notch [5]. Nowadays CSRRs have come up drastically because of their attractive characteristics. CSRRs are etched on the radiating patch to provide band notches. By etching two square CSRRs with proper dimensions in the mushroom-shaped radiating patch, band notches at WLAN, and WiMAX can be obtained [2]. CSRRs can be of square or ring. Etching a ring-shaped CSRR on the slanted feed line helps to achieve a band notch [6], CSRR can also be placed in the ground plane to provide notches [7, 8].

In this, an UWB antenna integrated with CSRRs is designed. Here, UWB is obtained by using tree-shaped radiating patch and by providing defects in ground. DGS not only provides wide frequency spectrum but also reduces antenna size [9, 10]. Two CSRRs are etched to provide band notches. The validation of proposed antenna is done using simulated and fabricated results.

2 Antenna Structure Without Dual Band-Notches

2.1 Simple Rectangular Patch Antenna

The geometry of simple rectangular patch antenna is given in Fig. 1 Rectangular patch is of dimension 20 mm × 27 mm. Antenna is printed on FR4 substrate with dielectric permittivity 4.4 and thickness 0.8 mm. Some portions of ground are etched. This defect helps to enhance impedance bandwidth. Defects in ground plane cause a capacitive and inductive effect this results additional resonating frequencies.

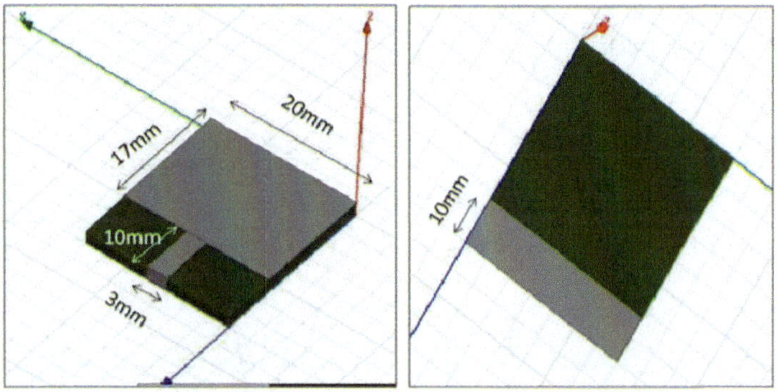

Fig. 1 Geometry of rectangular patch antenna

2.2 Antenna with Step-Shaped Patch

To improve the impedance bandwidth, step-shaped radiating patch can be used. Each step provides a resonant frequency. DGS and these step-shaped structures enhance the bandwidth. Resonating frequencies obtained are unified to obtain wider band. Figure 2 shows the top and bottom view of antenna with step-shaped radiation patch. Ground plane is of truncated shape. This truncation helps in impedance matching (Table 1).

Figure 3 shows the S_{11} characteristics of antenna with rectangular radiating patch and step-shaped radiating patch. Here, a bandwidth of 2.2 GHz (3.8–6 GHz) is obtained for simple rectangular patch antenna, and a bandwidth of 8.7 GHz from (4.2–12.9 GHz) is obtained for step-shaped patch antenna.

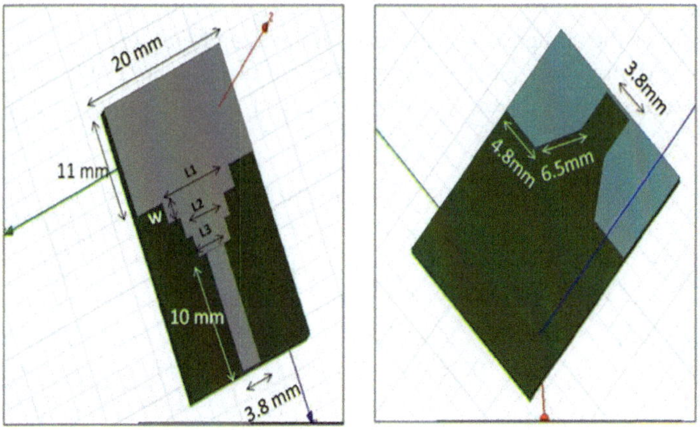

Fig. 2 Antenna with step-shaped radiating patches

Table 1 Dimensions of step-shaped radiating patch

$L1$ (mm)	$L2$ (mm)	$L3$ (mm)	W (mm)
11.5	7.5	5.5	2

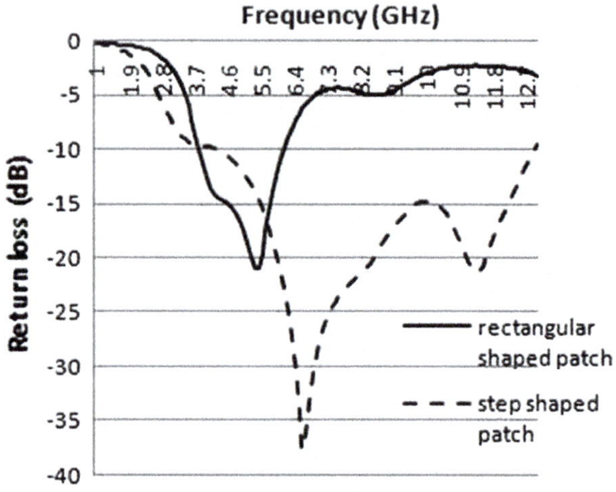

Fig. 3 S_{11} of rectangular and step-shaped patch

2.3 *Antenna with Tree-Shaped Patch*

Figure 4 shows the geometry of antenna with tree-shaped radiating patch and the photograph of fabricated antenna. The sudden discontinuities in Fig. 2 are made into smoothen and enhance the current distribution.

2.3.1 Comparison of Simulated and Measured (S_{11}) Characteristics of UWB Antenna

Figure 5 shows the comparison between simulated and measured S_{11} characteristics of tree-shaped radiating patch antenna. By simulation, an UWB of 8.9 GHz is obtained from 3.1 to 12 GHz. An enhancement of 1.1 GHz in lower frequency can

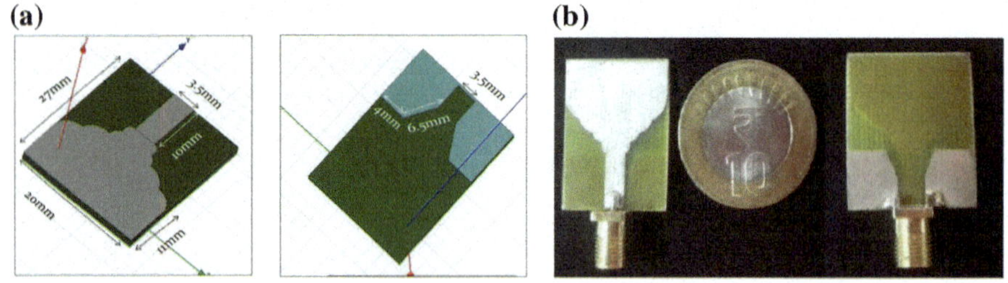

Fig. 4 Antenna with tree-shaped patch, a geometry, b photograph of fabricated antenna

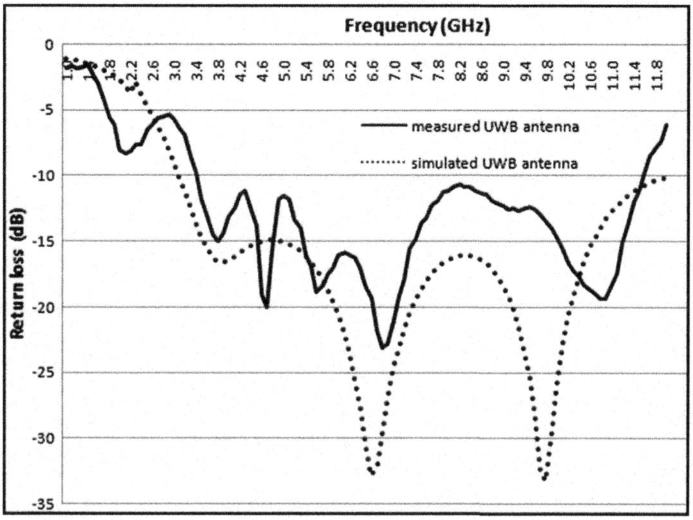

Fig. 5 Simulated and measured results of S_{11}

be seen when compared to antenna with sharp discontinuity. While measuring the fabricated antenna, an UWB from 3.4 to 11.5 GHz is obtained with bandwidth of 8.1 GHz.

2.3.2 Analysis Between Simulated and Measured VSWR of UWB Antenna

Figure 6 shows the comparison between measured and simulated VSWR of UWB antenna with tree-shaped radiating patch. Here, all the pass bands have a value less than 2, thus antenna radiates with less return loss.

3 Antenna Structures with Band Notches

UWB technology causes interference to coexisting users such as WLAN (5.1–5.9 GHz) and WiMAX (3.3–3.6 GHz). Thus, it is necessary for UWB antenna to perform band notched operations to avoid interferences. Complementary split-ring resonators are used for these. CSRRs are complementary to SRRs. These are designed by using babinet principle to split ring.

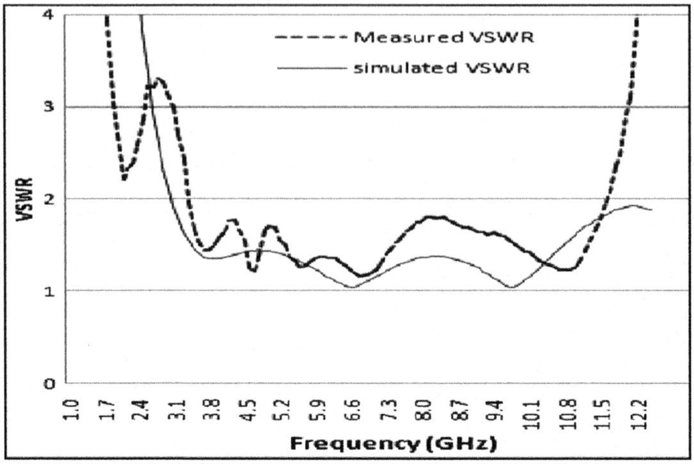

Fig. 6 Simulated and measured VSWR

3.1 Antenna with Dual Band-Notches

By etching square CSRR in the feed, a stop-band notch at WLAN frequency range is obtained, and a single CSRR in the patch provides a notch at WiMAX frequency range. The dimensions are taken as shown in Table 2 and Figs. 7, 8.

The mechanism of stop-band notches can be realized by using Fig. 9. At 3.5 GHz, surface currents are mainly concentrated on the square split ring etched in the patch, whereas at 5.5 GHz surface currents are distributed on the square split rings on the feed (Table 3).

Figure 10 shows the simulated and measured VSWR plots. Good agreement between measured and simulated results can be seen. All the pass bands have VSWR value less than 2 and greater than 2 in the vicinity of notched bands. Figure 11 shows the gain plot of UWB antenna and dual band-notched antenna. A maximum gain of 5.5 dB at 10.6 GHz and minimum gain of 2.2 dB at 4 GHz is obtained for UWB antenna. In the case dual band-notched antenna, maximum of gain of 5 dB at 9 GHz and minimum of 0.4 dB at 4 GHz is obtained. A variable positive gain is obtained for the pass bands and negative gain in the vicinity of notched bands, thus antenna achieves higher gain which is higher than that in the antenna presented in [1, 2].

Table 2 Dimensions of dual band-notch antenna (all dimensions are in mm)

A	*B*	*S*	*L*1	*W*1	*S*1	*L*2	*W*2	*S*2	*W*
7.5	9.5	2	7	3	0.5	4.5	1.25	1.5	0.5

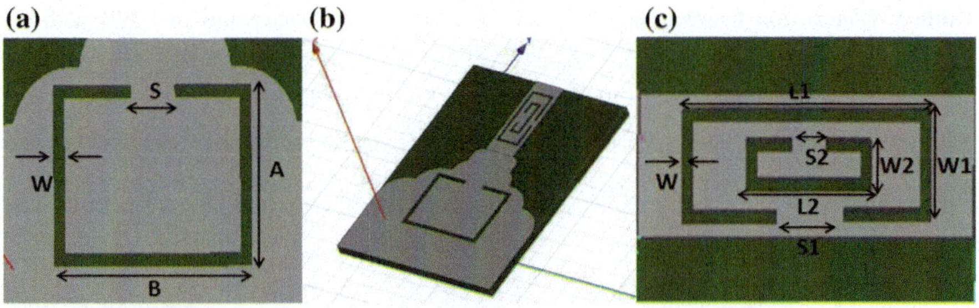

Fig. 7 **a** CSRR on patch, **b** top view, **c** CSRR on feed

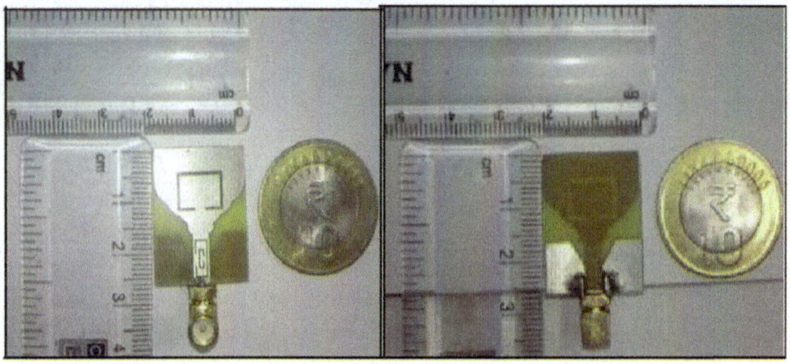

Fig. 8 Geometry of slotted section

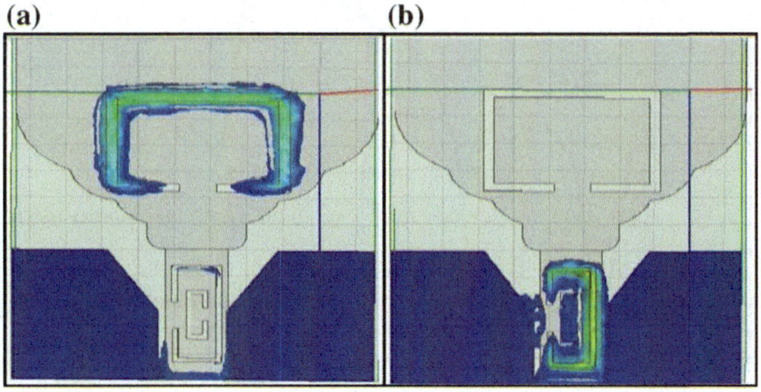

Fig. 9 **a** 3.5 GHz, **b** 5.5 GHz

Table 3 Comparison between simulated and measured S_{11} characteristics of UWB with dual notches

	Start freq (GHz)	Stop freq (GHz)	WiMAX notch (GHz)	WLAN notch (GHz)
Simulated	2.7	11.6	3.29–3.68	5.1–6.1
Measured	2.4	11.3	3.23–3.68	5.1–6.2

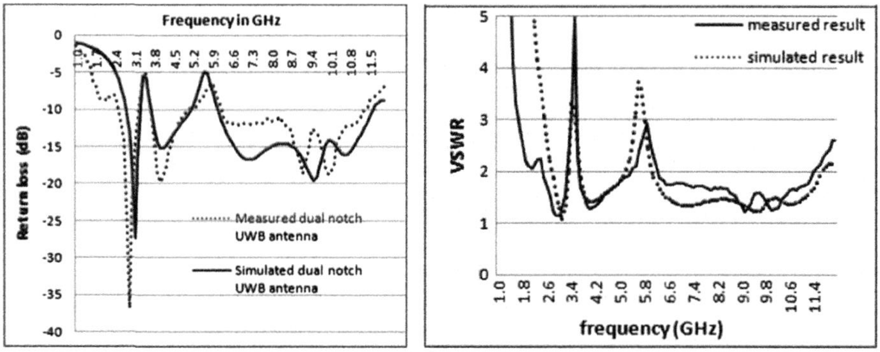

Fig. 10 Simulated and measured S_{11} and VSWR results of UWB antenna with dual band-notch

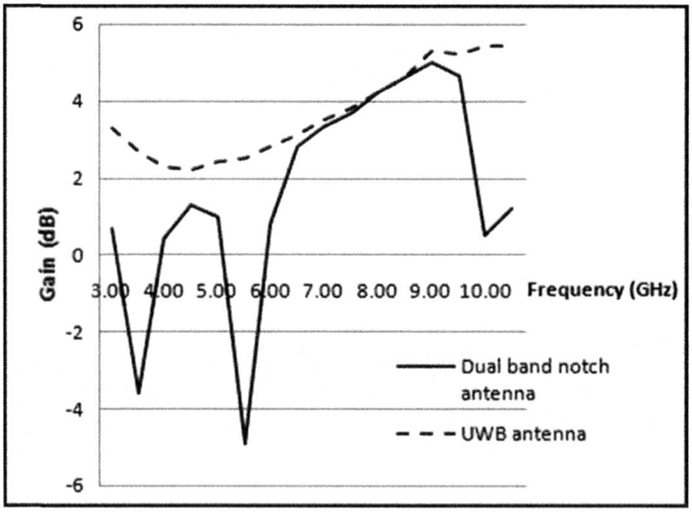

Fig. 11 Gain plot of antenna with dual band-notch

Table 4 Comparison

	Antenna		
	Reference antenna [1]	Reference antenna [2]	Proposed antenna
Size (mm^3)	14 × 16 × 1	30 × 34 × 1	20 × 27 × 0.8
Bandwidth (GHz)	2.8–19.2	3–11	2.8–11.6
Gain (dB)	<3	<4	<5.5
WiMAX (GHz)	3.4–3.6	3.4–3.6	3.29–3.68
WLAN (GHz)	5.1–5.6	5.1–5.9	5.1–6.1

4 Results

The comparison of proposed antenna with [1, 2] is given in Table 4. Proposed antenna has an area of 20 mm × 27 mm × 0.8 mm which is less than that of antenna presented in [2]. Highest gain obtained for the antenna is 4.78 dB which is higher than the presented antennas [1, 2].

5 Conclusion

In the case of simple rectangular patch, sharp and sudden discontinuity in the connection point between the micro-strip feed line, and the patch causes lowering of bandwidth. Above problem can be mitigated by using step-shaped radiating patch and a modified ground structure, which acts as an impedance matching element. Thus, desired ultra-wideband is obtained. By using tree-shaped patch, bandwidth and gain of the antenna are increased. An UWB ranging from 3.1 to 12 GHz is obtained with variable gain less than 6 dB; UWB antenna causes interferences to existing wireless technologies. So, stop-band notches are provided in these frequency ranges. Complementary split-ring resonators can be used for providing band notches at WLAN (3.3–3.6 GHz) and WiMAX (5.1–5.9 GHz) frequencies. Designed antenna is fabricated and measured using VNA. Measured results show a good agreement with simulated results with an acceptable variance in frequency.

References

1. Karimian R, Oraizi H, Fakhte S (2014) Design of a compact ultra-wide-band monopole antenna with band rejection characteristics. IET Microw Antennas Propag 8(8):604–610
2. Lai H-Y, Lei Z-Y, Xie Y-J, Ning G-L, Yang K (2011) UWB antenna with dual band rejection for WLAN/WiMAX bands using CSRRs. Prog Electromag Res Lett 26:69–78
3. Jung J, Lee H, Lim Y (2008) Compact band-notched ultra-wideband Antenna. Electron Lett 44(6)

4. Chung K, Hong S, Choi J (2007) Ultra wide-band printed monopole antenna with band-notch filters. IET Microw Antennas Propag 1(2):518–522
5. Movahedinia R, Azarmanesh MN (2010) Ultra-wideband band-notched printed monopole antenna. IET Microw Antennas Propag 4(12):2179–2186
6. Kamma A, Shrikanth Reddy G, Mishra SK, Mukherjee J (2012) Printed band notched U shaped ultra wide band antenna with modified split ring resonator. In: 2012 IEEE Asia-Pacific conference on antennas and propagation, 27–29 Aug 2012
7. Zhang M, Zhou X, Guo J, Yin W (2010) A novel ultra-wideband planar antenna with dual band-notched performance. Microw Opt Technol Lett 52(1):90–92
8. Xiaoyang C, Long J, Ming-Yang L, Qiang-Qiang Z, Qing-Ling Y (2015) Compact triple-band-notched UWB planar monopole antenna based on modified CSRR. In: IEEE 5th Asia specific conference on antenna and propagation 2015, pp 271–272. doi:https://doi.org/10.1109/APCAP.2016.7843202
9. Anuja Raj N, Dwivedi RP (2015) Design and simulation of high gain antenna for X Band applications. Int J Appl Eng Res 10(6). ISSN 0973-4562 (© Research India Publications)
10 Dwivedi RP, Usha Kiran K, Veeramani (2015) Design and simulation of wideband antenna for wireless application. IEEE SPIN 2015, pp 15–18

SOA Parameters Optimization for High Data Rate Operation

K. Swetha, R. Manohari and Shanthi Prince

Abstract To overcome the bandwidth constraint and speed limitation possessed by the electrical interconnect processing, all-optical networks are regarded as probable for future of ultra high-speed switching. Using the nonlinear effects in semiconductor optical amplifier (SOA), the all-optical logic gates operating at 100 Gbps can be realized by optimizing the parameters of SOA. In this work, injection current and input powers are optimized so as to obtain the characteristics of practical SOA enabling the operation of all-optical circuits at high data rate.

Keywords All-optical networks · Nonlinear effects · Semiconductor optical amplifier · Cross-gain modulation

1 Introduction

To realize the need of ever-growing demand for a high-speed networking and to deal with bandwidth starvation, optics has distinguished its credibility in arithmetic and logic operations due to its high operational bandwidth. However, the electronic gates pose major limitation in high-speed communication because of the intermediate electrical processing involved in these networks. For the better performance in speed, it is essential to replace prevailing digital circuits by all-optical gates. Here, the nonlinearity of SOA is exploited to realize the all-optical logic gate to work at 100 Gbps.

All-optical digital devices have been implemented using the nonlinear effects in semiconductor optical amplifier (SOA) [1]. All-optical gates are divided into two

K. Swetha (✉) · R. Manohari · S. Prince
Department of ECE, SRM University, Chennai 603203, India
e-mail: swethakumarasamy93@gmail.com

R. Manohari
e-mail: manohari.r@ktr.srmuniv.ac.in

S. Prince
e-mail: shanthi.p@ktr.srmuniv.ac.in

G. Gnanagurunathan et al. (eds.), *Optical and Microwave Technologies*,
Lecture Notes in Electrical Engineering 468,
https://doi.org/10.1007/978-981-10-7293-2_27

basic structures which are with SOA and without SOA. The advantages of using SOA-MZI-based devices are that they use less power, low latency, high stability, and integration potential [2]. For the realization of logic gates, ultrafast nonlinear properties of SOAs are utilized, in single SOA structure and in interferometric structures by exploiting cross-gain modulation (XGM) [3–6]. SOA-MZI structure using XPM is ideal due to its features like low energy requirement, high BER, high-speed operation, simplicity, capability, and stability [7, 8]. Here, we optimize the SOA parameters for the all-optical logic gates to work at 100 Gbps.

After Introduction, Sect. 1 introduces the basic characteristics of SOA; Sect. 2 explains the optimization of SOA parameters for high data rate operation, and Sect. 3 concludes the improved results briefly.

2 Theory of SOA

The nonlinear optical material is essential for achieving all-optical processing functions where different light beams can interact. SOA is made from semiconductor materials such as GaAs, AlAs, InAs, InP, GaSb, HgTe, and CdTe. It is electrically pumped to create a population inversion for the generation of optical gain by stimulated emission. The parameters of SOA are length, width, height, differential gain, loss, optical confinement factor, initial carrier density, line width enhancement factor, recombination coefficient *A*, recombination coefficient *B*, and recombination coefficient *C*.

The interdependence of the gain of SOA *G*, the injection current *I*, and amplified spontaneous emission power *P* are explained as follows: With injection current density *n*, the material gain *g* is given by

$$g = a(n - n0) \tag{1}$$

Equation above (1) implies that the material gain depends on injection current density $n0$ and gain constant a; the amplified spontaneous emission power P and the injection current I are given by

$$G = \exp[(\tilde{A}g - \alpha)L] \tag{2}$$

$$P = h\acute{\imath}fWdBn^2(G - 1)/\left(\tilde{A}g - \alpha\right) \tag{3}$$

$$I = eWdL(Bn^2 + Cn^3) \tag{4}$$

where *B* and *C* are radiative and Auger recombination constants, respectively [9]. After electron–hole recombination, the additional energy is released in the form of photon. This is known as radiative recombination. In non-radiative recombination, energy is released in the form of phonons. It is due to defects in lattice structure.

Auger recombination involves three particles. In this, the additional energy after recombination is transferred as the kinetic energy of the third particle in the lattice.

From Eq. (2), it is implied that the gain depends on confinement factor Γ, optical absorption α, and length l, and Eq. (3) implies that the power depends on gain and photon energy *hí*. Equation (4) implies that the injection current I depends on charge of electron e and physical parameters such as width W, thickness d of SOA, and recombination coefficients. So these parameters of SOA should be optimized to get appropriate signal gain.

3 Gain Saturation

The parameters that affect the gain of SOA are the input signal power and the internal noise. There is a decrease in the amplifier gain when the carriers in active region get depleted due to increase in signal power. Values in the range of 8–12 dB are typical gain of practical devices.

The dependence of material gain on optical power is that the gain is higher for low optical power and is lower for high optical power. The output saturation power $P_{s,\ out}$ is the output power when the saturated gain G reduces to half of its unsaturated gain value $G0$, i.e., $G = G0/2$. $P_{s,out}$ is given by

$$P_{s,out} = (G0/(G0 - 2Ps)) \ln 2 \tag{5}$$

$P_{s,out}$ is smaller than P_s where the quantity P_s is the saturation power of the amplifier; the amplifier gain is $G = P_{out}/P_{in}$. Equation (5) implies that $P_{s,out}$ is depended on P_s and gain.

4 Optimization of Parameters

All the physical parameters and main parameter injection current of InGaAs-based SOA are tabulated in Table 1. In this table, normal value and range of value for SOA are given.

Using the value listed in Table 1 for SOA parameters, the gain of SOA is calculated for varying input power within a range (−35 to 35 dBm). The output power and calculated gain values for a data rate of 2 Gbps are tabulated in Table 2.

Based on the values in Table 2, a graph is plotted. The plot of gain saturation is shown in Fig. 1, and the output saturation power of 7.025 dBm is obtained.

Optimization of injection current parameter is done for different data rates till 100 Gbps for a gain range of 8–12 dB. This gain range is observed as the real-time gain in practical components. This optimized injection current range for various data rate is shown in Table 3 for an input power of −5 dB.

Table 1 Normal values of SOA parameters

Name	Normal value	Units	Range
Injection current	0.14	A	[0–1]
Length	0.0008	m	[0, 1e−003]
Width	3e−006	m	[0, 500e−006]
Height	8e−008	m	[0, 10e−006]
Differential gain	3.5e−020	m^2	[0, 50e−020]
Loss	2500	1/m	[0, 10e−004]
Optical confinement	0.3	–	[0–1]
Initial carrier density	1e+024	m^{-3}	
Line width enhancement	5	–	[−30 to 30]
Recombination coefficient *A*	1.43e+009	1/s	[0.1 to 4.5e+009]
Recombination coefficient *B*	3e−016	m^3/s	[1 to 9e−016]
Recombination coefficient *C*	3e−040	m^6/s	[1.97e–040]

Table 2 Power optimization

Input power (dBm)	Output power (dBm)	Gain (dB)
−35	−18.404	16.596
−30	−13.431	16.569
−25	−8.511	16.489
−20	−3.748	16.252
−15	0.626	15.626
−10	4.277	14.277
−5	7.025	12.025
0	8.962	8.962
5	10.288	5.288
10	11.201	1.201
15	11.966	−3.034
20	13.098	−6.902
25	15.301	−9.699
30	18.825	−11.175
35	23.230	−11.77

By using the values tabulated in Table 3, the graph of gain versus injection current for data rates till 100 Gbps is plotted in Fig. 2. These optimized values of SOA parameters can be used in all-optical logic gates to work for 100 Gbps data rate which can be used in designing the flip-flops and counters for optical networking (Table 4).

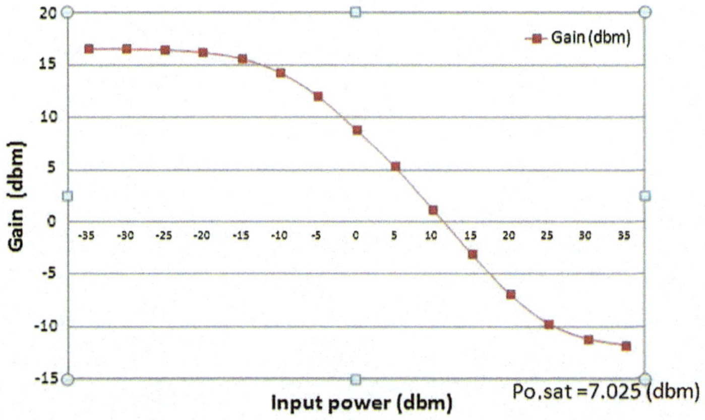

Fig. 1 Plot of gain saturation

Table 3 Optimization of injection current

Injection current (A)	Gain (dB)						
	2 Gbps	10 Gbps	20 Gbps	40 Gbps	60 Gbps	80 Gbps	100 Gbps
	Data rate						
0.12	3.895	4.47	4.529	4.557	4.433	4.572	4.441
0.13	7.148	7.737	7.795	7.824	7.719	7.839	7.227
0.14	9.847	10.455	10.514	10.542	10.458	12.557	10.465
0.15	12.025	12.655	12.713	12.742	12.676	12.756	12.683
0.16	13.773	14.426	14.484	14.513	14.402	14.527	14.469
0.17	15.192	15.866	15.925	15.453	15.913	15.959	15.921
0.18	16.367	17.057	17.116	17.144	17.113	17.108	17.121
0.19	17.35	18.061	18.12	18.148	18.124	18.162	18.131
0.2	18.21	18.921	18.98	17.007	18.989	19.022	18.996

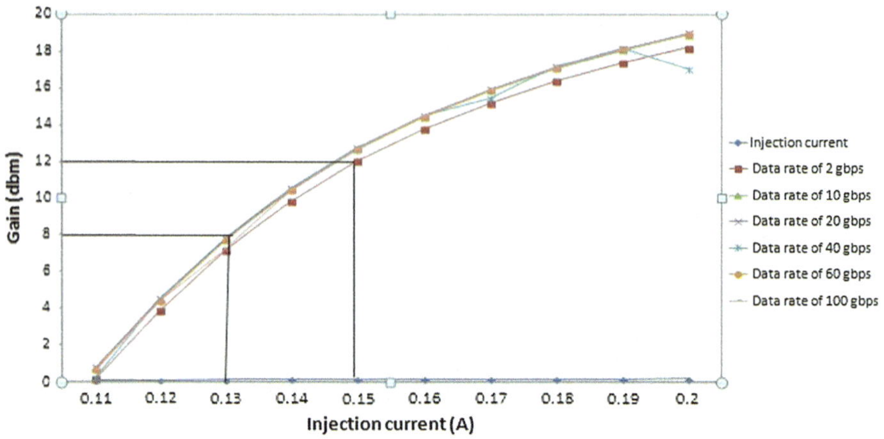

Fig. 2 Plot of gain versus injection current

Table 4 Optimized injection current range

Input power (dBm)	Injection current (A)
−10	0.13–0.14
−5	0.14–0.15
0	0.17–0.2
5	0.18–0.2

5 Conclusion

In this paper, optimization of SOA parameters such as input power and injection current has been achieved. First, input power is optimized as −5 dBm to have a gain saturation of 8–12 dB. Then, optimized injection current range for data rates till 100 Gbps with a gain range of 8–12 dB is obtained. Using this optimized parameter values, all-optical gates can be designed for higher data rates up to 100 Gbps.

References

1. Singh P, Tripathi DK, Jaiswal S, Dixit HK (2014) All-optical logic gates: designs, classification and comparison. Zoran Ikonic Academic Edition, 19 Mar 2014
2. Vagionas C, Fitsios D, Kanellos GT (2012) Optical RAM and flip-flops using bit-input wavelength diversity and SOA-XGM switches 30(18), 3003–3009
3. Wang J, Meloni G, Berrettini G (2010) All optical clocked flip flops and binary counting operation using SOA-based SR latch and logic gates. IEEE Top Quantum Electron 16
4. Wong WM, Blow KJ (2006) Design and analysis of an all-optical processor for modular arithmetic. Opt Commun 265:425–433

5. Wu Y-D, Shih T-T, Chen M-H (2008) New all-optical logic gates based on the local nonlinear Mach-Zehnder interferometer. Opt Express 16(1):248–257
6. Houbavlis T, Zoiros KE, Kalyvas M, Theophilopoulos G, Bintjas C, Yiannopoulos K, Pleros N, Vlachos K, Avramopoulos H, Schares L, Occhi L, Guekos G, Taylor JR, Hansmann S, Miller W (2005) All-optical signal processing and applications within the Esprit project DO_ALL. J Lightwave Technol 23(2):781–801
7. Kumar A, Kumar S, Raguwanshi SK (2013) Implementation of all optical logic gate using SOA-MZI. STM J. ISSN: 2231-0401
8. Jeong J-M, Marhic ME (1991) All-optical logic gates based on cross-phase modulation in a nonlinear fiber interferometer. Opt Commun North-Holland 85:430–436
9. Dutta NK, Wang Q (2013) Low reflectivity facet designs. Semiconductor Opt Amplif

All-Optical 3R Regenerator of Design and Simulation

Althi Bhavya Bindu and Shanthi Prince

Abstract The ever ending need for high data rate and speed for the communication leads to the evolution of optical communication. A communication network in current scenario must provide seamless and errorless connectivity. One of the major components used for this purpose is regenerator for long-haul communication. The latest evolution of 3R generator has a bottleneck of OEO conversion; therefore, there is a need for an all-optical 3R generator. The 3R performs reshaping, retiming, and reamplification on data pulse. This paper deals with techniques and implementations of reshaping, retiming, and reamplification as one unit module for point-to-point link establishment. This above integration results in reducing jitter, high data rate support, and improvement in bit error rate. The work will be supported by simulation and practical implementation result using optiwave system software.

1 Introduction

Communication using light to carry information with fiber cable as a medium is known as fiber-optic communication. An optical communication system consists of a transmitter, receiver, and channel. Formatted electrical bits are encoded using a laser in transmitter for transmission; later, bits are sent to the receiver using optical fiber channel. The receiver encodes optical signal into electrical using photodetector for further processing. Need for optical communication over electrical is the ever ending requirement of high data rate, large bandwidth, low loss capacity, avoidance of crosstalk, and security in communication. The current scenario for point-to-point link establishment in optical communication is constrained by many major factors. The electrical bottleneck optical-electrical-optical conversion constraints data rate of the link. This conversion requirement is due to unavailability of all-optical network components. Long-distance attenuation of fiber over link length and optical

A. B. Bindu (✉) · S. Prince
Department of Electronics and Communication Engineering, SRM University, Chennai, India
e-mail: althibhavyabindu@gmail.com

G. Gnanagurunathan et al. (eds.), *Optical and Microwave Technologies*, Lecture Notes in Electrical Engineering 468,
https://doi.org/10.1007/978-981-10-7293-2_28

losses in components constraints long-haul communication. Need for 3R regenerator attenuation, dispersion, and shape of the waveform must be compensated for long-haul communication. Power efficiency requirement of power is increased due to optical-electrical-optical conversion in the link. Complex integration due to unavailability of all-optical components, electrical and optical components in the network becomes essential. This, in turn, increases the cost, power, and complexity of the system. The compensation of all the above constraints can be practically nearly done using 3R regenerator. This leads to the motivation of this paper, and formulated problem statement of this paper is all-optical 3R regenerator to be realized with reamplification, reshaping, and retiming features for power efficient and simple optical integrated network component. This system must lead to high data rate, less jitter, and improved BER optical signal transmission.

The paper is organized as follows:

Section 1—Introduction of optical fiber communication with current scenario and need for this proposal.
Section 2—Link establishment with real-time parameters available in the market and its effects.
Section 3—1R regenerator design, parameters, and working.
Section 4—2R regenerator need, working theory, real-life feasible techniques implementation.
Section 5—Results and observations.
Section 6—Conclusion and possible future work.

2 Link Establishment

The design of an optical link involves many interrelated variables and the fiber, source, and photodetector operating characteristics, so that the actual link design and analysis may require several iterations before they are completed satisfactorily. Since performance and cost constraints are very important factors in fiber-optic communication links, the designer must carefully choose the components to ensure that the desired performance level can be maintained over the expected system lifetime without over-specifying the component characteristics. The key system requirements needed for analyzing a link are a channel, bit rate, bit error rate, and transmission distance, i.e., link length. The type of channel determines the environment of communication and its possible bit error rate. Bit rate and bit error rate are interconnected to evaluate performance and stability of the system established. For stable and realizable optical communication system, the min BER is 10^{-9} with quality factor 6. Link length affects the communication between source and destination due to various loss and dispersion effects. Since transmission length and data rate are inversely proportional, a trade off must be optimized among these factors. To fulfill the above requirements, the designer has to choose components with their associated parameters by an iteration process [1].

The choice of components and parameters for this paper is as below.

Optical fiber	Optical source: Laser	Photodetector
Core size: 8.2 μm	Emission wavelength: 1550 ± 30 nm	Responsivity: 0.65 a/W
Relative index difference: 0.36	Spectral line width: 5–10 nm	Operating wavelength: 1550 nm
Dispersion: 16.75 ps/nm/km	Output power: 10 dbm	Dark current: max 200 nA
Attenuation: 0.2 dB		
Numerical aperture: 0.14		

3 1R Regenerator

3.1 Line Losses

Attenuation is the reduction in optical power as light travels through the fiber. The main causes of optical attenuation in fibers are as follows: coupling loss, splice loss, optical fiber loss and connector loss, and also scattering, absorption of the light, irregularities in the glass structure. Apart from actual losses suffered, while designing the system, it is also important to incorporate a margin of 6–8 dB to account for losses from splices or other components that may have to be added at a future date and also to allow for any deterioration of components due to aging. Amplification of light can be performed by several techniques. Many optical amplifiers were designed for same. Doped fiber amplifiers utilize stimulated emission in the active medium for amplification. In SOAs, electron and hole recombination occurs. In Brillouin and Raman amplifiers, scattering of incoming light and change of wavelength in photons cause amplification of the signal. Parametric amplifiers use parametric amplification. For the given link, which has attenuation loss of 0.2 dB, the fiber loss is calculated as follows:

A [dB] = $\alpha * L$ = 0.2 dB/km * 10 km = 2 dB (for the whole link).

The number of amplifier required (N) = L/LA, where L = distance for the link.

The loss levels show the requirement of deploying amplifiers in the link to make up for the lost power. This lost power must be recovered so that the output power should be high. Multiple combinations of amplifiers are required based on link length.

Amplifier spacing (L_A [km]): This is the space between two adjacent amplifiers in the link.

L_A [km] = Pt [dBm] − Pr [dBm]/α [dB/km], where Pt = transmitted power, Pr = received power.

4 2R Regenerator

Loss of information due to loss of envelope of the signal is an undesirable effect. Therefore, it is essential to maintain the shape of optical signal; hence, it raises the needed for reshaping along with reamplification for high performance over longer transmission distance.

4.1 Dispersion

Dispersion is the time distortion of an optical signal that results from the time of light differences of different components of that signal, typically resulting in pulse broadening. Figure 1 shows dispersion of signal over length transverse.

Usage of single mode fiber mitigated large-scale dispersion. The major chromatic and polarization mode dispersion has to be further analyzed. Dispersion has different effects for digital and analog transmission. The envelope distortion leads to the unacceptable threshold level for composite second-order distortion in analog, whereas it affects the overall performance of data rate, link length for digital.

4.2 Dispersion Compensation

The overall chromatic dispersion of a system by adding is controlled by optical elements with a suitable amount of dispersion. Dispersion management essentially means mitigation of chromatic dispersion. This process is performed by dispersion compensation technique. It avoids overlapping of signal waveforms due to broadening of the pulse. Dispersion compensation is majorly implemented for long link length optical communication system and telecommunication using fiber cable as its channel. Even minute variation in envelope due to broadening at high data rates results in distortion in transmitted signal. Fiber Bragg grating is generally used for dispersion compensation. Figure 2 is the layout schematic of 2R regenerator using fiber compensation technique. Figure 2 shows FBG compensated 2R regenerator.

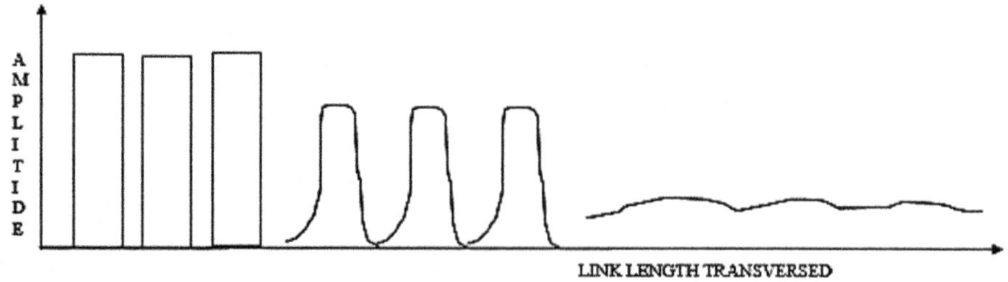

Fig. 1 Dispersion of signal over link length

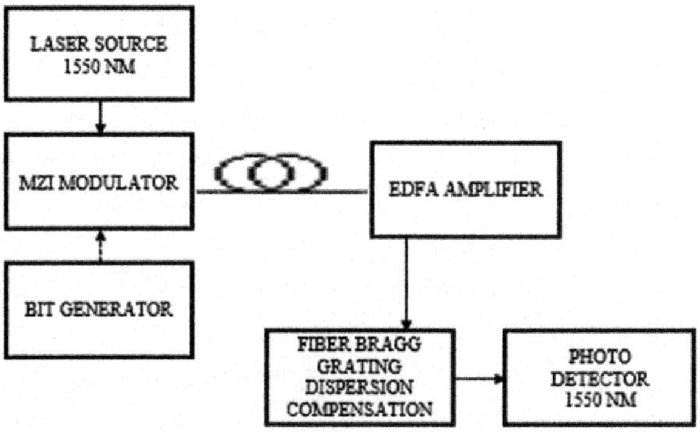

Fig. 2 Dispersion-compensated 2R regenerator using fiber

Figure 3 shows the implementation of 2R regenerator in Optisystem. Even though dispersion can be compensated using dispersion-compensated fiber Bragg grating, the time jitter must be compensated in the link, therefore, there is a need for clock recovery for the transmitted signal. Decision circuit implemented using a nonlinear optical digital gate (NODG) is utilized for reshaping a digital signal. The 2R output is given as input to optical gate realized using semiconductor optical amplifiers. The mechanism of operation is similar to an analog gate giving output when both gates are on and off at same time. The off state represents digital 0 symbol and on state represents digital 1 symbol. The nonlinear optical data gate design utilized in this paper in Fig. 4 was taken from the literature of optical gates design. This design uses optical couplers, WDM, and simple amplifiers making the design completely optical and thus mitigating the requirement of signal processing at this block which was the major weakness for initial optical regenerators [2].

4.3 Clock Recovery

Sampling clock is required for retiming of a signal. Therefore, recovery of sampling clock becomes essential. Receiver requires a reference clock of approximately same frequency for recovery purpose. The reference clock is aligned to the transition of an incoming data stream to generate recovery clock. On the usage of recovered clock for sampling on incoming data stream, data is recovery. Together, this is called clock data recovery. This process is required in absence of any accompany clock signal to recover data. Frequency detection and phase alignment are the main functions for recovery of clock and data.

Frequency detection: The time differences between consecutive edges of the data stream are estimated to retrieve locked frequency for regenerating the transmitted data.

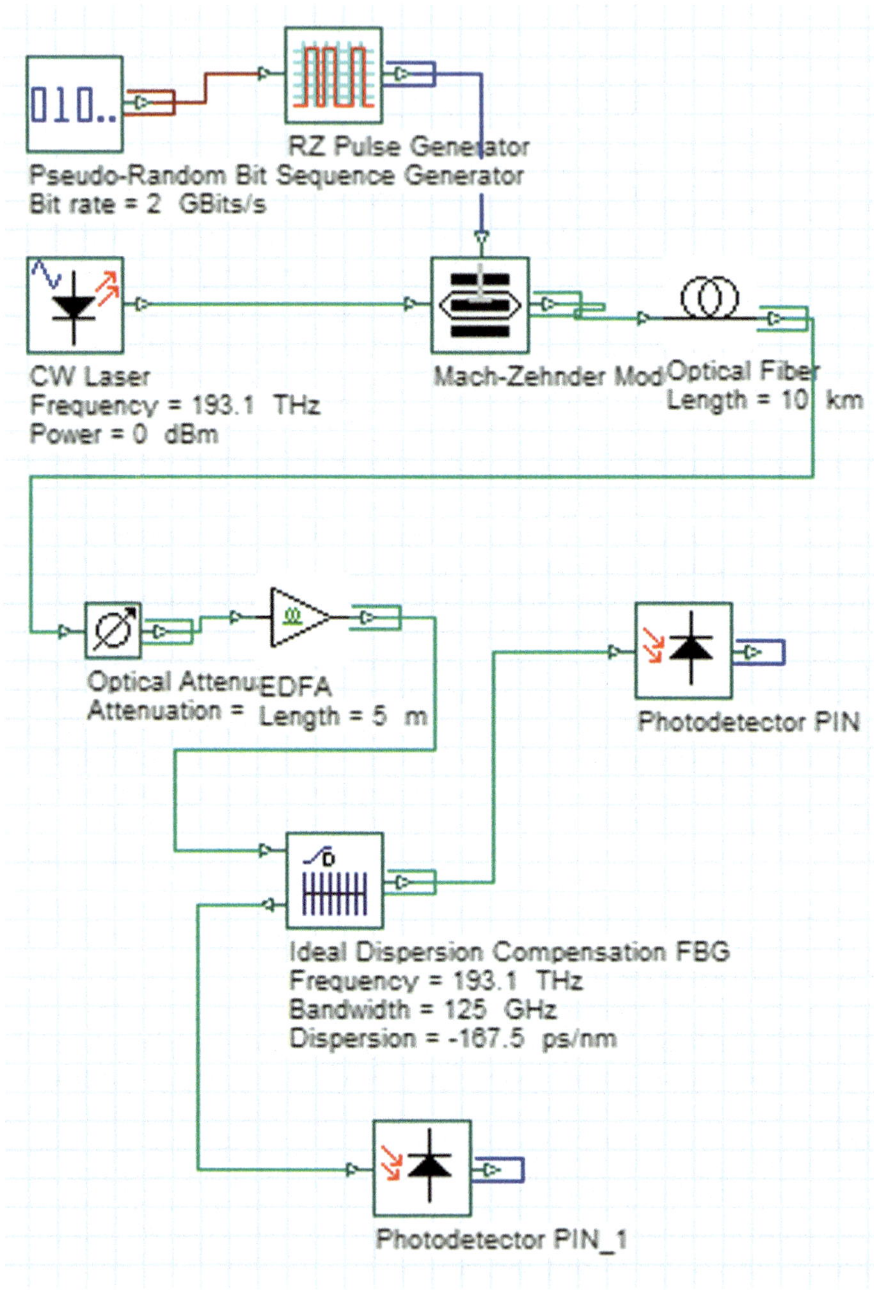

Fig. 3 The dispersion-compensated 2R regenerator using fiber layout using Optisystem

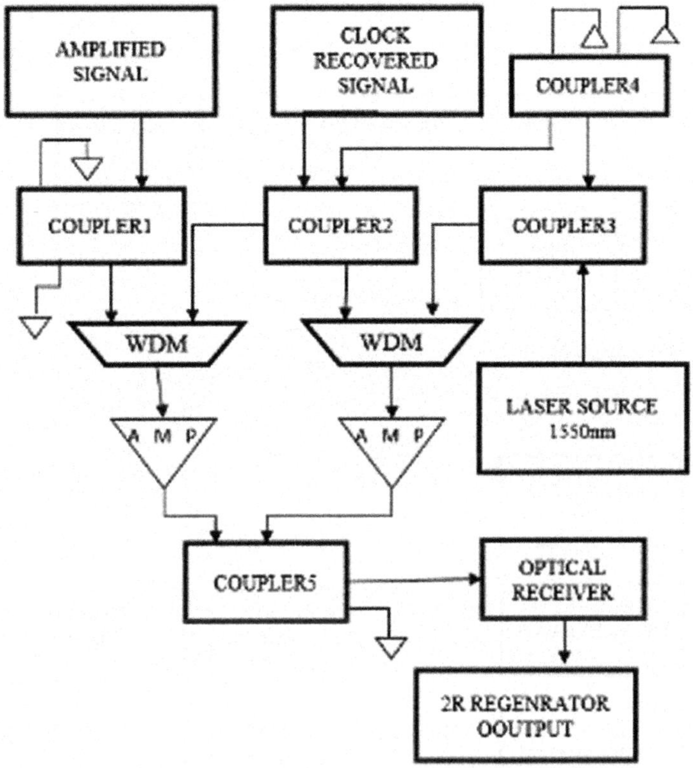

Fig. 4 Nonlinear optical data gate

Phase alignment: The phase of the incoming data stream and recovered clock is matched for alignment of phase [3–5].

Simple clock recovery layout schematic for a communication link using reference signal generated locally, input signal from the fiber at receiver side, is a given in below Fig. 5.

Figure 6 depicts designed block diagram of 2R regenerator using clock recovery. Dispersion effects over large link length are compensated to a large extent using above designed clock recovery unit by aligning the input signal according to trained digital pulses [6–8].

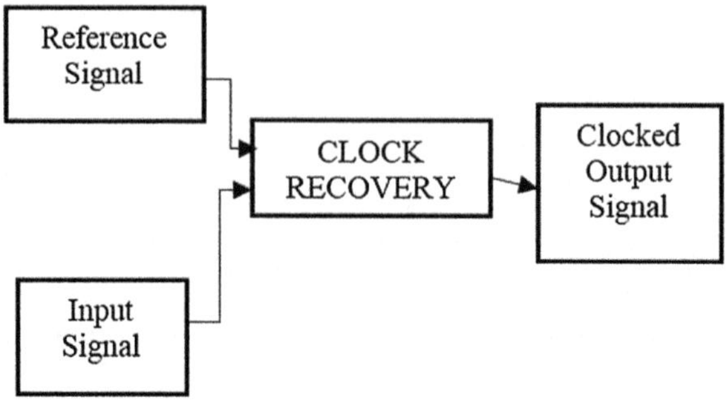

Fig. 5 Clock recovery circuit

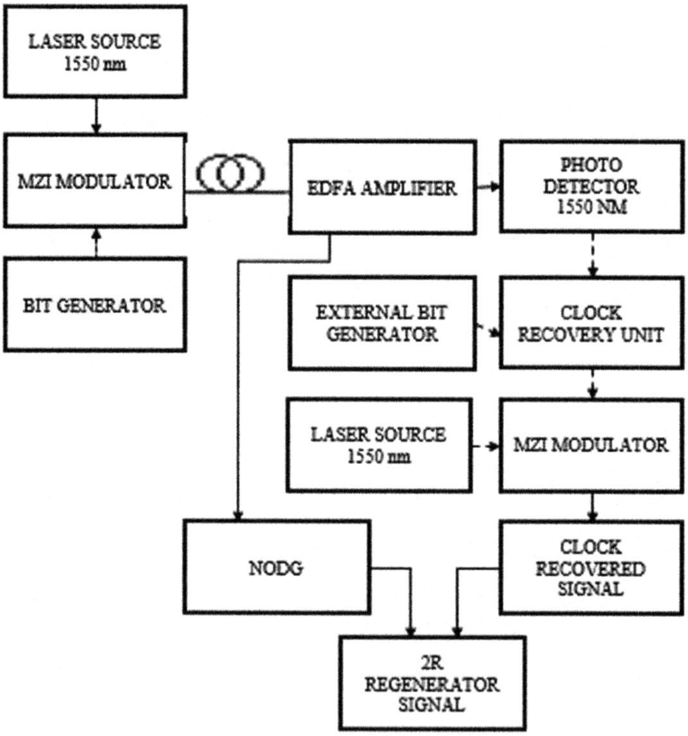

Fig. 6 2R regenerator layout schematic

5 Results and Inferences

Figure 7 is the simulation result for simplex link stability against loss over link length. We can conclude that at high-frequency operations trivial, attenuation loss leads to enormous information loss; hence, the attenuation must be mitigated using repeater or amplifiers. The min Q factor of 6 is achieved at:

Loss = 23.2 dBm for 1 Gb/s with BER = 7.74485e−10.
Loss = 18.2 dBm for 5 Gb/s with BER = 1.06549e−11.
Loss = 17.2 dBm for 10 Gb/s with BER = 2.09343e−11.

For current scenario, multimedia high data rate 40 Gb/s and above bit rate system must be realized. Some compensation technique must be employed for achieving min Q factor for transmission link of such high data rate.

Figure 8 is the simulation result for amplified point-to-point link showing inclusion of amplifier in the communication link on considering repeater spacing, improved BER of the system. The system is loss resistant for longer link length, i.e.,

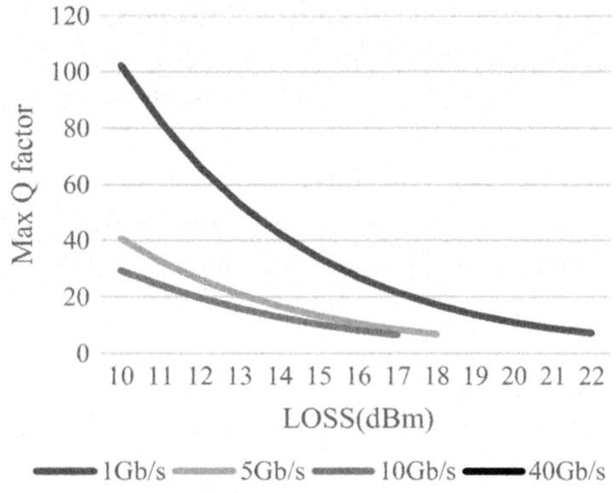

Fig. 7 Loss versus Max Q factor for simplex link (1R regenerator)

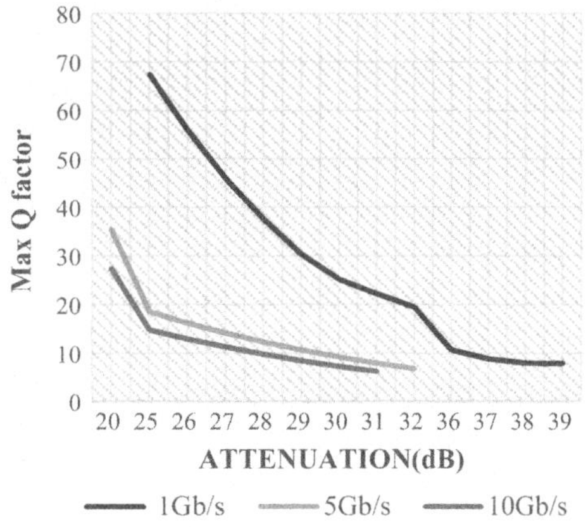

Fig. 8 Loss versus Max Q factor for amplified transmission link

initially, the communication for 1, 5, and 10 Gb/s data rate became unrealizable at 22, 18, and 17 dBm loss, but on amplification, it was stable up to 39, 32, and 31 dBm. Since we should also consider the shape of the received signal for decoding information, optical time domain visualizer was used to view it as shown in Figs. 9 and 10. Figures 9 and 10 are pulse shape of transmitted and received signal for FBG compensated 2R regenerator The dispersion is very less compared to the pulse size, but that also must be mitigated since loss of some nanosecond or picosecond information is also a high loss of information data at high data rates.

The major factor that effects optical communication is dispersion, which is cumulative in nature, i.e., as length of the link increases, its effect increases. This effect must also be considered as it leads loss of information due to overlapped symbols. In 1R system, this is not mitigated hence leading to requirement of 2R, another being higher data rates. On implementation of 2R regenerator using FBG fiber compensation in the transmission line, higher data rates are achieved for same link parameters as shown in Fig. 10, i.e., at 40 Gb/s, dispersion-compensated transmission link can be established with min Q factor of 11.8329. For reshaping of signal, time jitter must also be compensated. Therefore, there is need for clock recovery in the transmission link. Higher data rate system can be achieved on clock recovered transmission link as shown in Fig. 10, but there is a constraint that all bit generators must have same frequency of operation.

Figures 11, 12, and 13 above shows eye diagram and min BER at min Q factor for each realized optical link in Optisystem before 1R, after 1R, and 2R, respectively. We can infer that the performance is improved as reamplification and reshaping process is utilized in the optical communication link. This increases complexity for operation because if the frequencies are not aligned, then system

Fig. 9 Generated signal for 2 Gb/s

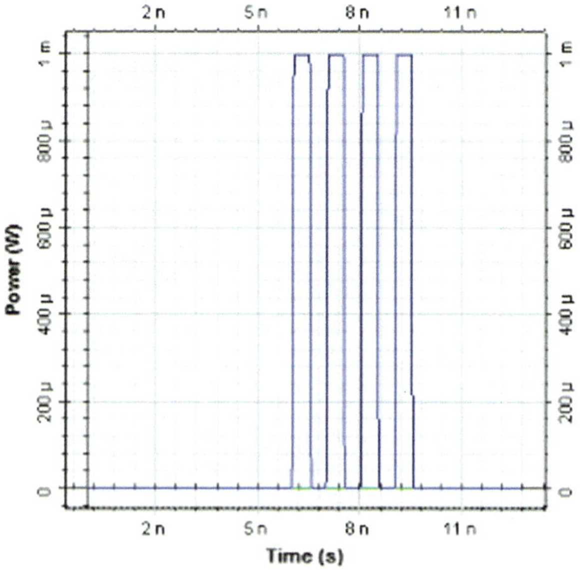

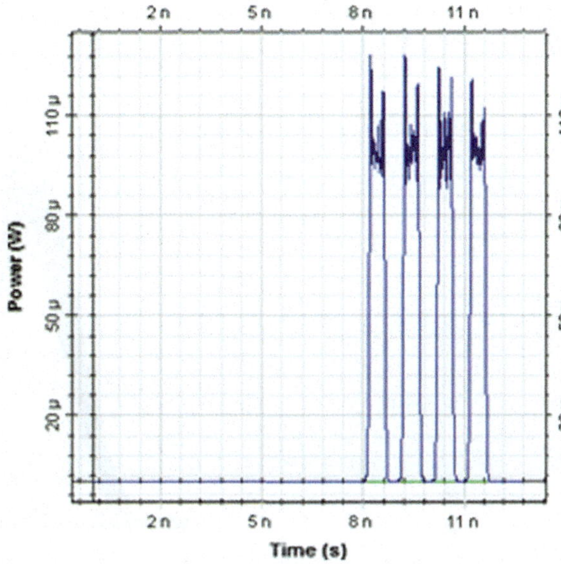

Fig. 10 Received signal after FBG compensation

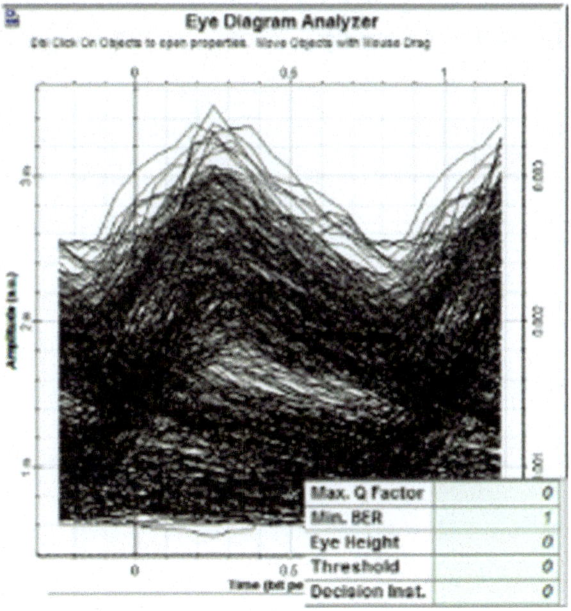

Fig. 11 Eye diagram before 1R regenerator

becomes unstable. For clock recovery, the locally generated signal frequency is estimated by assumption; hence, there is always a probability of misalignment in the transmission link. This leads to the need of 3R regenerator [9].

O-E-O conversion requires more cost and power; hence, design of all-optical regenerator system mitigates this disadvantage [10].

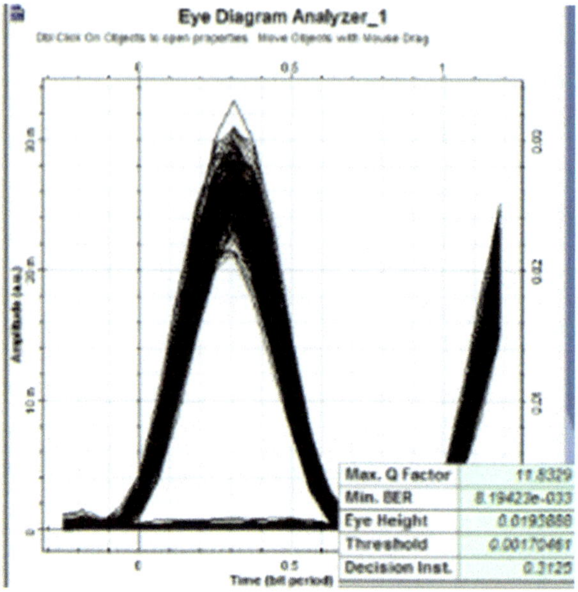

Fig. 12 Eye diagram after 1R regenerator

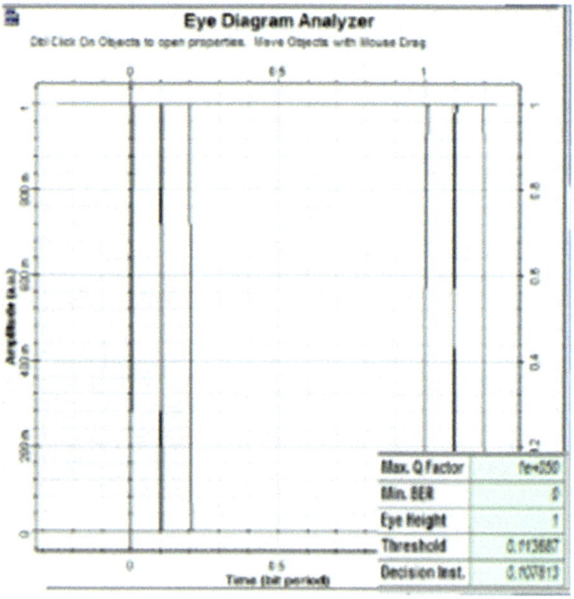

Fig. 13 Eye diagram after 2R regenerator

6 Conclusion

The optical communication point-to-point link establishment is dependent on bit rate, transmission link length, and bit error rate. The development of all-optical 3R regenerator along with reamplification, reshaping, and retiming as its features become essential. The proposed paper established point-to-point link with real-time parameters. The amplified transmission link compensated the losses in the network. Dispersion-compensated link reduced the effect of distortion hence further increasing the data rate of transmission. Clock recovery was implemented to reduce time jitter of the signal which leads to flexibility of link length and data rate.

References

1. Romero Cortés L, Azaña J (2014) All-optical reconfigurable regenerative RZ to NRZ format converter based on a Mach-Zehnder interferometer and a temporal photonic integrator. IEEE Photonics 314–315. doi:www.doi.org/10.1109/IPCon.2014.6995370
2. El-Sayed Y, Wageeh A, Ismail T, Mostafa H (2015) All-optical clock and data recovery using self-pulsating lasers for high-speed optical networks. IEEE Photonics. doi:www.doi.org/10.1109/ICEAC.2015.7352193
3. Kamatani O, Kawanishi S (1996) Ultrahigh-speed clock recovery with phase lock loop based on four-wave mixing in a traveling-wave laser diode amplifier. J Lightwave Technol 14 (8):1757–1767
4. Sua G, Choi WS, Hanumola PK (2016) Digital clock and data recovery circuits for optical links. In: Compound semiconductor integrated circuit symposium (CSICS), IEEE. doi:www.doi.org/10.1109/CSICS.2016.7751036
5. Raupach SMF, Grosche G (2014) Chirped frequency transfer: a tool for synchronization and time transfer. IEEE Trans Ultrason Ferroelectr Freq Control 61(6):972
6. Gomez Agis F, Ware C, Erasme D, Ricken R, Quiring V, Sohler W (2006) 10-GHz clock recovery using an optoelectronic phase-locked loop based on three-wave mixing in periodically poled lithium niobate. Photonics Technol Lett 18(13):1460–1462
7. Xin H, Wenmeng L, Daonong Z, Song Y, Qiwei D (2014) Smart substation IEC61588 time synchronization system and security evaluation. In: Proceedings of IEEE symposium on precision clock synchronization for measurement control and communication (ISPCS), pp 97–101
8. Ingram DME, Schaub P, Campbell DA, Taylor RR (2012) Evaluation of precision time synchronisation methods for substation applications. In: Proceedings of IEEE symposium on precision clock synchronization for measurement control and communication (ISPCS), pp 1–6
9. Stojanovci N, Xu C (2015) Clock recovery in coherent optical receivers. OSA, March 2015, paper 060.1660, 060.2230
10. www.optiwave.com. 2014

Analysis of Dispersion Compensation Methods in WDM Systems

S. Sesha Sai Srikar and N. Subhashini

Abstract Objectives: WDM systems are high-performance systems. Light signals can be transmitted to longer distances by using WDM systems. The efficiency of light signals is reduced by optical effects such as dispersion and attenuation. Attenuation can be reduced by using powerful optical amplifiers. Dispersion causes errors in optical communication systems. So dispersion has to be minimized to obtain better performance of WDM systems. This paper analyzes the different types of dispersion compensation techniques that can be used in WDM Systems to minimize dispersion. **Methods/Statistical analysis**: In dispersion compensation technique, dispersion compensation fiber (DCF) is placed at different positions of optical fiber to reduce dispersion. The negative dispersion in DCF removes the dispersion in SMF. The three dispersion compensation techniques pre-compensation, post-compensation, and symmetric compensation techniques are analyzed with Optisystem. The structure of FBG also helps to compensate dispersion in optical systems. Dispersion compensation by FBG is also analyzed with Optisystem. **Findings**: Pre-compensation technique shows good performance for very shorter distances. Performance of pre-compensation is reduced for long distances. Post-compensation also do not show good quality factor for longer distances. Symmetric compensation shows the best performance and quality factor for longer distances and is the best method out of the three compensation techniques. FBG also compensates the dispersion very effectively in long-distance communication. **Application/Improvements**: Symmetric compensation is the best technique to reduce dispersion in all three dispersion compensation techniques. By using FBG, dispersion can be reduced effectively. These methods can be used in WDM systems to minimize errors and to improve performance.

S. Sesha Sai Srikar · N. Subhashini (✉)
School of Electronics Engineering, VIT University, Chennai, Tamil Nadu, India
e-mail: subhashini.n@vit.ac.in

S. Sesha Sai Srikar
e-mail: saisrikar59@gmail.com

G. Gnanagurunathan et al. (eds.), *Optical and Microwave Technologies*,
Lecture Notes in Electrical Engineering 468,
https://doi.org/10.1007/978-981-10-7293-2_29

Keywords Dispersion compensation · Dispersion compensation fiber
Fiber Bragg grating · Single mode fiber · Symmetrical compensation
Wavelength division multiplexing

1 Introduction

In the modern era, there is high need for long-distance transmission in the optical communication. The advancement in the fiber optics leads to development of long-distance communication. WDM networks are having high capacity, high speed, and high bandwidth compared to other optical networks. WDM networks can be used to send multiple signals from transmitter to receiver. The intensity of light is reduced when light is traveling through the fiber. This reduction of light intensity is due to dispersion and other nonlinearities. The use of fiber in optic transmission is increasing rapidly. Optical power amplifiers like EDFA can easily transmit the signals to very high distances. These types of amplifiers reduce the requirement of electric format signal conversion into optic format signal conversion and vice versa. Optical power amplifiers like EDFA, semiconductor optical amplifier, hybrid amplifier, and Raman amplifier can be used in long-distance transmissions to reduce power loss. Hybrid optic amplifier provides great bandwidths to optical systems [1]. It is very useful in long-distance transmissions [2]. In very high speed WDM networks, change of channel bitrate also changes the nonlinear impairments [3]. As channel number and transmission distance increases, light signals in fiber become vulnerable to optical nonlinear effects. The optical signal transmission may be affected by optical nonlinear effects like chromatic dispersion and waveguide dispersion. Various techniques have been proposed to reduce these optical impairments which include optical phase conjugation [4, 5], dispersion compensating devices [6, 7], and fiber Bragg grating. The usage of dispersion compensation fiber is one of the easiest and useful ways to improve the already existing structure of the normal single mode fiber (SMF). Some amount of negative dispersion can be used for compensating positive dispersion in the normal single mode fiber. Eye margins are affected by compensation fiber lengths as power in compensating fiber increases when signal-to-noise ratio increases, and therefore, system performance is also improved. The performance can be improved by placing one dispersion compensated fiber (DCF) before the single mode fiber (SMF) at every amplifier section [8]. Fiber Bragg gratings are used to reduce the dispersion in optical fibers. Fiber Bragg grating is a distributed type reflector in a small portion of fiber. FBG reflects some wavelengths and passes remaining wavelengths as shown in Fig. 1. The light that satisfies Bragg condition is first resonated by the fiber Bragg grating structure, and then, it is reflected. A fiber Bragg grating can be used like filter for blocking of wavelengths. FBG has several advantages like quality improvement and cost reduction. FBG has very low internal loss and cost efficient. In papers [9, 10] they designed pre-compensation scheme, post-compensation scheme, and symmetric dispersion compensation scheme for NRZ modulation

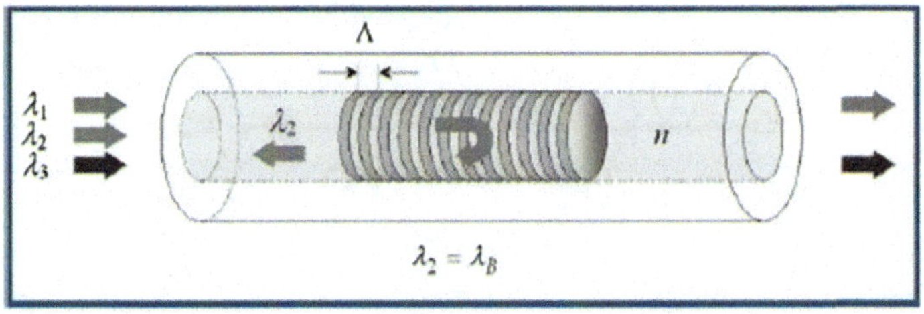

Fig. 1 Reflection of wavelengths in FBG

schemes with the help of the standard fibers and compensation fibers. Erbium doped fiber amplifier (EDFA) compensates the losses in the optical signals. Similarly the paper also analysed the dispersion compensation schemes for carrier suppressed return-to-zero (CSRZ) systems for systems with fiber nonlinearities and obtained that symmetric compensation technique gives good outputs compared to other systems. But capacity of these techniques is low and not sufficient to be used in large speed systems. In [11] dispersion compensation and power compensation were achieved at the same time by using pumped DCF with Raman amplification. In [12–14], methods were analysed to reduce dispersion and FWM effects. All these models cannot be used in high-capacity WDM system because these are having low speed and low number of channels. Here, we compare all the three dispersion compensation techniques in terms of their quality factors. The rest of the paper is designed as follows: Simulation setup is described in Sect. 2. Result analysis has been shown in Sects. 3 and 4 conclude the paper.

2 Simulation Setup

The three dispersion reduction methods are structured as shown in Fig. 3. All the three compensation schemes are designed using Optisystem software as shown in Fig. 2. In pre-compensation, the system first undergoes pre-amplification process, and then, dispersion in the system is eliminated with the negative dispersion produced by DCF that cancels the dispersion in SMF as shown in Fig. 2. Post-compensation scheme is designed such that system undergoes post-amplification process. Then, dispersion is eliminated by negative dispersion produced by DCF that cancels the dispersion in SMF as shown in Fig. 2. In symmetrical compensation scheme, the system undergoes symmetrical amplification process, and then, dispersion is eliminated by the two compensation fibers. In Fig. 3, the transmitter includes one data source which generates the data signals, laser that generates the light signals, one electrical driver for driving the electrical signals, and amplitude modulator that modulates the signals. Here, data source is in return-to-zero modulated format. The required data format is generated by electrical driver, which converts the binary signal to electrical format. The continuous wave

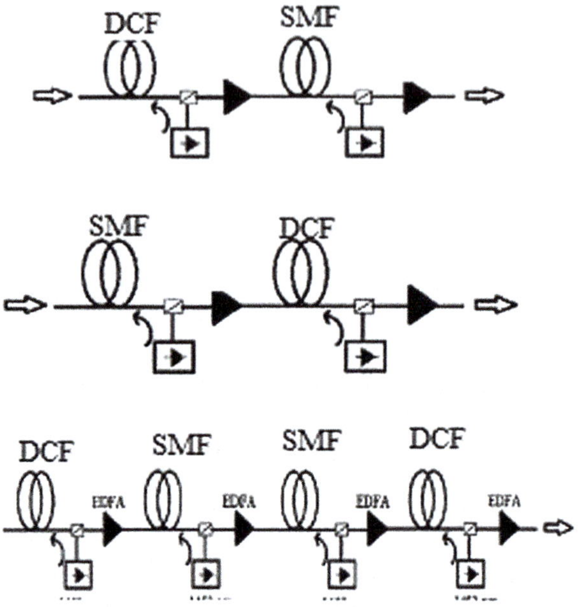

Fig. 2 Pre-compensation, post-compensation, and symmetrical compensation techniques

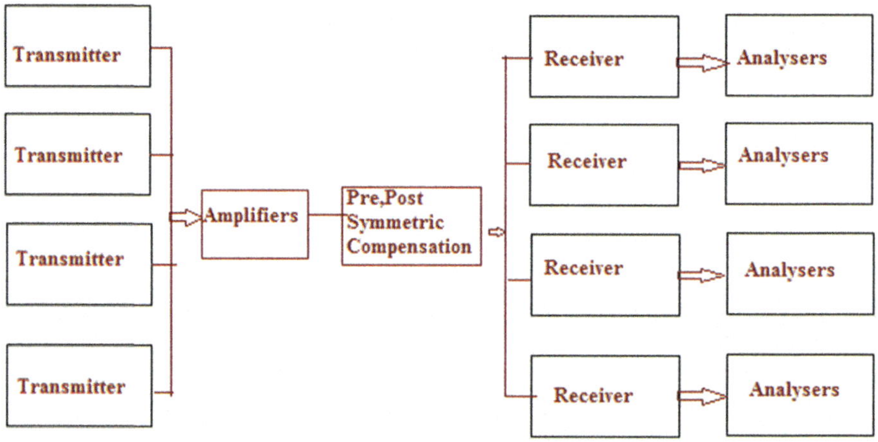

Fig. 3 Simulation setup for three compensation schemes

(CW) laser generates laser beam with required bandwidth. The data source is modulated by a modulator. The modulated signals are amplified by using one power amplifier like EDFA, then passed into optical fiber, so that the signal can be transmitted without any loss. EDFA amplifiers are placed at the DCF and SMF to reduce loss. The transmitted signal is received at the detector with the PIN diode,

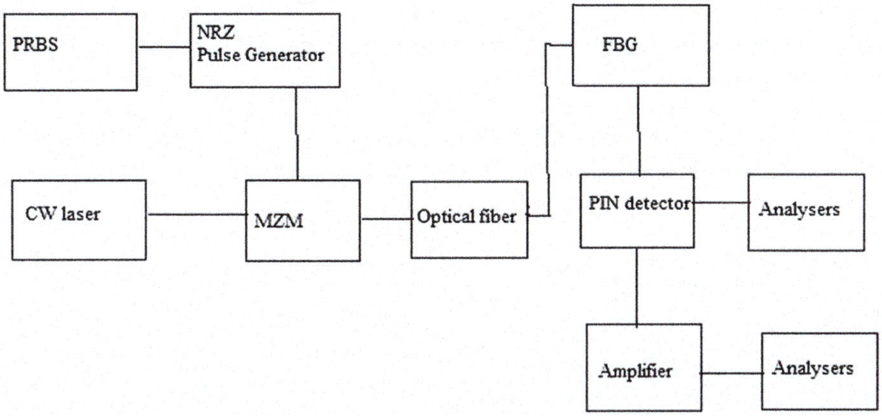

Fig. 4 Simulation setup for dispersion reducing by FBG

and received signal is passed into Bessel filter. BER analyzer is used to observe the received signal. The bit error rate and Q factor are obtained from BER analyzer. Bessel filter is used as an electrical filter. PIN detector detects the signal and its responsivity is 0.875. The simulation setup for reducing the dispersion by using fiber Bragg grating is as shown in Fig. 4. FBG is very useful equipment, and FBG can be used in filtering of different wavelengths and also in sensing of different smart devices. FBG is the very easiest technique in compensating the dispersion in WDM systems. The structure of FBG helps to compensate the dispersion of the system. Dispersion can be reduced by introducing different types of FBG compensators. By using Optisystem software, the circuit is constructed. A CW laser with 193.5THz frequency and power of 1mw is used to transmit the signal. Transmitted signal is modulated by MZM and passed through the fiber. The transmitted signal is given to the FBG, and then, it is received with PIN detector.

3 Result Analysis

All the results are taken at 1550 nm wavelengths. The optical fiber length is varied in accordance with the different lengths. The equation for obtaining SMF lengths is given as

$$L_2 = -\left(\frac{D_1}{D_2}\right) * L_1$$

Here, L_1 is normal SMF length, and L_2 is DCF length. D_1 is dispersion of SMF, and D_2 is dispersion of DCF. Different lengths of DCF and SMF for different cases are shown in Table 1. The eye diagrams of pre-compensation, post-compensation, and symmetric compensation are shown in Figs. 5, 6, and 7. The performance of the system is analyzed by graph between signal quality factor and the different lengths of fibers as shown in Fig. 8. From the results, it is seen that signal quality is reduced as lengths of both SMF and DCF is increased. Pre-compensation shows good performance for short distances. But for long distances, the performance of

Table 1 Several cases for different lengths of SMF and DCF

Case	DCF length (kms)	SMF length (kms)
1	8	80
2	16	160
3	24	240
4	32	320
5	40	400
6	48	480
7	56	560
8	64	640
9	72	720
10	80	800

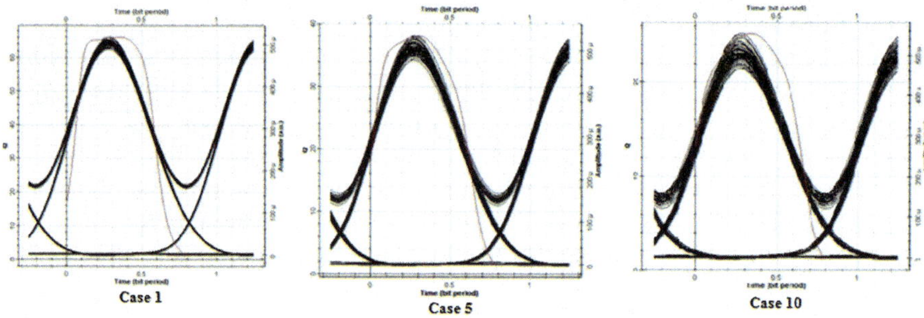

Fig. 5 Eye diagrams for different cases of pre-compensation

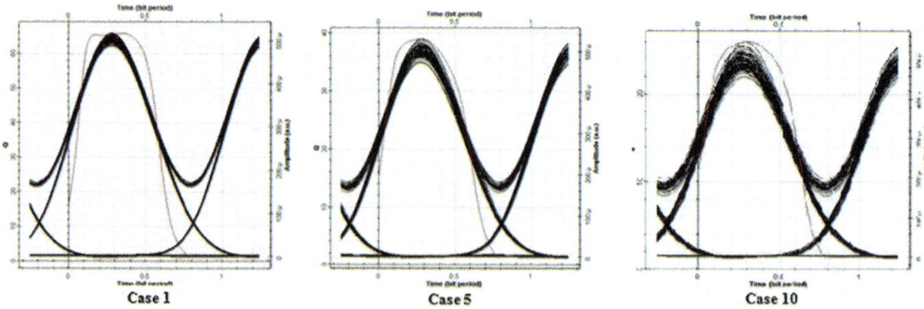

Fig. 6 Eye diagrams for different cases of post-compensation

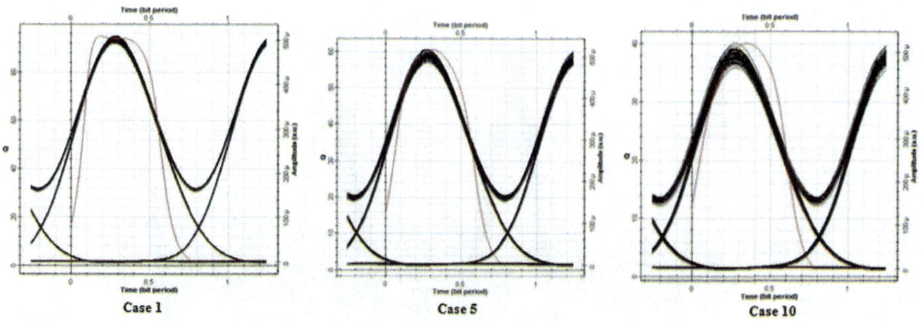

Fig. 7 Eye diagrams for different cases of symmetric compensation

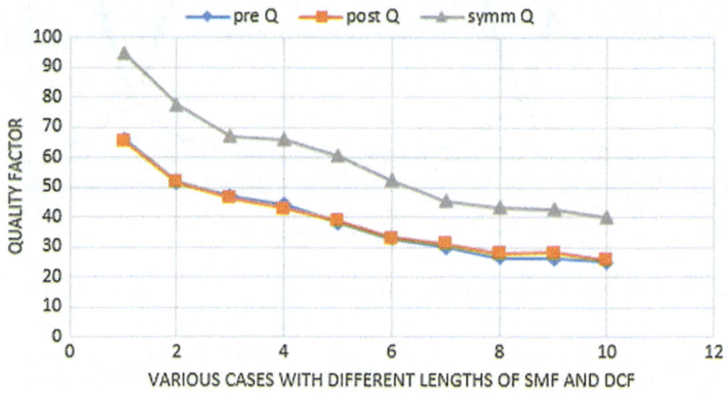

Fig. 8 Graph for pre-compensation, post-compensation, and symmetric compensation in terms of quality factor

the pre-compensation is worst. In the post-compensation scheme, the quality of the signal is better showing good performance. But increasing the lengths of DCF and SMF reduces the quality of the signal. For very large distances, symmetrical compensation is better than post-compensation, and the quality of pre-compensation is worst. Dispersion compensation by using FBG is designed, and eye diagrams are obtained for different apodization function like tanh, Gaussian, and uniform apodization function by using FBG. These are shown in Figs. 9, 10, and 11.

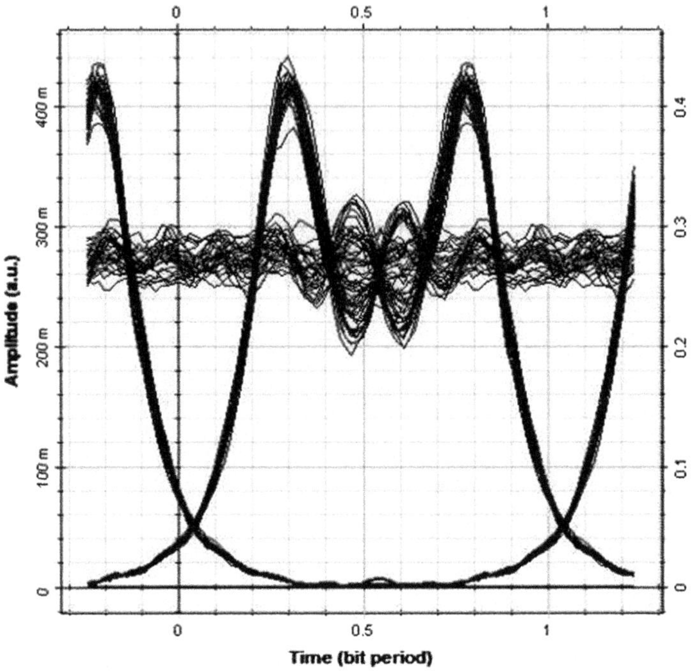

Fig. 9 Eye diagram for Gaussian apodization function

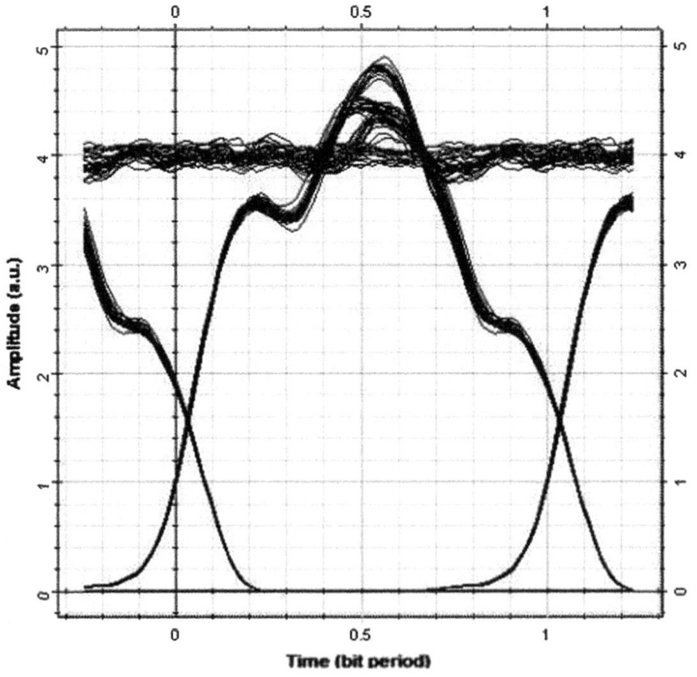

Fig. 10 Eye diagram for Tanh apodization function

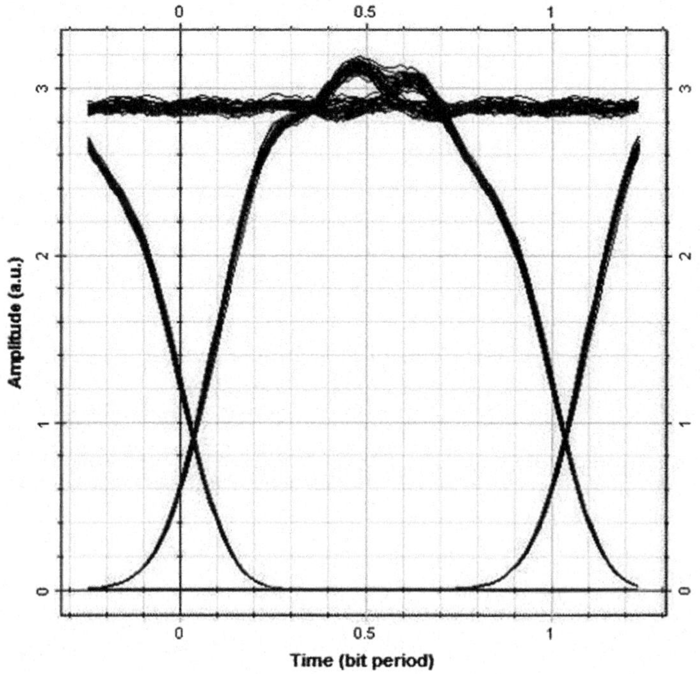

Fig. 11 Eye diagram for uniform apodization function

4 Conclusion

Three dispersion compensation methods such as pre-dispersion compensation, post-dispersion compensation, and symmetric dispersion compensation techniques in a WDM System are compared in this paper. All the three compensation schemes were simulated for a WDM system for various distances, and the quality factor was measured at the output. Symmetrical compensation technique was found to be a better technique when compared to the other two techniques. Dispersion compensation by using FBG is also analyzed for different apodization function like tanh, Gaussian, and uniform apodization function. Of all the three apodization functions, tanh apodization function has the best performance.

References

1. Seo HS, Chung WJ, Ahn JT (2005) A novel hybrid silica wide-band amplifier covering S + C + L bands with 105-nm bandwidth. IEEE Photonics Technol Lett 17(9):1830–1832
2. Chung HS, Han J, Chang SH, Kim K (2005) A Raman plus linear optical amplifier as an inline amplifier in a long-haul transmission of 16 channels × 10 Gbit/s oversingle-mode fiber of 1040 km. J Opt Commun 244(1–6):141–145

3. Garg AK, Kaler RS (2008) Novel optical burst switching architecture for high speed networks. Chin Opt Lett 6(11):807–811
4. Watanable S, Naito T, Chikma T (1993) Comparison of chromatic dispersion in a single-mode fiber by optical phase conjugation. IEEE Photonics Technol Lett 5(1):92–95
5. Poole CD, Wieserfed JM, DiGiovanni DJ, Vengsarkar AM (1994) Optical Fiber based dispersion compensation using higher order modes near cut off. J Lightway Technol 12 (10):1746–1758
6. Kaler RS, Kamal TS, Sharma AK (2002) Approximate and exact small signal analysis for single mode fiber near zero dispersion wavelength with higher order dispersion. Fiber Integr Opt 21(15):391–415
7. Randhawa R, Sohal JS, Kaler RS (2010) Pre-post and hybrid dispersion mapping techniques for CSRZ optical networks with nonlinearities. Opt 121(14):1274–1279
8. Rothnie DM, Midwinter JE (1996) Improved standard fibre performance by positioning the dispersion compensating fibre. Electron Lett 32(20):1907–1908
9. Isa N, Ahmet A (2005) Design of a chirped fiber Bragg grating for use in wideband dispersion compensation. In: The fourth International Conference on Electrical and Electronics Engineering ELECO. pp 1–5, Sep 2005
10. Kaler RS, Sharma AK, Kamal TS (2002) Comparison of pre-, post- and symmetrical dispersion compensation schemes for 10 Gb/s NRZ links using standard and dispersion compensated fibers. J Opt Commun 209:107–123
11. Tiwari U, Rajan K, Thyagarajan K (2008) Multi-channel gain and noise figure evaluation of Raman/EDFA hybrid amplifiers. Opt Commun 281(6):1593–1597
12. Ajmani Manisha, Singh Preeti (2015) Comparative analysis of DCF and OPC as means to minimize FWM in WDM system. Indian J Sci Technol 8(27):1–12
13. Barza Badar, Anisha AP (2015) Simulative analysis and compensation of dispersion in WDM optical systems. Int J Eng Res Technol 4(1):721–725
14. Kahlon NK, Kaur G (2014) Various dispersion compensation techniques for optical system: a survey. Open J Commun Softw 1(1):64–73

Wideband Antenna for Medical Application

Drishti Gandhi and Niraj Kumar

Abstract In this paper, return loss (S_{11}) of the antenna is discussed. Several antenna size miniaturization techniques are discussed, and few are implemented to design a medical antenna for neural data transfer operating in the Medical Implant Communication Service (MICS) band as assigned by the FCC. The antenna also operates in bands near ISM band which makes the antenna to operate for other medical application as well. The current path is lengthened by using combination of spiral-slot and meandered slot antenna. Shorting pin is used to lower down the frequency, and use of superstrate is made to isolate the antenna from getting in direct contact with the human body. The design of the antenna is done using High-Frequency Simulation Software (HFSS), and the simulated results are discussed in the paper.

Keywords Implantable medical device (IMD) · Miniaturization
Meandered slot · Metamaterials · Shorting pin · Spiral-slot

1 Introduction

The European Telecommunications Standards Institute (ETSI) has standardized the Medical Implant Communication System (MICS). The first one is for telecommunication between a base station and an implanted device. The second one is for telecommunication between medical implants within the same body. The frequency band allocated for Medical Implant Communications Systems (MICS) is 402–405 MHz. The 433.1–434.8 MHz, 868–868.6 MHz, 902.8–928 MHz, and 2400–2500 MHz industrial, scientific, and medical (ISM) bands are also suggested for implantable medical device biotelemetry. The advantage of using MICS band is its

D. Gandhi · N. Kumar (✉)
School of Electronics Engineering, VIT University, Chennai, India
e-mail: niraj.kumar@vit.ac.in

D. Gandhi
e-mail: drishtigandhi@gmail.com

G. Gnanagurunathan et al. (eds.), *Optical and Microwave Technologies*,
Lecture Notes in Electrical Engineering 468,
https://doi.org/10.1007/978-981-10-7293-2_30

worldwide availability and being feasible with low-power and low-cost circuits, reliably supporting high data rate transmissions, falling within a relatively low-noise portion of the spectrum, lending itself to small antenna designs, and acceptably propagating through human tissue [1].

Biotelemetry provides wireless communication from outside the body to inside it, or vice versa. For medical implants, higher frequency telemetry links are being developed. Dipole antenna and spiral or serpentine microstrip antenna as well as circumferential antenna, and waffle-type antenna have been designed for implantation into the shoulder for cardiac telemetry. Implantable antennas are of two types microsystems and stimulators. The microsystems are implants done within the human body like temperature monitoring, electrocardiograms ECG, blood glucose sensor. The stimulators receive information from the external unit which is managed by the doctor; it is used for stimulating specific nerve, pacemaker, functional electrical stimulators (FES) and for retinal implants. The major concern for an implantable antenna is miniaturization and its biocompatibility with the human body. There are several techniques for minimizing the antenna size, but the techniques involved results in decrease of antenna's efficiency and narrower impedance bandwidth. For bandwidth enhancement mode, coupling between proximity elements and addition of vertical lines to the meander strip is done. Antenna size miniaturization is based on decrease in the phase velocity. Biocompatibility achieved by coating material like polymethylsiloxane, polymethylmethacrylate (PMMA).

2 Proposed Methodologies

Methodologies to reduce the antenna size includes the use of high dielectric substrate materials, adding ground pins, planar inverted F antenna (PIFA), and spiraling the conductor shape (planar helix). In the 402–405 MHz MICS band both spiral and serpentine designs were found to be effective radiators for communication with medical implants. The antenna must be compatible with the human body causing no damage to the human tissues. SAR is a measure of power absorbed by the tissue, and it is evaluated for safety concerns caused by the implanted device on the human body. The standards of SAR are regulated by IEEE. Several design issues of implantable antennas have been discussed in [2, 3]. Miniaturization proposed for implantable patch antennas is achieved by using high-permittivity dielectric (substrate/superstrate) materials. Dielectrics having high permittivity like ceramic alumina ($\varepsilon r = 9.4$) and Rogers 3210 ($\varepsilon r = 10.2$) are used for implantable patch antennas to utilize their property of miniaturization of antenna because of shortening the effective wavelength which result in lower resonance frequencies. However, even with such high-permittivity dielectrics, the superstrate layer still insulates the antenna from the high-permittivity tissue. Hence for miniaturization purpose, a high-permittivity dielectric material is used for substrate and superstrate [4].

The operating frequency increases with increase in thickness of superstrates but it requires enhanced physical dimensions.

3 Design Methodology

3.1 Meander Slot Antenna

Meander line antenna is one type of microstrip antenna which allows designing antenna with a small size and provides wideband performance. We have chosen the slotted meander line antenna because it reduces the size of the antenna. Smaller size makes it very easy to be shifted. Factors including positions of slots and number of turns of the meander line affect the performance of the slotted meander line antenna. Number of meander slots also affects the result of meander line antenna. Meandered antenna increases the effective current flow path length on the patch surface. These phenomena are responsible for reducing the resonance frequency and can achieve a more compact size for the implantable antenna [5, 6].

3.2 Spiral-Slot Antenna

The main purpose behind using the spiral design is that spiral antenna has many turns which help in lengthening the current path. So, as per the theory, by lengthening the current path, the operating frequency decreases.

3.3 Shorting Pin

A shorting pin is also introduced between the ground and patch plane for miniaturization purpose. The effective size of antenna increases due to the shortening pins, and hence it reduces the required physical size or dimensions of the antenna at particular operating frequency. To obtain the resonance at the frequency of interest, the location of the coaxial feeding and the location of the shorting pin are varied. Significant reduction of volume is observed by using a shorting pin at appropriate location. The position and the diameter of the shorting pin affect the resonant frequency; as we decrease the diameter of the shorting pin, the resonant frequency decreases [7–12].

4 Proposed Antenna Design

4.1 Use of High Dielectric Substrate and Superstrate

In this method, the antenna patch is placed in-between two substrates, the upper substrate is called "superstrate." It is used to lower the frequency of the antenna; it also helps in size reduction of the antenna [13].

4.2 Change of Feeding Point

In this case, the feeding point of the antenna is moved from center of the antenna to the center of the joining of the spirals [14].

4.3 Use of Shorting Pin

The shorting pin on either side of the spiral is used; basically shorting pin lowers the antenna's operating frequency [15].

4.4 Separated Spiral with Change of Feed Point

In this case, the spirals are separated by introducing slot on the joining of spirals, and the feed point is moved [16].

4.5 Feed Location

The choice of the microstrip feed depends on its ease of fabrication and stability and impedance matching.

In this section, a different antenna design is explained with its effect on the antenna performance. In Fig. 1, simple spiral antenna is designed with feed location as the center. The size of the antenna is of ($6.4 \times 15.2 \times 1.27$) mm. The main purpose behind using the spiral design is that spiral antenna has many turns which help in lengthening the current path, so by the principle lengthening the current path decreases the operating frequency. The result of the same is shown in Fig. 2 as it

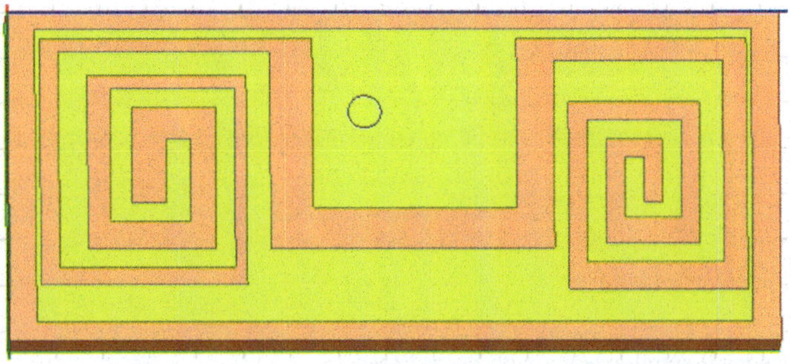

Fig. 1 Spiral antenna with center feed

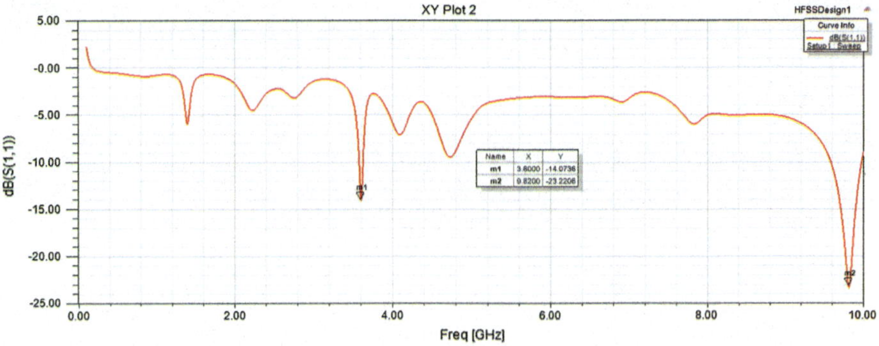

Fig. 2 S_{11} of antenna with center feed

can be seen that the antenna has many resonating frequencies and the lowest resonating one is 3.6 GHz. The operating frequency of the antenna is low but it is still not desired. So, further improvement is required.

As the part of improvement, the meandering of the antenna is done. Figure 3 shows the meandered antenna in which the spiral antenna is connected to the meander slot strip. The new dimension of the antenna is (8.8 × 15.2 × 1.27) mm. Here, the design lengthens the current path further, and the expected resonating frequency is very low. The result of the same is shown in Fig. 4. Now the lowest resonating frequency is 1.4 GHz. It is very clear that the meandering of the antenna does help in lowering the operating frequency but it is still not sufficient.

Figure 5 below shows the further improvement in the design to lower down the operating frequency. The feed location has been changed here. The new result is shown in Fig. 6.

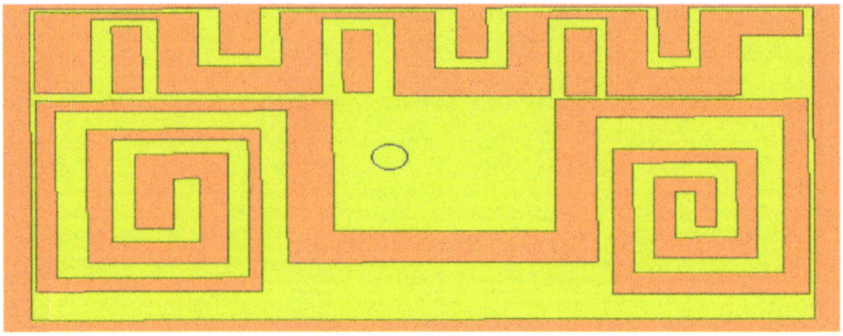

Fig. 3 Meandered spiral antenna with central feed

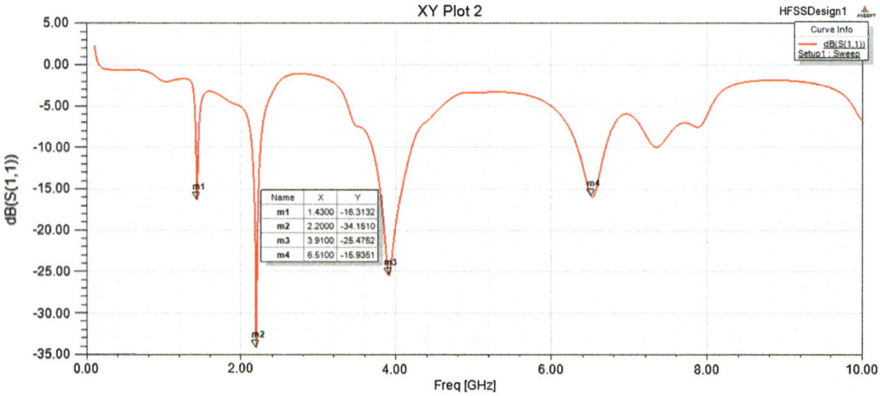

Fig. 4 S_{11} of meandered antenna

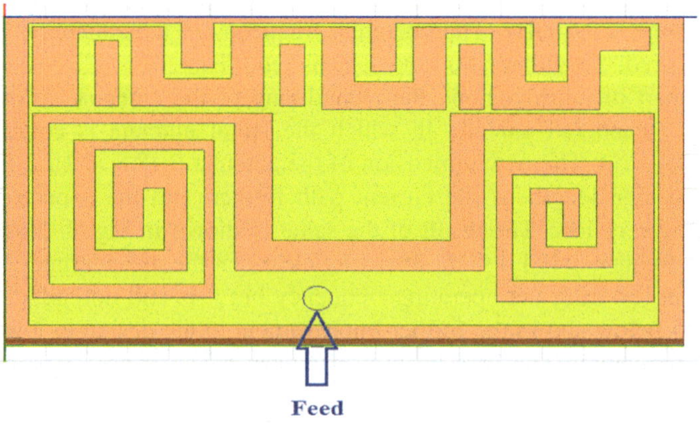

Fig. 5 Antenna with change of feed location

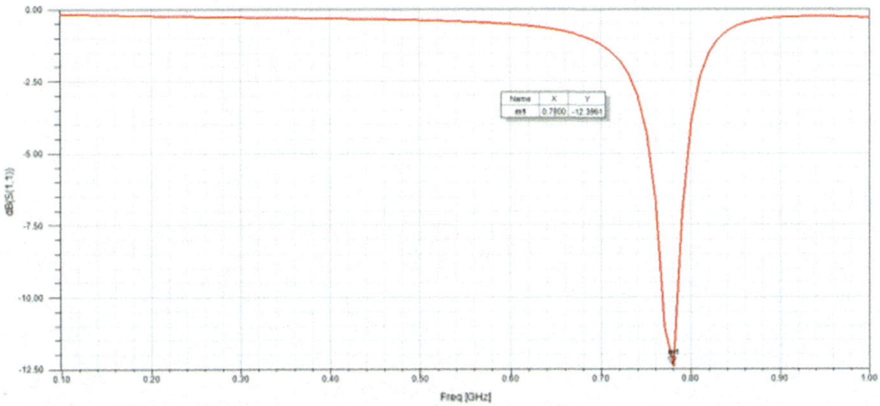

Fig. 6 S_{11} of antenna with change of feed location

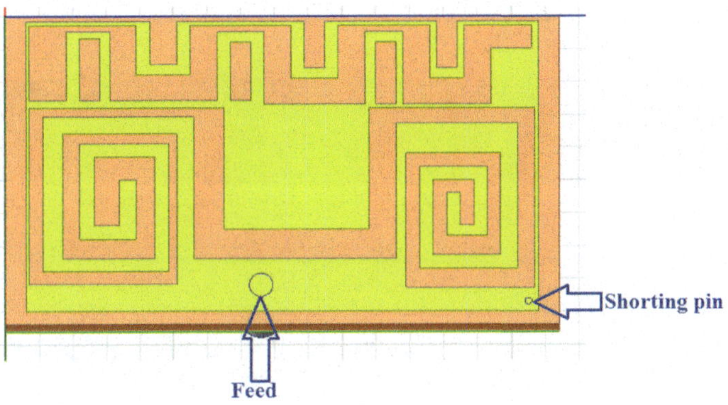

Fig. 7 Antenna with shorting pin

After the change of feed location, antenna shows very great improvement, and new resonating frequency is very low as much as 780 MHz with $S_{11} = -12$ dB. So, the change of feed location does help in lowering the operating frequency but the desired frequency band is still not achieved. So, further technique of lowering the operating frequency is used here. Figure 7 shows the use of the shorting pin; here, shorting pin creates the perturbation of current flow on the antenna patch leading to the lower resonating frequency. As it can be noticed, new resonating frequency now is 600 MHz. The return loss graph of the same is shown in Fig. 8.

The change of feed location and addition of shorting pin does help in lowering the operating frequency of the antenna, so far the lowest operating frequency achieved is about 600 MHz with $S_{11} = -15$ dB. But the desired operating frequency is still not achieved which is 405 MHz. So, again the feed location has been changed, and now the feed is moved on the upward side of the antenna. The arrangement is shown below in Fig. 9; the S_{11} of the same is shown in Fig. 10, which shows the new lowest operating frequency possible is 410 MHz.

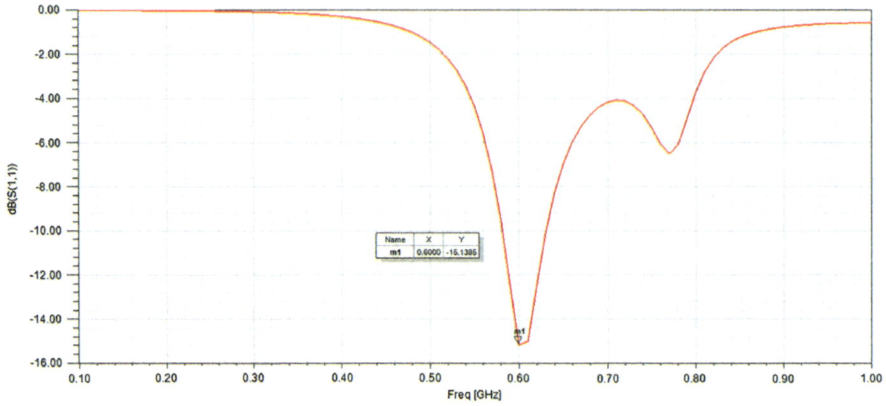

Fig. 8 S_{11} of antenna with shorting pin

Fig. 9 Antenna with shorting pin and change of feed

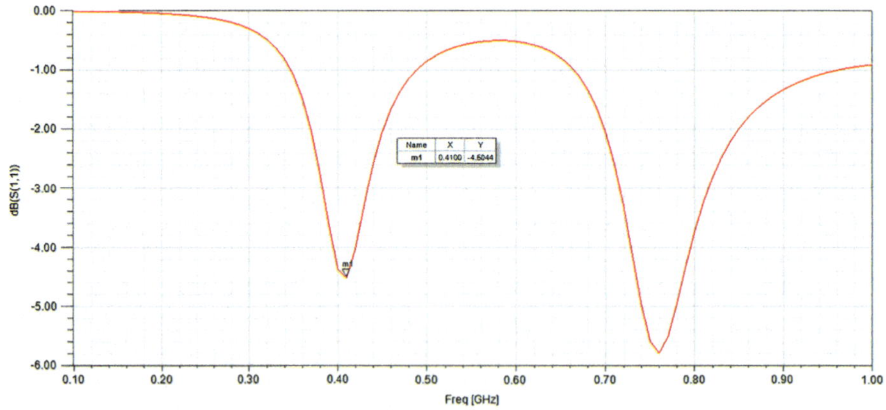

Fig. 10 S_{11} of antenna with shorting pin and change of feed

Here, $S_{11} = -4$ dB which is not sufficient for antenna to operate practically. Further improvement of S_{11} is required to make the antenna operate efficiently at 403–405 MHz band.

5 Conclusion

A wideband implantable antenna is proposed for neural data monitoring. The antenna operates in the MICs band. Combination of spiral-slot and meander slot is used. The results with two-spiral joined are analyzed with antenna of size (6.4 × 15.2 × 1.27) mm. For miniaturizing the antenna, meander slot is introduced in the design. Now the antenna size is (8.8 × 15.2 × 1.27) mm. The current path is increased which lowers down the frequency range and makes it operate in the lower medical frequency ranges in MHz For better tuning of the antenna, a shorting pin is introduced. Coaxial feed is used to feed the antenna which makes the design easy. Hence, summarizing the content into tabular form is shown in Table 1.

Table 1 Performance of spiral antenna with different configuration

S. No.	Design	Resonating frequency	Comment
1	Spiral antenna	3.6 and 9.8 GHz	Higher resonating frequency
2	Spiral with meander slot using Roger 6010	1.4, 2.2, 3.9, and 6.51 GHz	Lower resonating frequency
3	Spiral with meander slot using FR-4	2.1, 3.2, and 5.9 GHz	Shifting of resonating frequency on higher side
4	Use of high dielectric substrate and superstrate	780 MHz	Frequency lowered but $S_{11} > -10$ dB
5	Change of fed point	780 MHz	$S_{11} = -12.4$ dB
6	Use of shorting pin	600 MHz	$S_{11} = -15$ dB
7	Separate spiral and change of fed point	410 MHz	$S_{11} = -4.5$ dB

References

1. Rahimi M, Zarrabi FB, Ahmadian R, Mansouri Z, Keshtkar A (2014) Miniaturization of antenna for wireless application with difference metamaterial structures. Progress Electromagnet Res 145:19–29
2. Singh N, Singh AK, Singh VK (2015) Performance of wearable ultra wide band textile antenna for medical applications. Microwave Opt Technol Lett
3. Jacob N, Sam NA, George NA (2015) Design of wearable antenna. In: Proceedings of 4th IRF international conference, 19th April 2015
4. Ur-Rehman M, Abbasi QH, Akram M, Parini C (2014) Design of band-notched ultra wideband antenna for indoor and wearable wireless communications. IET Microwaves Antennas Propag 9(3):243–251
5. Agarwal K, Guo YX (2015) Interaction of electromagnetic waves with humans in wearable and biomedical implant antennas. In: APEMC
6. Liu C, Guo YX, Xiao S (2016) A review of implantable antennas for wireless biomedical devices. In: Forum for electromagnetic research methods and application technologies (FERMAT)
7. Liu H, Liu XY (2015) A miniaturized differentially fed implantable Hilbert curve fractal antenna at MICS band. IEEE
8. Rao S, Llombart N, Moradi E, Koski K, Bjorninen T, Sydanheimo L, Rabaey JM, Carmena JM, Rahmat-Samii Y, Ukkonen L (2014) Miniature implantable and wearable on-body antennas: towards the new era of wireless body-centric systems. IEEE Antennas Propag Mag 56(1)
9. Chamaani S, Akbarpour A (2015) Miniaturized dual-band omni directional antenna for body area network base stations. IEEE Antennas Wireless Propag Lett 14
10. Xue HG, Yang XX, Ma Z (2015) A novel micro strip-CPW fed planar slot antenna with broadband and circular polarization. IEEE Antennas Wireless Propag Lett 14
11. Basari, Sirait DC, Zulkifli FY, Rahardjo ET (2013) A helical folded dipole antenna for medical implant communication. In: IEEE MTT-S international microwave workshop series on RF and wireless technologies for biomedical and healthcare applications (IMWS-BIO)
12. Duan Z, Guo YX, Je M, Kwong DL (2014) Design and in vitro test of a differentially fed dual-band implantable antenna operating at MICS and ISM bands. IEEE Trans Antennas Propag 62(5)
13. Sarva E, Sridevi PV (2016) A novel triple band planar microstrip patch antenna with defected ground structure. Indian J Sci Technol 9(3)
14. Mezaal YS (2016) New compact microstrip patch antennas: design and simulation results. Indian J Sci Technol 9(12)
15. Rajaraman G, Anitha M, Mukerjee A, Sood K, Jyoti R (2015) Dual-band, miniaturized, enhanced-gain patch antennas using differentially-loaded metastructures. Indian J Sci Technol 8(1):11–16
16. Lak A, Parhizgar N, Lak M (2015) Effect of presence of human body on antenna gain. Indian J Sci Technol 8(30)

Mitigation of Cross-Phase Modulation in Multiband Radio Over Fiber Systems

Ch. Venkata Dharani and A. Brintha Terese

Abstract Cross-phase Modulation (XPM) is third-order nonlinear optical effect appears in wavelength division multiplexing (WDM) channel. XPM converts the power or intensity fluctuation in one channel to phase fluctuation in another channel. Due to this, the total refractive index seen by a particular channel is influenced by the power of that channel as well as optical intensity of the neighboring channels. The effect of XPM in WDM for radio over fiber (RoF) systems has been analyzed, and a mitigation process has been presented in this paper. The effect of XPM with and without RF signals is analyzed for different power levels at fixed dispersion value 20 ps/nm/km dispersion rate itself.

Keywords Cross-phase modulation · RF signal · BER · Dispersion Predistortion

1 Introduction

The rising demand by the users for video calls and vast data transmission had led to the emergence of wireless data transmission [1, 2]. The advantages of larger bandwidth, low attenuation loss, ease of maintenance, and installation made radio over fiber the flexible medium for the optical and wireless systems [3, 4]. The effect of XPM on wave length division multiplexed systems (WDM), in two different distances with nonzero dispersion value without RF signal is analyzed and observed that the lesser distance have better BER than longer distance [5]. The research had been done by varying the dispersion value from −4 to 4 ps/nm/km in order to reduce the effect of cross-phase modulation which resulted in the nonlinear curve of

Ch.Venkata Dharani · A. Brintha Terese (✉)
Department of Communication Engineering, School of Electronics Engineering, VIT University, Chennai, India
e-mail: abrinthatherese@vit.ac.in

Ch.Venkata Dharani
e-mail: chillarigae.venkatadharani2014@vit.ac.in

G. Gnanagurunathan et al. (eds.), *Optical and Microwave Technologies*, Lecture Notes in Electrical Engineering 468,
https://doi.org/10.1007/978-981-10-7293-2_31

the Q factor [6]. The intermodulation3 (IM3) nonlinear effect for producing the optical dual-side bands (ODSB) by introducing fiber dispersion for the RoF systems was analyzed [7].

The light wave traveling through the optical fiber will have little interaction with each other when the capacity of the channel is less and the interaction increases if the channel capacity increases. This nonlinearity leads to the distortion of the original signal and degradation of system performance [8]. The nonlinearities are of two different types. First kind depends on the variation in refractive index of the medium. They are self-phase modulation (SPM), cross-phase modulation (CPM), and four-wave mixing (FWM). Another category depends on inelastic scattering phenomenon, they are stimulated Raman scattering (SRS) and stimulated Brillouin scattering (SBS). The effect of nonlinearity is not negligible. There are various techniques proposed, and even more research is going on to compensate the nonlinearities. There are different techniques in which the RoF technology can be used such as RF over fiber (RoF-F), IF over fiber (IFoF), and baseband over fiber (BBoF). The main advantages of the RoF are ease of maintenance, installation, provides larger bandwidth, and flexible. The disadvantage is noise and distortion [9].

Along with high data rate transfer, coverage of longer distances is also a big issue for the information transfer. The longer distances are being limited up to 15–20 km due to the presence of the intermodulated3 (IM3) nonlinearity in the RoF systems. Many predistortion techniques had been proposed to suppress the IM3 and sub-carrier nonlinearities. A vector control theory was proposed in order to suppress the IM3 even after 20 km [10]. A predistortion circuit was introduced for linearizing the system for the efficient performance of the broadband RoF links. This circuit is responsible for the improvement of the system inferred power up to 2 dB [11]. A predistortion IC circuit had been introduced in front of the MZM in order to eliminate the second- and third-order intermodulated nonlinearity and also to linearize the MZM output. But the drawback of this technique is, it is bit costlier to implement and merit is it can also be applicable to linearize the higher modulation index [12]. A polynomial predistortive signal is given in order to linearize the RF power amplifiers so that the effect of IM3 nonlinearity in the ROF systems can be eliminated [13].

In the present work, analysis and mitigation of cross-phase modulation have been done for different power levels between 0 and 30 dBm at a dispersion rate of 20 ps/nm/km, whereas the mitigation process is done at 0 ps/nm/km dispersion.

2 Cross-Phase Modulation

The intensity-dependent characteristics give rise to the nonlinearity in the optics. The nonlinearities arise in the fiber due to the change in the refractive index of the medium. The main nonlinear effects due to this reason are self-phase modulation (SPM), cross-phase modulation (CPM), and four-way mixing (FWM). In WDM channel, intensity of one beam changes the phase of the other beam due to Kerr effect.

The cross-phase modulation is caused due to the interaction of the two signals passing through the same medium which results in the occurrence of error or distortion in the output signal. The cross-phase modulation is bit similar to the self-phase modulation, which means the refractive index of pulse not only depends on its own refractive index but also on the pulse traveling simultaneously in another channel. Cross-phase modulation introduces cross talk in the system, produces amplitude and time jittering, and causes pulse broadening. The effect of the cross-phase modulation in WDM system is analyzed.

3 Methodology

Figure 1 shows the block diagram to analyze the cross-phase modulation without RF signal and Fig. 2 shows the block diagram to analyze the XPM with RF signal.

The Mach–Zender modulator (MZM) is used to modulate the signals given to it. Continuous wave (CW) at 1552.52 nm with line width of 10 MHz is given to the MZM as the light source and the pseudorandom generator gives the data source in the form of NRZ pulses to the MZM as shown in Fig. 1. For the normal WDM systems, the change in refractive index occurs due to the Kerr effect for which the RF signal is not given whereas to the RoF systems, the RF signal acts as carrier wave and it is responsible for the change in the refractive index in the system which causes the cross-phase modulation. The sinusoidal RF signal with 28 GHz along with bias circuit is given to the MZM as shown in Fig. 2. The modulated and multiplexed data are transmitted through the optical fiber of a length 50 km to the remote access unit (RAU). In the RAU, the obtained optical signal will be converted to the electrical signal with the help of photodetector and filtered with help of a Bessel filter and then transmitted to the end users.

Fig. 1 Multiband WDM system

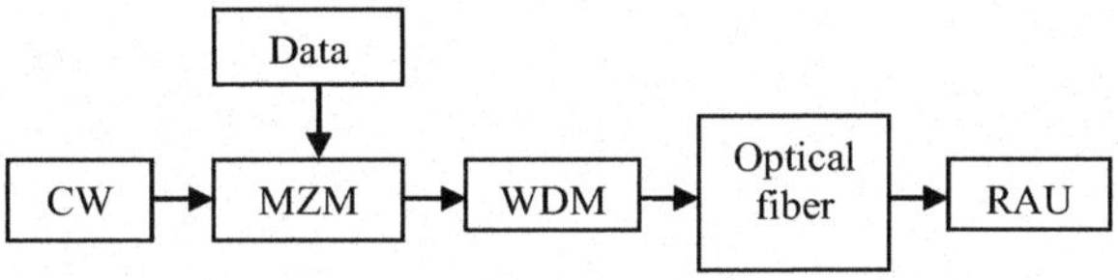

Fig. 2 Multiband radio over fiber system

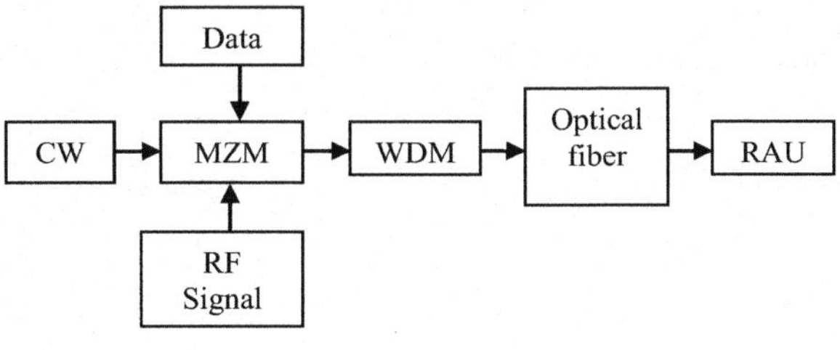

Phase shift keying (PSK) and offset quadrature pulse shift keying (OQPSK) are the two modulation techniques used in order to have better performance of the system. Here, the pseudorandom generated data will be given to the PSK and OQPSK modulators and the modulated data are then given to the MZM.

4 Performance Evaluation

Let us now see the performance of the system with different power levels.

4.1 NRZ Pulse Train

The data source from the pseudorandom generator is given to the NRZ generator in order to get the NRZ pulse train to give as input data source to the MZM. The effect of cross-phase modulation observed is more as the power level increases. The power range is varied between 0 and 30 dBm. The output obtained in the form of eye diagram with and without RF signal are shown in Figs. 3 and 4.

The performance of WDM system for different power levels is tabulated in Tables 1 and 2.

The observation from Tables 1 and 2 says that as the power increases the effect of cross-phase modulation also increases which results in increase of number of error bits. The Q factor value also reduces due to the effect of XPM.

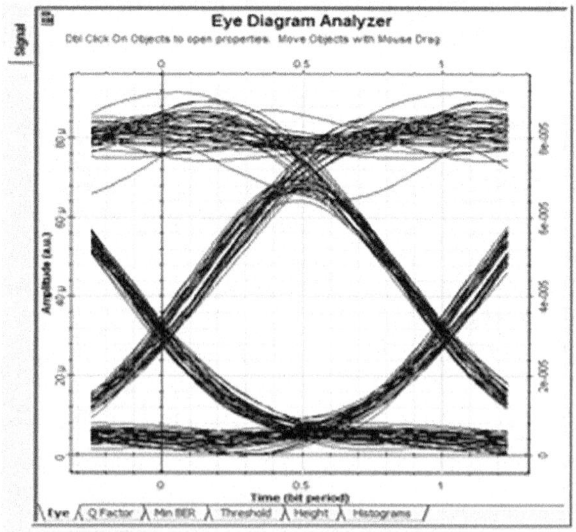

Fig. 3 BER for power level 0 dBm for a dispersion of 20 ps/nm/km without RF signal

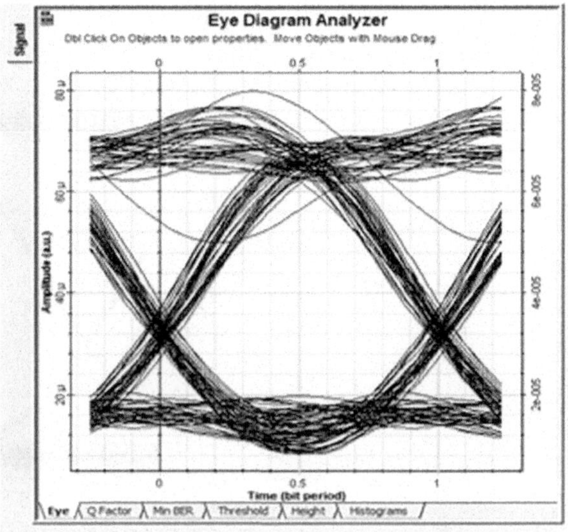

Fig. 4 BER for power level 0 dBm for a dispersion of 20 ps/nm/km with RF signal

Table 1 System performance without RF signal at 20 ps/nm/km

Power in dBm	Q factor	BER
0	10.8071	1.37932e−27
10	10.4268	8.5001e−27
20	7.6522	6.1807e−15
30	–	Infinite

Table 2 System performance with RF signal at 20 ps/nm/km

Power in dBm	Q factor	BER
0	10.598	1.47736e−26
10	8.21723	1.0357e−16
20	4.39158	4.69772e−6
30	2.10893	0.0126024

4.2 PSK Pulse Train

The data source from the pseudorandom generator is given to the PSK modulator in order to get the PSK pulse train to give as input data source to the MZM. The effect of cross-phase modulation observed is more as the power level increases. The power range is varied between 0 and 30 dBm. The output obtained with and without RF signal are shown in Figs. 5 and 6.

The performance of WDM system for different power levels are tabulated in Tables 3 and 4.

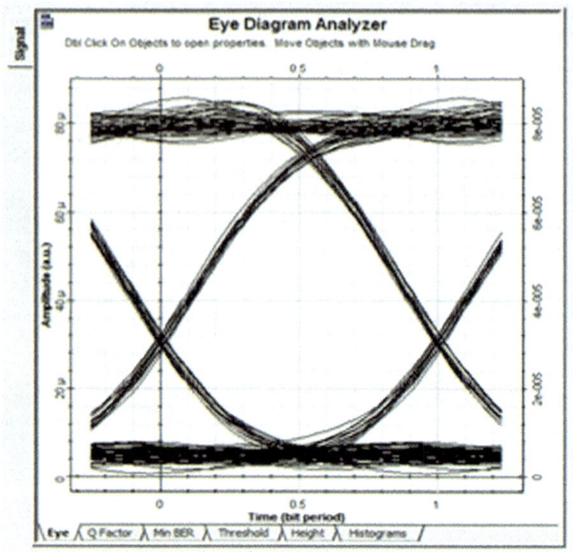

Fig. 5 BER for power level 0 dBm for a dispersion of 20 ps/nm/km without RF signal

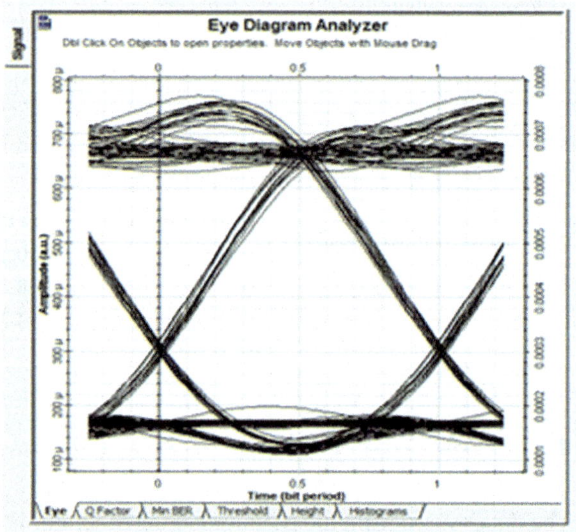

Fig. 6 BER for power level 0 dBm for a dispersion of 20 ps/nm/km with RF signal

The observation from Tables 3 and 4 says that as the power increases the effect of cross-phase modulation increases which results in the increase of number of error bits and decrease of Q factor. When compared to the normal NRZ pulse train the number error bits were less in the input side in PSK modulation but as the power increases the number of error bits in the output became infinite.

Table 3 System performance without RF signal at 20 ps/nm/km

Power in dBm	Q factor	BER
0	13.5015	6.45579e−42
10	9.35212	3.66651e−21
20	3.88114	4.25477e−5
30	–	Infinite

Table 4 System performance with RF signal at 20 ps/nm/km

Power in dBm	Q factor	BER
0	15.0052	3.32482e−51
10	11.9595	2.87404e−33
20	6.74627	6.62565e−12
30	–	Infinite

4.3 OQPSK Pulse Train

The data source from the pseudorandom generator is given to the OQPSK modulator in order to get the OQPSK pulse train to give as input data source to the MZM. The effect of cross-phase modulation observed is more as the power level increases. The power range is varied between 0 and 30 dBm. The output obtained with and without RF signal are shown in Figs. 7 and 8.

The performance of WDM system for different power levels are tabulated in Tables 5 and 6.

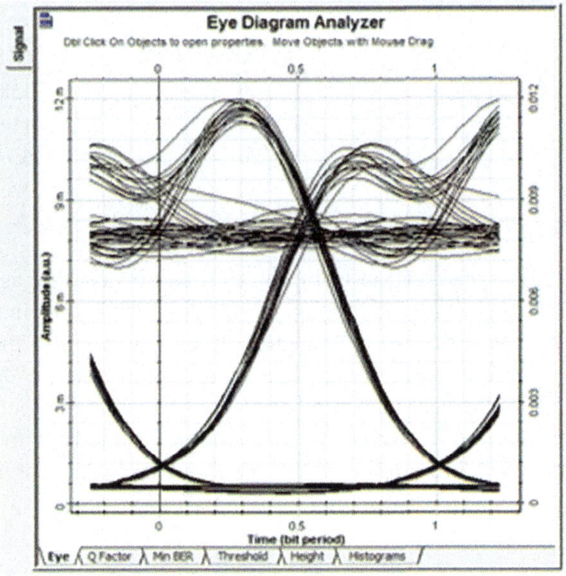

Fig. 7 BER for power level 0 dBm for a dispersion of 20 ps/nm/km without RF signal

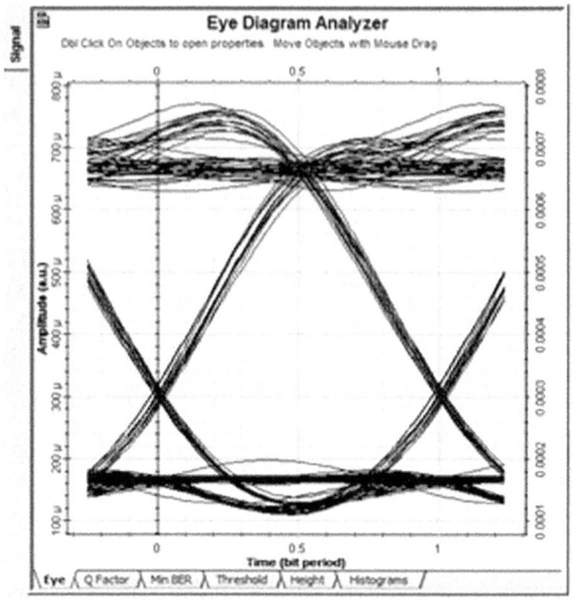

Fig. 8 BER for power level 0 dBm for a dispersion of 20 ps/nm/km with RF signal

Table 5 System performance without RF signal at 20 ps/nm/km

Power in dBm	Q factor	BER
0	14.7126	2.19104e−49
10	9.50215	8.6787e−22
20	3.93764	3.3892e−5
30	2.53731	0.00539969

Table 6 System performance with RF signal at 20 ps/nm/km

Power in dBm	Q factor	BER
0	15.0052	3.32482e−51
10	11.9595	2.87404e−33
20	6.74627	6.62565e−12
30	–	Infinite

The observation from Tables 5 and 6 says that as the power increases the effect of cross-phase modulation increases which results in the increase of number of error bits and decrease of Q factor. When compared to the normal NRZ pulse train, the number error bits were less in the input side in OQPSK modulation, but as the power increases the number of error bits in the output became infinite for the system containing RF signal. When compared to the PSK modulation, OQPSK modulation has very little difference in the input BER and output remains the same.

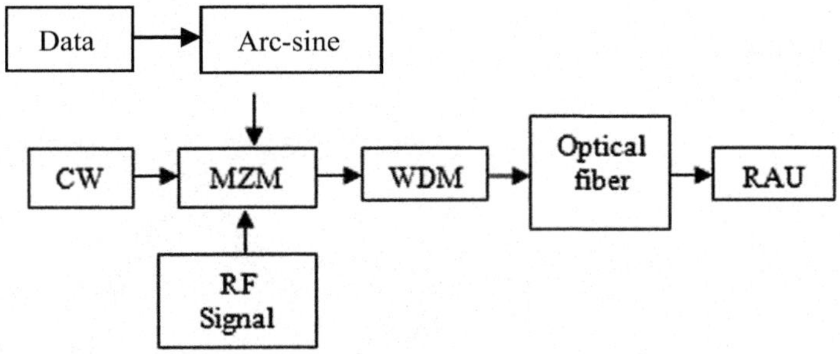

Fig. 9 Block diagram of predistortion technique

5 Predistortion

The mitigation of cross-phase modulation is done with the help of the predistortion block with an arcsine signal which compensates the nonlinearities present in the signal. The block diagram of the predistortion technique is as shown in Fig. 9.

The arcsine signal is given before the MZM in order to linearize signal and reduce the effect of nonlinearities present in the system. The simulation results of the NRZ, PSK, and OQPSK pulse train were given below. The Q factor and BER values are improved at 0 ps/nm/km dispersion when compared to the values at 20 ps/nm/km dispersion.

6 Results and Analysis

The results of the WDM systems with and without RF signal and also with predistortion are discussed below.

6.1 NRZ Pulse Train

The effect of XPM modulation has been reduced due to the precompensation of the signal with an arcsine signal. The number of error bits has been reduced when compared to the normal systems without predistortion block. The eye-diagram in Fig. 10 shows the analysis of the WDM system with RF and predistortion for NRZ pulse train.

The performance analysis for different power levels is tabulated in Table 7.

The observation from Table 7 shows that the performance of the system has been improved when compared to the systems with and without RF signals. The performance of the systems in the absence of predistortion is observed to be good at

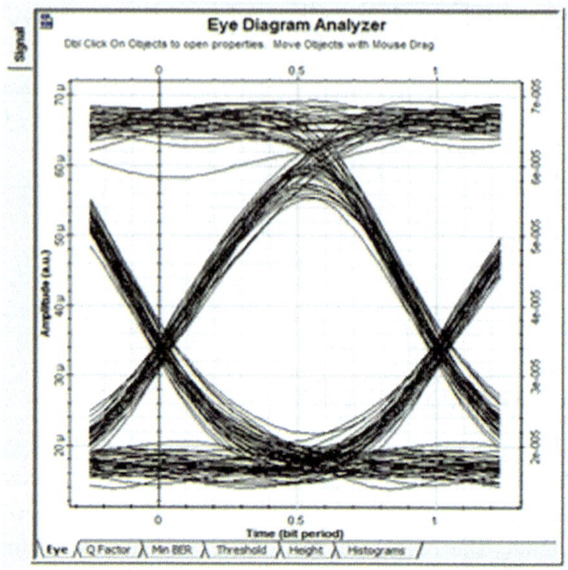

Fig. 10 BER for power level 0 dBm for a dispersion of 0 ps/nm/km with RF signal

Table 7 System performances with RF signal at 0 ps/nm/km

Power in dBm	Q factor	BER
0	9.30512	6.15425e−21
10	8.47233	1.04171e−17
20	7.01478	9.89993e−13
30	2.51413	0.00456827

20 ps/nm/km dispersion rate, but in the presence of predistortion the system performance is effective at 0 ds/nm/km dispersion itself.

6.2 *PSK Pulse Train*

The effect of XPM modulation has been reduced due to the precompensation of the signal with an arcsine signal. The number of error bits has been reduced when compared to the normal systems without predistortion block. The eye-diagram in Fig. 11 shows the analysis of the WDM system with RF and predistortion for PSK pulse train.

The performance analysis for different power levels is tabulated in Table 8.

The observation from Table 8 shows that the performance of the system has been improved when compared to the systems with and without RF signals. The performance of the systems in the absence of predistortion is observed to be good at 20 ps/nm/km dispersion rate, but in the presence of predistortion the system

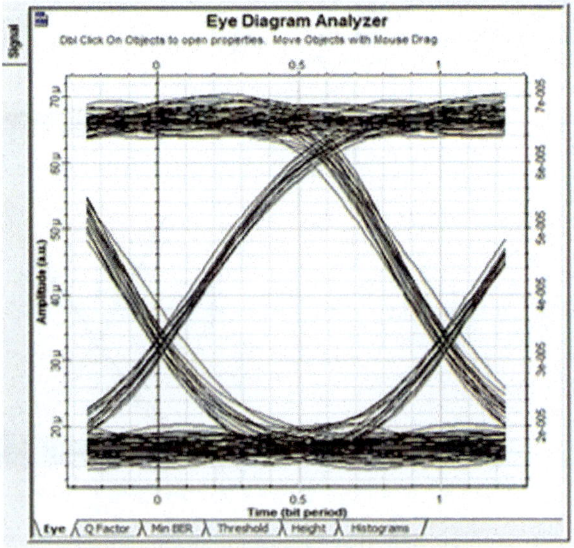

Fig. 11 BER for power level 0 dBm for a dispersion of 0 ps/nm/km with RF signal

Table 8 System performances with RF signal at 0 ps/nm/km

Power in dBm	Q factor	BER
0	12.1543	2.45246e−34
10	10.3757	1.32406e−25
20	7.80719	2.36799e−15
30	3.03815	8.76635e−3

performance is effective at 0 ds/nm/km dispersion itself. The performance of the PSK is much better than the NRZ pulse train.

6.3 *OQPSK Pulse Train*

The effect of XPM modulation has been reduced due to the precompensation of the signal with an arcsine signal. The number of error bits has been reduced when compared to the normal systems without predistortion block. The eye-diagram in Fig. 12 shows the analysis of the WDM system with RF and predistortion for PSK pulse train.

The performance analysis for different power levels is tabulated in Table 9.

The observation from Table 9 shows that the performance of the system has been improved when compared to the systems with and without RF signals. The performance of the systems in the absence of predistortion is observed to be good at 20 ps/nm/km dispersion rate, but in the presence of predistortion the system performance is effective at 0 ds/nm/km dispersion itself. The performance of the

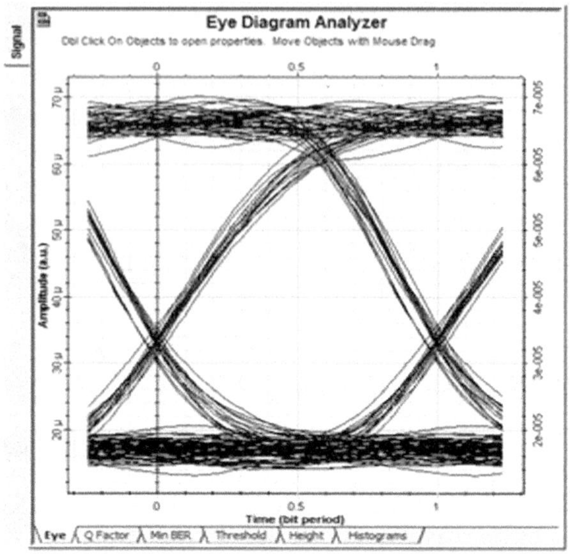

Fig. 12 BER for power level 0 dBm for a dispersion of 0 ps/nm/km with RF signal

Table 9 System performances with RF signal at 0 ps/nm/km

Power in dBm	Q factor	BER
0	12.0753	6.43944e−34
10	10.2423	5.51746e−25
20	7.69301	6.3367e−15
30	3.14533	7.005e−3

OQPSK is much better than the NRZ pulse train and nearly equal to the performance of the PSK.

7 Conclusion

Hence, the effect of cross-phase modulation has been analyzed and performance difference between the normal NRZ pulse train and the PSK, OQPSK pulse train has also been observed. The offset QPSK gives better performance than the other two pulse trains, even though it has infinite error at the maximum power level. But the offset QPSK gives limited error rate at higher power levels in the presence of arcsine signal which compensates the nonlinearity. The performance of the systems has been improved by using the predistortion technique and can be implemented for the micrometer wave applications.

References

1. Smulders P (2002) Exploiting the 60 GHz band for local wireless multimedia access: prospects and future directions. IEEE Commun Mag 40(1):140–147
2. Park C, Rappaport TS (2007) Short-range wireless communications for next-generation networks: UWB, 60 GHz millimeter-wave WPAN, and ZigBee. IEEE Wireless Commun 14 (4):70–78
3. Jia Z, Yu J, Ellinas G, Chang G-K (2007) Key enabling technologies for optical-wireless networks: optical millimeter-wave generation, wavelength reuse, and architecture. J Lightw Technol 25(11):3452–3471
4. Chang G-K, Chowdhury A, Jia Z, Chien H-C, Huang M-F, Yu J, Ellinas G (2009) Key technologies of WDM-PON for future converged optical broadband access networks. IEEE/ OSA J Opt Commun Netw 1(4):C35–C50
5. Saini G, Gupta D, Maurya N (1956) Effect on WDM due to cross phase modulation. Int J Electron Comput Sci Eng. ISSN-2277-1956
6. Kashyap K, Singh H, Singh P, Gupta C. Effect of cross phase modulation (XPM) on optical fiber using two wavelength division multiplexed (WDM) channels. Int J Emerg Technol Comput Appl Sci (IJETCAS)
7. Cho TS, Kim K (2006) Optimization of radio-on-fiber systems employing ODSB signals by utilizing a dual-electrode Mach–Zehnder modulator against IM3. IEEE Photon Technol Lett 18(9)1076–1078
8. Thing VLL, Shum P, Rao MK (2004) Bandwidth-efficient WDM channel allocation for four-wave mixing-effect minimization. Trans Commun 52(12):2184–2089
9. Pooja M, Saroj S, Manisha B (2015) Advantages and limitation of radio over fiber system. IJCSMC 4(5):506–511. ISSN 2320–088X
10. Park S-H, Choi Y-W (2005) Significant suppression of the third intermodulation distortion in transmission system with optical feedforward linearized transmitter. IEEE Photon Technol Lett 17(6):1280–1282
11. Hraimel B, Zhang X (2012) Low-cost broadband predistortion-linearized single-drive x-cut mach-zehnder modulator for radio-over-fiber systems. IEEE Photon Technol Lett 24 (18):1571–1573
12. Sadhwani R, Jalali B (2003) Adaptive CMOS predistortion linearizer forfiber-optic links. J Lightw Technol 21(12):3180–3193
13. Shah AR, Jalali (2005) Adaptive equalisation for broadband predistortion linearisation of optical transmitters. IEE Proc Optoelectron 152:16–32

The Design of High Gain Substrate Integrated Waveguide Antennas with FR4 and RT Duroid

Caroline Sebastian, V. J. Amirtha Vijina and R. Ramesh

Abstract Background: The design and implementation of circularly polarized substrate integrated waveguide (SIW) antenna in *X* band using the FR4 and RT Duroid is presented in this paper. The substrate integrated waveguides represent the family of substrate integrated circuits which is planar in nature and used for the transmission of electromagnetic energy. **Methods/Statistical Analysis**: In this paper, the design techniques for substrate integrated waveguide using slots with the integration of microstrip with the substrates FR4 and RT Duroid are presented, which emphasize the broad perspective of the substrate integrated waveguide in circular polarization. **Findings**: The design has simulated in HFSS Software. From the simulation results, it is observed that gain has increased in SIW structure with slots for circular polarization as compared to SIW structure without slots also by comparing the substrate structure with RT Duroid shows good gain than the structure with FR4.

Keywords Substrate integrated waveguide · RT Duroid · Vias
Microstrip tapering

1 Introduction

The transmission of electromagnetic energy is accomplished conventionally by two conductor lines such as parallel line, coaxial cables and two wire cables. The energy will not travel through normal electrical wire without great losses [1]. It can encounter large losses such as radiation loss, copper loss, dielectric loss and skin effect. Also the transmission lines encounter difficulty in interconnecting with the planar circuits due to its non-planar nature. Waveguides are the better option for the transmission of electromagnetic energy in comparison with transmission lines [2].

C. Sebastian · V. J. Amirtha Vijina · R. Ramesh (✉)
School of Electronics Engineering (SENSE), VIT University Chennai Campus,
Chennai, India
e-mail: ramesh.R@vit.ac.in

G. Gnanagurunathan et al. (eds.), *Optical and Microwave Technologies*,
Lecture Notes in Electrical Engineering 468,
https://doi.org/10.1007/978-981-10-7293-2_32

Since waveguides are non-planar and bulky, substrate integrated waveguide can be a substitute option for the transmission of the electromagnetic energy. But it has the benefits of conventional waveguides such as high power handling capability and high-quality factor [1].

Communication systems design makes use of either linear (vertical or horizontal) or circular polarization. Due to the advancement of signal propagation characteristics, circularly polarized antennas play a major role. The axial ratio is the important parameter to determine whether the polarization is circular [3]. It is the ratio of the E-field orthogonal components. A circularly polarized field consists of two orthogonal electric field components of equal amplitude and phase difference of 90. The axial ratio for circularly polarized antenna is preferably 1, since the E-field components are equal in magnitude [3].

2 SIW Antenna Design

Substrate integrated waveguides (SIW) is basically dielectric integrated waveguide structure which has the advantages of both waveguides and planar transmission lines. SIW structures are fabricated by connecting two parallel metal plates which contain two rows of conducting cylinders and slots embedded in a dielectric substrate. The SIW has propagation characteristics similar to rectangular waveguides. Like rectangular waveguides, the fundamental mode in the SIW is TE^{10} mode [4]. The design constraints in rectangular waveguides are length and width due to the relation with cut-off frequency. The physical parameters that play importance in design are '*d*' the diameter of the vias (holes), '*p*' the centre to centre distance between the vias, '*h*' the height of the substrate, '*a*' spacing between the centre to centre distance of two rows of holes, 'aequ' width of the dielectric-occupied rectangular waveguide.

The width of dielectric-filled rectangular waveguide is related to the width of conventional waveguide by the dielectric constant.

$$\text{aequ} = \text{acon}/€_{\text{r}} \tag{1}$$

In our proposal, we concentrated on the *X* band application, so the antenna is designed to operate on 10 GHz. The cut-off frequency related to velocity of light '*c*' and 'aequ' the width of the dielectric-occupied rectangular waveguide.

$$fc = c/(2\text{aequ}\sqrt{€_r}) \tag{2}$$

The spacing between the centre to centre distances of two rows of holes '*a*' can be calculated by

$$a = \text{aequ} + \left(d^2/0.95p\right). \tag{3}$$

The centre to the centre distance between the holes 'p' and diameter of the vias 'd' is chosen such that the leakage radiation is minimum.

$$p > d \tag{4}$$

The phenomena of electromagnetic band gap can occur in periodic structures. The effects of bandstop in SIW can be tackled by considering the bandwidth of the operation and the centre to the centre distance between the holes.

$$(p/\lambda c) > 1/4. \tag{5}$$

The SIW shows the fundamental mode TE^{10} as similar to rectangular waveguide. TM modes cannot propagate through the SIW due to the presence of vias in the structure. The holes in the structure result in a large radiation for the surface current. This thwarts the propagation of TM modes in the SIW structure.

By means of above design constraints 'd' the diameter of the vias (holes), 'p' the centre to centre distance between the vias, 'h' the height of the substrate, 'a' spacing between the centre to centre distance of two rows of holes, 'aequ' width of the dielectric-occupied rectangular waveguide, a conventional waveguide can be adapted to the SIW.

FR4 is a grade assigned to a glass-reinforced epoxy laminate sheets, rods and printed circuit boards (PCB). It is nothing but a composite material that contains the materials such as woven fibreglass cloth with an epoxy resin binder which is flame resistant. In the design of SIW using the FR4 substrate of dielectric constant of 4.4 and the height of the substrate is 1.6 mm. RT Duroid high frequency circuit materials are basically filled PTFE (random glass or ceramic) composite laminates for use in high reliability, aerospace and defence applications while the design of SIW using RT Duroid the dielectric constant of 2.2 and the substrate height is chosen as 1.575.

3 Design of Microstrip Transition

The design of the SIW structure in planar circuits has to integrate with conventional transmission lines such as microstrip line. The portion of radiating surface in the microstrip line has to taper with the SIW structure for the proper impedance matching. Thus, the design of SIW with microstrip is needed.

The width to height ratio of the microstrip can be obtained by the equations such as

$$\begin{aligned} {}^{w}/_{d} &= {}^{8e^{A}}/_{e^{2A}} - 2; \quad {}^{w}/_{d} < 2 \\ &\frac{2}{\pi}\left\{B - 1 - \ln(2B - 1) + \frac{€_r - 1}{2€_r}\left\{\ln(B - 1) + 0.39 - \frac{0.61}{€_r}\right\}\right\} \end{aligned} \tag{6}$$

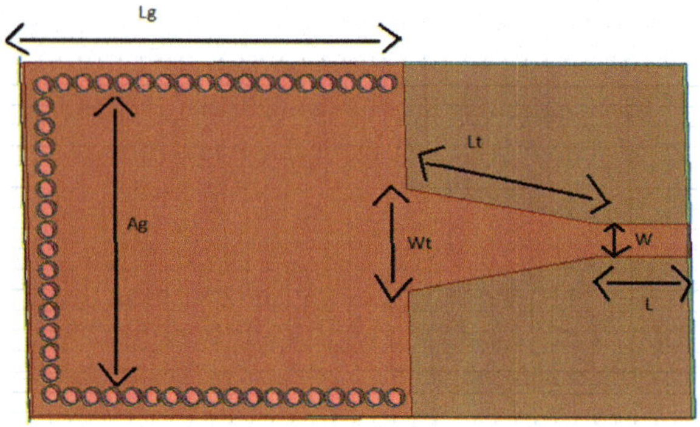

Fig. 1 Microstrip transition to the waveguide

where A and B in Eq. 6 are

$$A=\frac{Z_0}{60}\sqrt{\frac{€_r+1}{2}}+\frac{€_r-1}{€_r+1}\left(0.23+\frac{0.11}{€_r}\right) \quad (7)$$

$$B=\frac{377\pi}{2Z_0\sqrt{€_r}} \quad (8)$$

The tapering is essentially an interconnection with microstrip width to waveguide width. The microstrip width W_t and tapering width A_g is related such that

$$W_t/A_g=0.4. \quad (9)$$

Also the wavelength of waveguide and cut-off wavelength is related as

$$\lambda_c=c/f_c \quad (10)$$

$$\lambda_g=\lambda_c/€_r \quad (11)$$

See Fig. 1.

4 Design of SIW Slot Antenna for Circular Polarization

Our objective is the design of the slots in SIW structure for the circular polarization in X band applications. The length of slots can be achieved from equation as

$$L=\lambda_0/\sqrt{2}(€_r+1) \quad (12)$$

The length from the wall of waveguide to centre of slot is considered as $\lambda_g/4$, and distance between the slots is taken as $\lambda_g/2$.

Using the above equations for the dimensions of design slots can be obtained in circular polarization. Here, we illustrated a SIW and microstrip transition in circular polarization for *X* Band. The design shows good properties if the reflection coefficient is less than −15 dB. For the design purposes, the width of conventional waveguide is used as 18.71 mm.

The simulation results for SIW in circular polarization and its E-field pattern are shown in Figs. 3 and 4. The parameters for SIW and its microstrip transition design are given.

W_t (mm)	W (mm)	L_t (mm)	λ_c (mm)	$€_r$	λ_g (mm)	A_g (mm)
7.5	3.23	18.75	37.5	4.4	17.88	18.75

In the design parameters of SIW using FR4 and its microstrip transition, the length is indicated in millimetre and frequency is indicated in GHz (Figs. 2, 3, 4, 5, 6, 7, 8, 9, 10 and 11).

While the design of SIW using the RT Duroid for the dimensions of design slots in circular polarization, dielectric constant of 2.2 and height of substrate is chosen as 1.575 mm.

W (mm)	L_t (mm)	$€_r$	λ_g (mm)	A_g (mm)	Λ_c (mm)	W_t (mm)
3.126	5.066	2.2	20.22	10.11	30	4.044

In the design parameters of SIW using RT Duroid and its microstrip transition, the length is indicated in millimetre and frequency is indicated in GHz (Figs. 12, 13, 14, 15, 16, 17, 18, 19, 20 and 21).

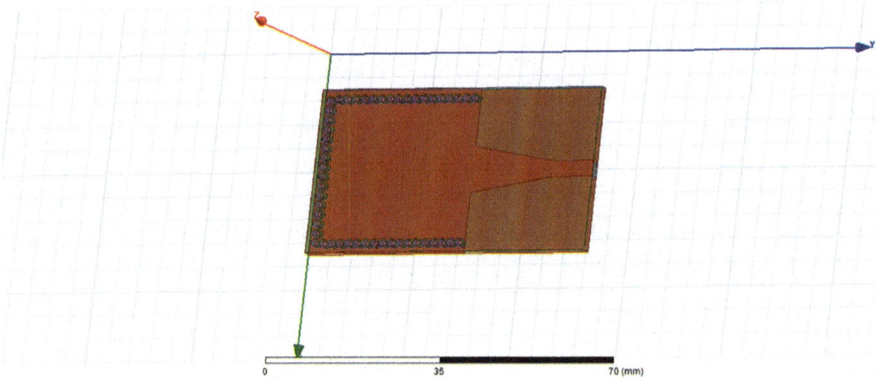

Fig. 2 Proposed structure of SIW antenna without slots using FR4 substrate

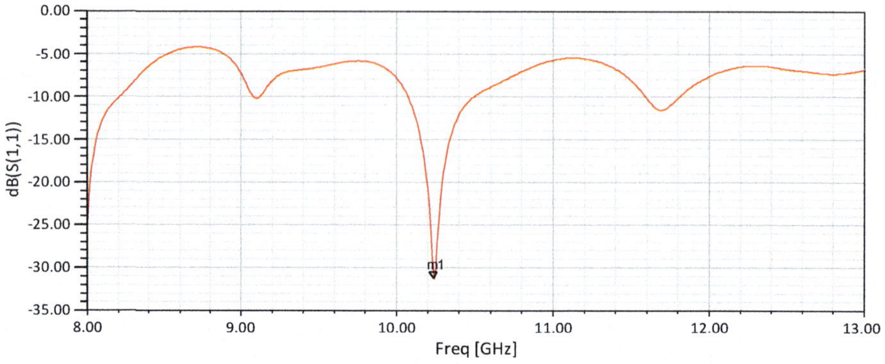

Fig. 3 Graph showing reflection coefficient and frequency of SIW structure without slots using FR4 substrate

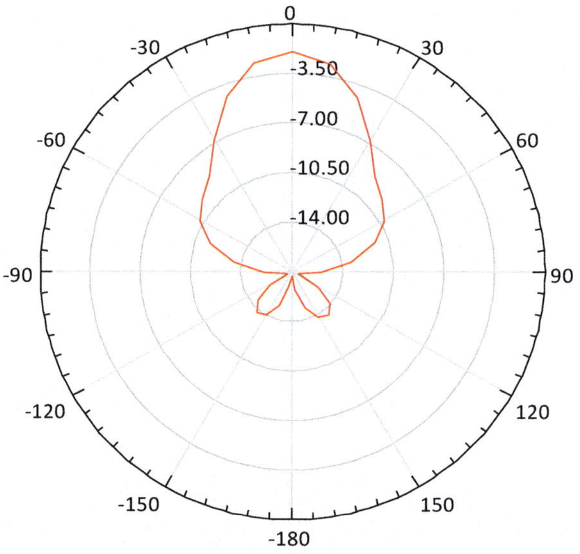

Fig. 4 Radiation pattern of SIW antenna without slots using FR4 substrate

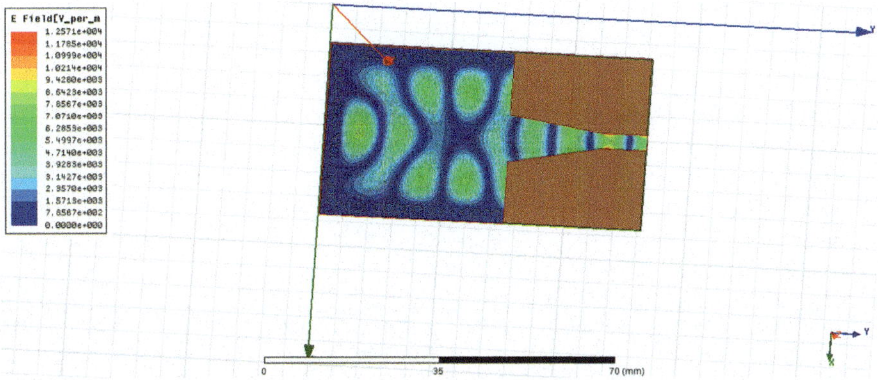

Fig. 5 E-field pattern for SIW without slots

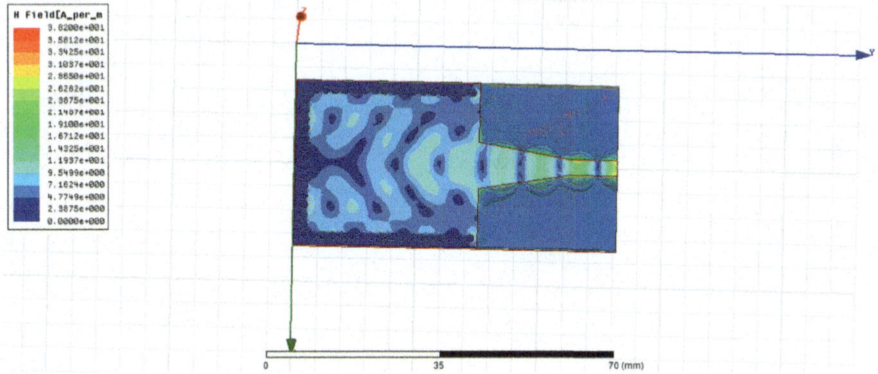

Fig. 6 H-field pattern for SIW without slots

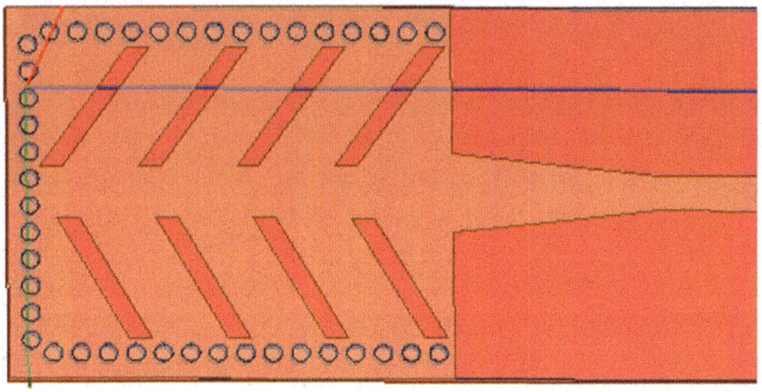

Fig. 7 Proposed structure of SIW antenna with slots

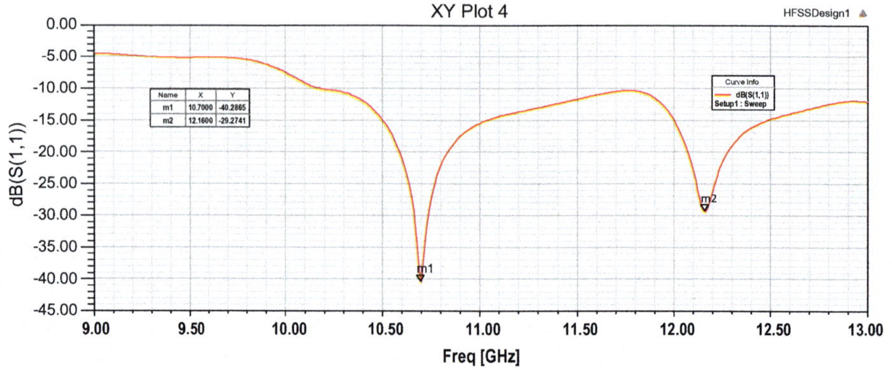

Fig. 8 Graph showing reflection coefficient and frequency in the circular polarization

Fig. 9 Radiation pattern of SIW antenna with slots in the circular polarization

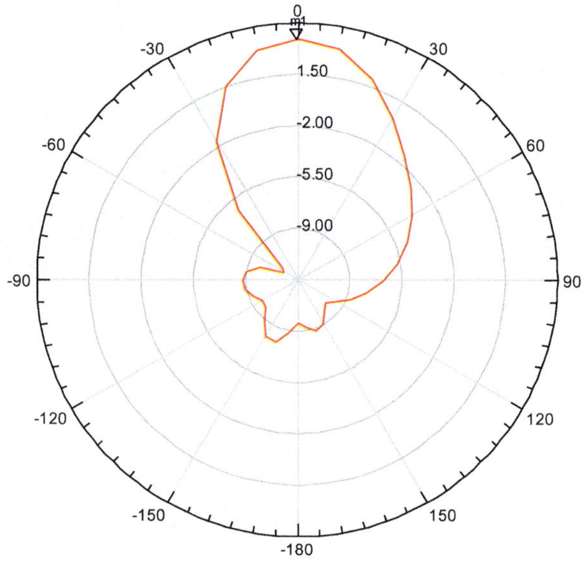

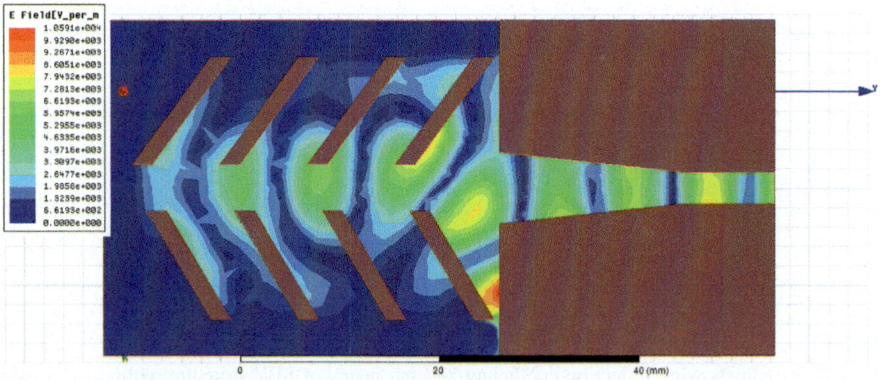

Fig. 10 E-field pattern for SIW with slots

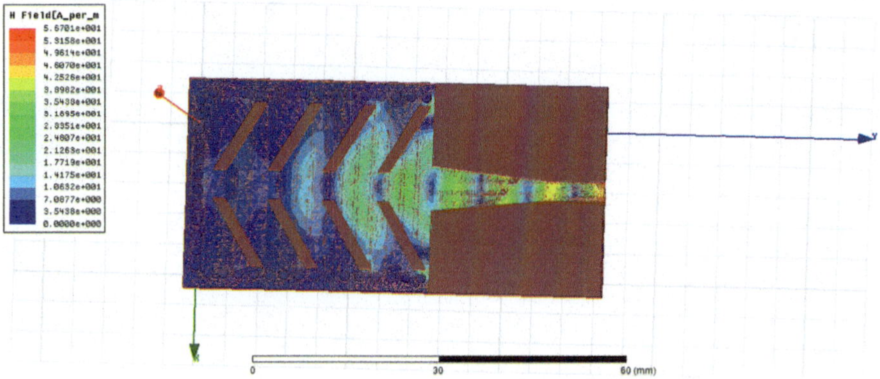

Fig. 11 E-field pattern for SIW with slots

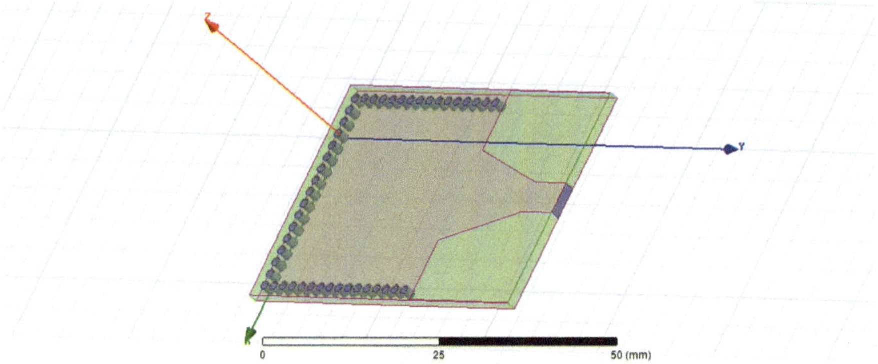

Fig. 12 Proposed structure of SIW antenna without slots using RT Duroid substrate

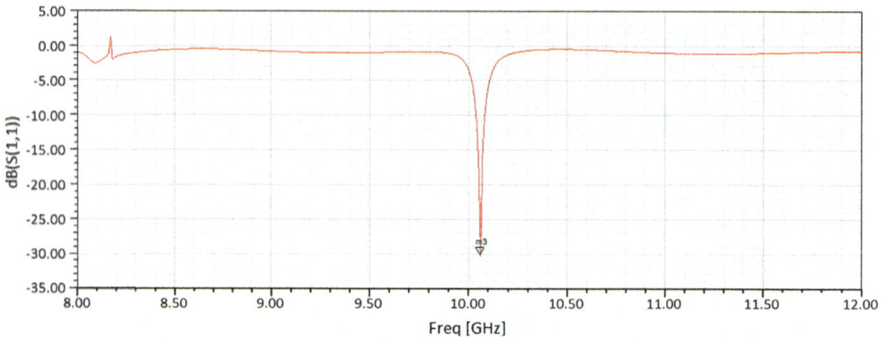

Fig. 13 Graph showing reflection coefficient and frequency of SIW structure without slots using RT Duroid substrate

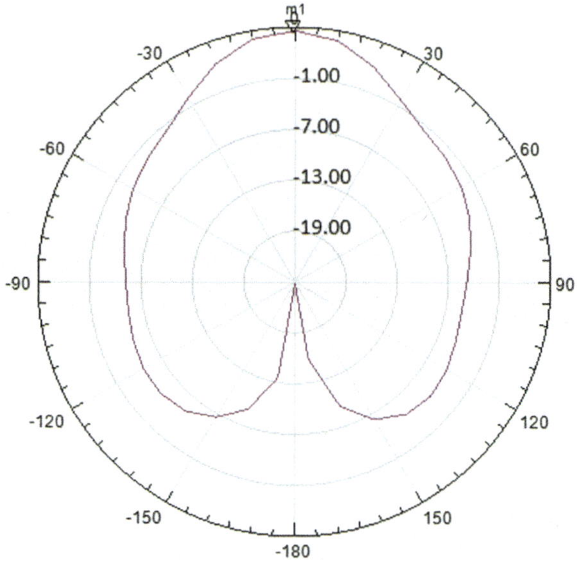

Fig. 14 Radiation pattern of SIW antenna without slots using RT Duroid

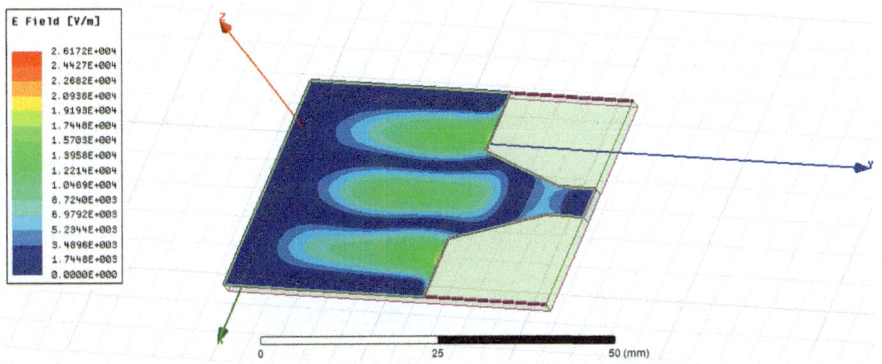

Fig. 15 E-field pattern for SIW without slots

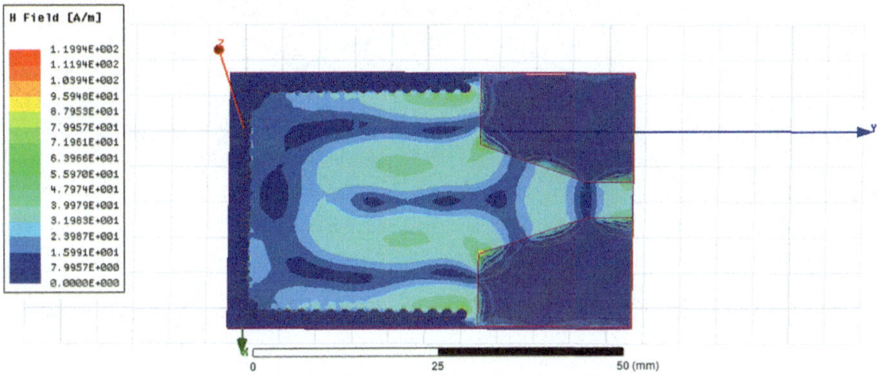

Fig. 16 H-field pattern for SIW without slots

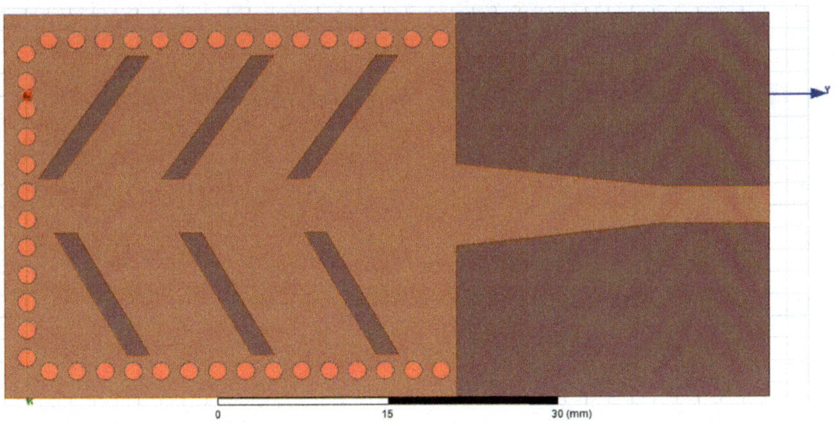

Fig. 17 Proposed structure of SIW antenna without slots using RT Duroid substrate

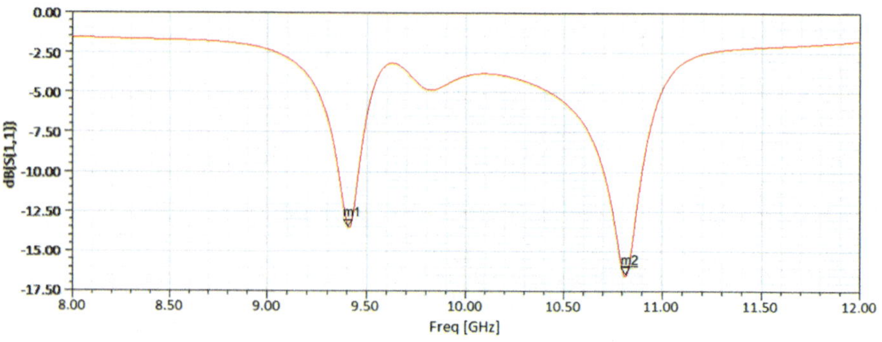

Fig. 18 Graph showing reflection coefficient and frequency of SIW structure with slots using RT Duroid substrate

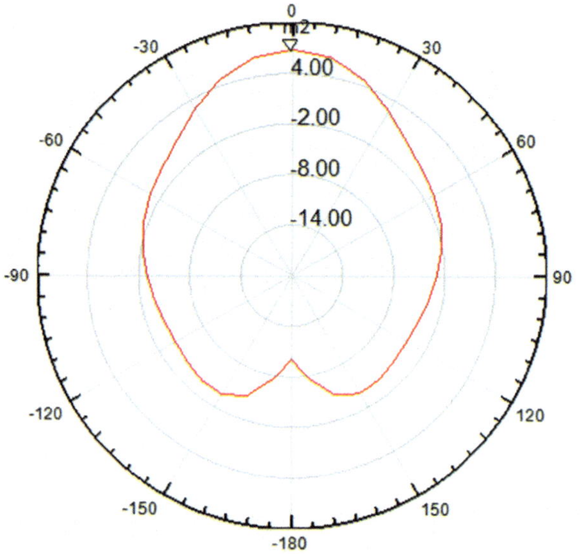

Fig. 19 Radiation pattern of SIW antenna with slots using RT Duroid substrate

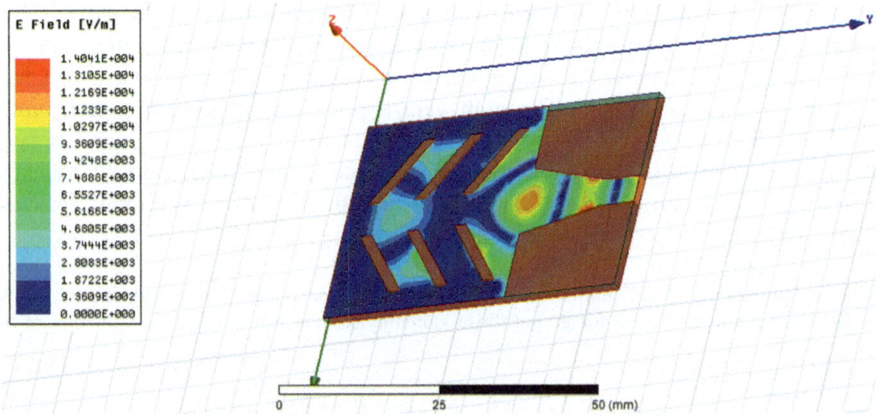

Fig. 20 E-Field pattern for SIW with slots

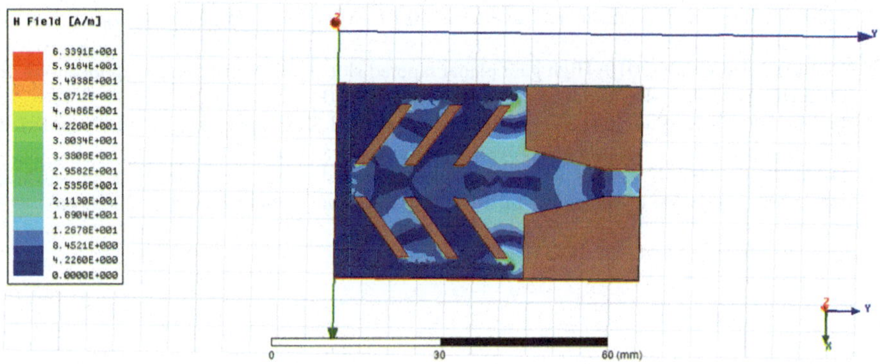

Fig. 21 H-field pattern for SIW with slots

5 Conclusion

The design of substrate integrated waveguide and its microstrip tapering is done in the *X* Band. From the results of simulation of SIW and its microstrip interconnection, the reflection coefficients are below −15 dB that is −38.58 from 10.24 GHz which shows in *X* Band. After the introduction of slanting slots in SIW, the structure resonates in two frequencies of 10.70 and 12.16 GHz with a reflection coefficient of −40.2865 and −29.2741 dB, respectively. From the radiation pattern for SIW having no slots, the gain is comparatively less. But with the introduction of slots, the gain has increased to a reasonable value of 4.0025 as compared to the structure of SIW without slots. While the design of SIW using RT Duroid substrate without slots the gain is 4.5961 dB which is higher as compared to the substrate with FR4, also the SIW structure with slots using RT Duroid has increased to 6.0389 dB. SIW is a promising technology to realize the waveguide in a substrate and offers an elegant way to integrate the waveguide with microwave and millimetre wave planar circuits. For high gain applications in circular polarizations, SIW with RT Duroid can be used. Using SIW technology, a waveguide has been designed on RT/Duroid 5880 commercial substrate at 10.62 GHz centre frequency. The filter has 55 mm × 35 mm dimensions, and the response has a return loss less than −15 dB.

Acknowledgements The Authors would like to extend the deep gratefulness and heartfelt thanks to Dr. Usha Kiran Kommuri for giving those valuable assistance towards the completion of the work.

References

1. Deslandes D, Perregrini L, Arcioni P, Bressan M, Wu K (2002) Dispersion characteristics of substrate integrated rectangular waveguide. IEEE Microwave Wireless Compon Lett 12 (9):333–335
2. Yoon DG, Hong YP (2010) High gain planar tapered slot antenna for Ku-band applications. In: Proceedings of Asia–Pacific microwave conference 2010
3. Liu C, Guo Y-X, Bao X, Xiao S-Q (2012) 60-GHz LTCC integrated circularly polarized helical antenna array. IEEE Trans Antennas Propag 60(3):1329–1335
4. Deslandes D, Wu K (2002) Design consideration and performance analysis of substrate integrated waveguide components. In: European microwave conference, Oct 2002, pp 1–4

The manufacturer's authorised representative in the EU is Springer Nature Customer Service Centre GmbH, Europaplatz 3, 69115 Heidelberg, Germany. If you have any concerns regarding our products, please contact ProductSafety@springernature.com

Printed and bound by CPI Group (UK) Ltd, Croydon, CR0 4YY

01/10/2026

02225871-0005